Microgrids for Commercial Systems

Scrivener Publishing
100 Cummings Center, Suite 541J
Beverly, MA 01915-6106

Publishers at Scrivener
Martin Scrivener (martin@scrivenerpublishing.com)
Phillip Carmical (pcarmical@scrivenerpublishing.com)

Contents

Acknowledgements

Foremost, thanks to the Almighty for his everlasting love throughout this endeavour.

Acknowledgement is always a phrase to appreciate the dedication, commitment, resources and timely solution either with the digital platform or real-time medium, timely support and bond of encouragement is the vital tool for teachers and researchers from their Institution.

In these regards, we editors express our sincere thanks to Mr. S. Muthukumaran, Director, TECH Engineering Services, Chennai, India; Mr. K. Balaji, Vertiv, Chennai, India; Mr. K. Sasikumar, Electrical Engineer, Mott MacDonald, Noida, India; Department of Electrical Engineering, Information Technology, and Cybernetics, University of South-Eastern Norway, Norway and Center for Electric Vehicle Technologies, Center for Energy Storage Systems, Department of Electrical and Electronics Engineering, College of Engineering Guindy, Anna University, Chennai, India.

We editors got the full support and executed the task promptly where our Institution devoted the time and liberty for enhancement with research in particular to make this book a great success.

I wish One and All for the devoted time frame effort for the grand success with the book.

Warm regards,
Editors

1

Smart Energy Source Management in a Commercial Building Microgrid

A. C. Vishnu Dharssini*, S. Charles Raja and P. Venkatesh

Department of Electrical and Electronics Engineering, Thiagarajar College of Engineering, Tamil Nadu, Madurai, India

Abstract

In the most recent times, renewable resource-based local power generation and effective utilization are evolving as a hot research topic, as it mainly focuses on ensuring a reliable energy supply to the consumers. It mitigates the challenges of the volatile nature of renewable sources and uncertainties in demand rise. To ensure the sustainable operation of microgrids specifically, a deep knowledge of demand and generation trends is required. The proposed research framework highlights the promising supply-side management approach of available renewable sources along with the existing grid connection, the optimal energy configuration of which is found with the help of Hybrid Optimization of the Multiple Energy Resources simulation software. The scope of the chapter is to assure the continual supply to meet the load of a particular department block in an educational institution, Thiagarajar College of Engineering (TCE), Madurai (9.8821° N, 78.0816° E). The feasibility study deals with prioritizing available sources such as solar photovoltaic generation, diesel generation, and backup facilities in TCE. Based on seasonal energy trends and availability, a significant demand-side management technique (DSM) technique is feasible for educational buildings: load shifting is implemented. The approach tracks the optimal energy configuration among four configurations and compares the configuration before and after the DSM incorporation. It also pinned out the probability of cost reduction by the optimal source allocation strategy.

***Keywords*:** Building energy management, feasibility analysis, demand side management, optimal source allocation, ToU tariff

**Corresponding author*: acvishnudharssini@gmail.com

P. Sivaraman, C. Sharmeela and P. Sanjeevikumar (eds.) Microgrids for Commercial Systems, (1–20) © 2024 Scrivener Publishing LLC

1.1 Introduction

In this modern era, the local power generation and utilization concept is considered an asset and efficacious approach to have a better energy management system [1]. This encourages the penetration of renewable energy resources and considerably reduces the usage of conventional fossil fuels [2]. The real-time challenge with the approach is adopting it to an existing grid-connected system without collapsing the system and also ensuring optimal source utilization [3, 4]. The alarming usage of conventional resources which are emitting greenhouse gas paves the way to emphasize the need of incorporating an appropriate demand-side management (DSM) strategy [5]. The DSM techniques encompass various kinds of strategies such as load shifting, peak clipping, valley filling, strategic load growth, strategic load conservation, etc. [6]. Some DSM implemented reduce loads in peak hours, resulting in a demand-response approach [7]. The applicability of DSM with respect to available resources in a location provides a clear insinuation on conventional sources [8]. Some strategies are also tariff-based, aimed to reduce additional charges and switch load in reference to the time of use principle [9]. The preliminary requirement for introducing building energy management is possible with smart energy meters which enable to model the demand pattern of that particular building. The process is generally termed modeling demand, involving the procedural gathering of available information as a numerical note and updating knowledge on the state of existing energy systems [10].

Modeling the demand pattern of smart buildings is essential based on economic and environmental impact on energy utilization trends and sustainable use of available resources [11]. Enormous studies and research articles plotted the problem by various approaches and techniques, but it has a limitation as it varies from system to system and also relies upon system considerations [12]. Each work proposes a different and unique supply and demand-side management approach according to the dataset type, i.e., whether univariate or multivariate. Some researchers focused on comparing supply–demand result patterns [13] accompanied by an analysis of the cost involved in bidding for supply from each kind of source in case of the availability of multiple sources [14].

The study presented in [15] put forth a stochastic programming model for staging the performance of a smart microgrid by reducing the operating cost and emissions with non-conventional resources. Plenty of work dealt with this problem using optimization techniques which are found to be generic and primordial. Vishnupriyan and Manoharan [16] adopted the

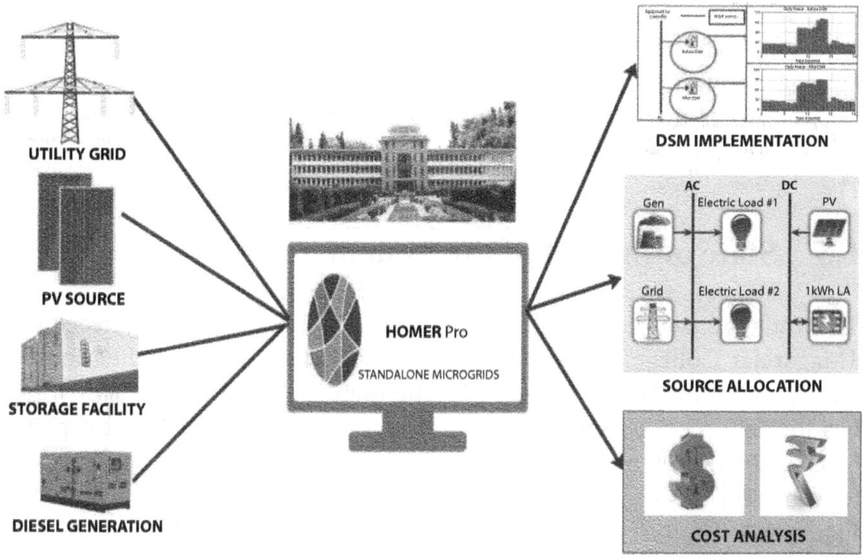

Figure 1.1 Overview of HOMER analysis.

study with a HOMER simulation software for Hybrid Renewable Energy Systems under six different climatic conditions, and the results were optimized to improve the renewable fraction in both before and after the DSM strategies. Similar works are presented in [17, 18] for residential and commercial buildings, promoting optimal energy sources based on computed net present cost (NPC) and cost of energy (COE).

From this extensive survey, it is clear that predictive DSM implementation by data pertaining to the demand load of higher education institutional buildings was not collected and approached to pick out optimum energy configuration with a standard analytical approach. This paves the way to the initiation of strategic DSM, i.e., load shifting and HOMER software solution as in Figure 1.1.

1.2 Motivations of the Study

- ✓ To ensure a balancing operation of the microgrid from both the supply end and the customer end.
- ✓ To prioritize the utilization of local generation with available renewable resources which, in turn, results in the reduction of transmission losses.

- ✓ To promote building energy management.
- ✓ To schedule the operation of each load and to trace corresponding source combinations to feed them.
- ✓ To estimate the lifetime expenditure, i.e., over 25 years for each source configuration.

1.3 State of the Art of the System

The test system—Department of Electrical and Electronics Engineering in Thiagarajar College of Engineering—is the study location in the Madurai district of Tamil Nadu, as shown in the map (Figure 1.2). The study area has four sources of supply which are solar PV panels with a net capacity of 18 kW from an overall 450-kW system, a 10-kVA diesel generator system, battery storage of 1 kWh lead acid, and grid supply as depicted in Figure 1.3. The test system consists of highly equipped laboratories with Elmeasure smart meters affixed to trace the energy usage pattern. The data gets continually stored in the server which is reclaimed in the centralized server system on demand using Elmeasure's ElNet software.

1.4 Overview of the Proposed Methodology

Figure 1.4 shows the overall process initiating from data fetching to optimal configuration selection. The data obtained from various smart meters will possess certain default shortcomings that can be offset by adopting proper data handling techniques. There exists a subsequent process to get

Figure 1.2 Location of Thiagarajar College of Engineering.

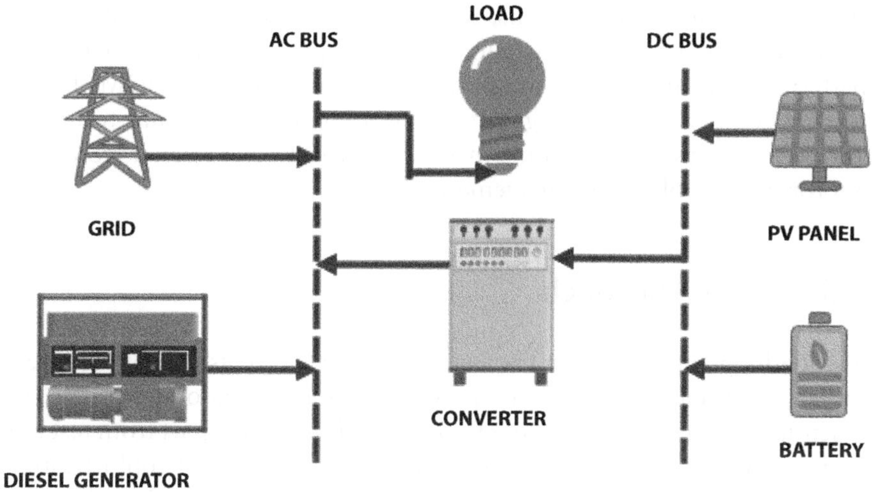

Figure 1.3 Block diagram of the test system.

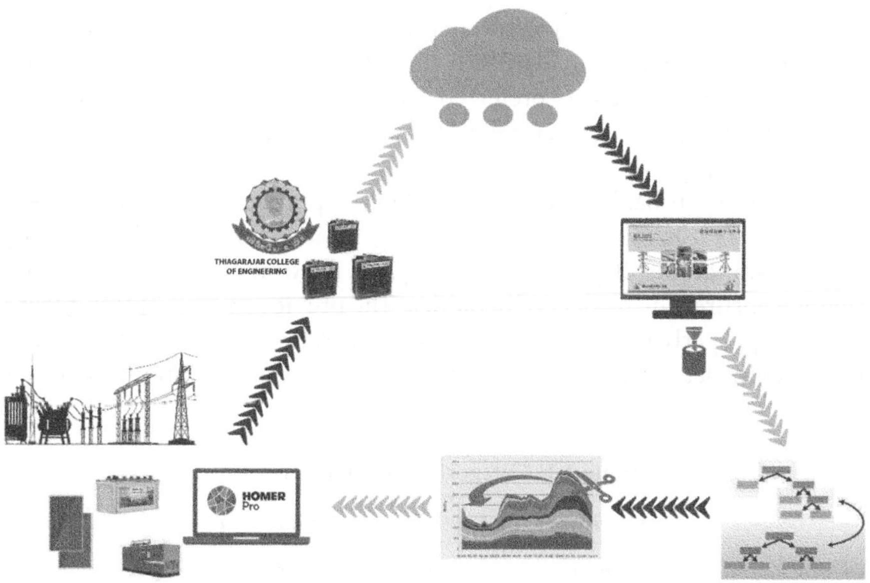

Figure 1.4 Process flow of the proposed methodology.

coherent worthwhile data sources at the end of the data pre-processing approach. These data catered to the analytics part to draw insights from the data. In the proposed methodology, data is fed to the decision tree algorithm which decides on loads to be operated at a particular time. In short, it schedules the primary load and secondary loads in a building with respect to time. Following this, it incorporates the most probable DSM strategy, i.e., load shifting. Next, for effective supply side management, load patterns before and after DSM are laid hold of and fed to the HOMER software. It compares each configuration in terms of environmental, technical, and economic aspects, resulting in a best approach (before/after DSM with appropriate supply-side management).

1.5 DSM Approach

The loads and appliances in the test system are tagged as primary and secondary loads based on customer preferences and utilization premises. The essential loads for the active functioning of laboratories apart from test kits and devices are tagged as primary loads, while the rest, for sophistication in utilization, are tagged as secondary loads. The categorization of loads in the building is presented in Table 1.1.

Lighting loads, fans, routers, projectors, and plug points are considered primary, while air conditioners, water dispensers, exhausters, printers, and coolants are secondary loads. With the knowledge of available loads and their importance, loads are preferred accordingly. Tracing the existing pattern of consumption is essential as it is one of the determining factors for implementing DSM.

The most probable and effective DSM approach—load shifting—is applicable as it involves the effective usage of loads without violating any kind of limitations. It prevents reaching the maximum demand limit and thereby ensures the neglect of penalties. The previous consumption record which is properly maintained in the server is retrieved and treated to feed the machine learning algorithm. The decision tree algorithm thus extracts

Table 1.1 Categorization of loads.

Primary load	Secondary load
Lights	Air conditioners
Fans	Water dispensers
Plug points	Exhausters
Projectors	Printers
Routers	Refrigerators

Table 1.2 Distinguished time slots.

Time slot	Period
11.15 a.m. to 01.15 a.m. and 02.15 p.m. to 04.15 p.m.	Peak hours
09.00 a.m. to 11.15 p.m.	Normal hours
01.15 p.m. to 2.15 a.m. and 04.15 p.m. to 09.00 a.m.	Off-peak hours

insights from data and makes decisions on switching appliances. Its decision relies entirely on energy patterns obtained from smart meters, maximum demand limit, the peak time of consumption, and, finally, on the time of the day. The algorithm is designed to have average monthly patterns on a 24-h time scale for incorporating the appropriate DSM technique, and it also varies with academic semester schedules. Thus, net demand patterns before and after DSM over each month are required to be called out to comparatively visualize the role of the ML algorithm.

The procedure adopted to train the ML algorithm is shown in Figure 1.5. The loads are shifted in correspondence to the time slot. The 24 h of a day are split into three distinguished time periods, namely, peak, normal and off-peak, as shown in Table 1.2. The algorithm checks the time period at the moment and cross-verifies the criteria as shown in the figure. Only primary loads are prioritized during peak hours, i.e., active laboratory sessions. During normal hours, loads are utilized based on the probability of maximum demand reach, and during off-peak hours, it is up to the consumer's choice of load usage, accounting for both norms in the prior two slots.

The patterns of before and after DSM visualizes that a maximum reach of 80 kWh/day falls in January and is considered to be the peak consumption month, while in March and October the lowest consumption of below 40 kWh/day occurs. It was also observed that it is quite challenging to reduce the peak by shifting, as the possible time of shifting is only around morning (7:30 a.m.) to evening (6:30 p.m.). Still the ML was found to be effective in modifying the demand curve by reducing the peak.

Based on the decision, loads are shifted, and the demand curve is traced back on approximation, as shown in Figure 1.6. It was seen that reducing the peak over a working time slot will aid in the quality of energy utilization. In addition to ensuring this, both demand traces before and after DSM are fed to HOMER software, which comparatively analyzes the system to have suitable and profitable energy configurations.

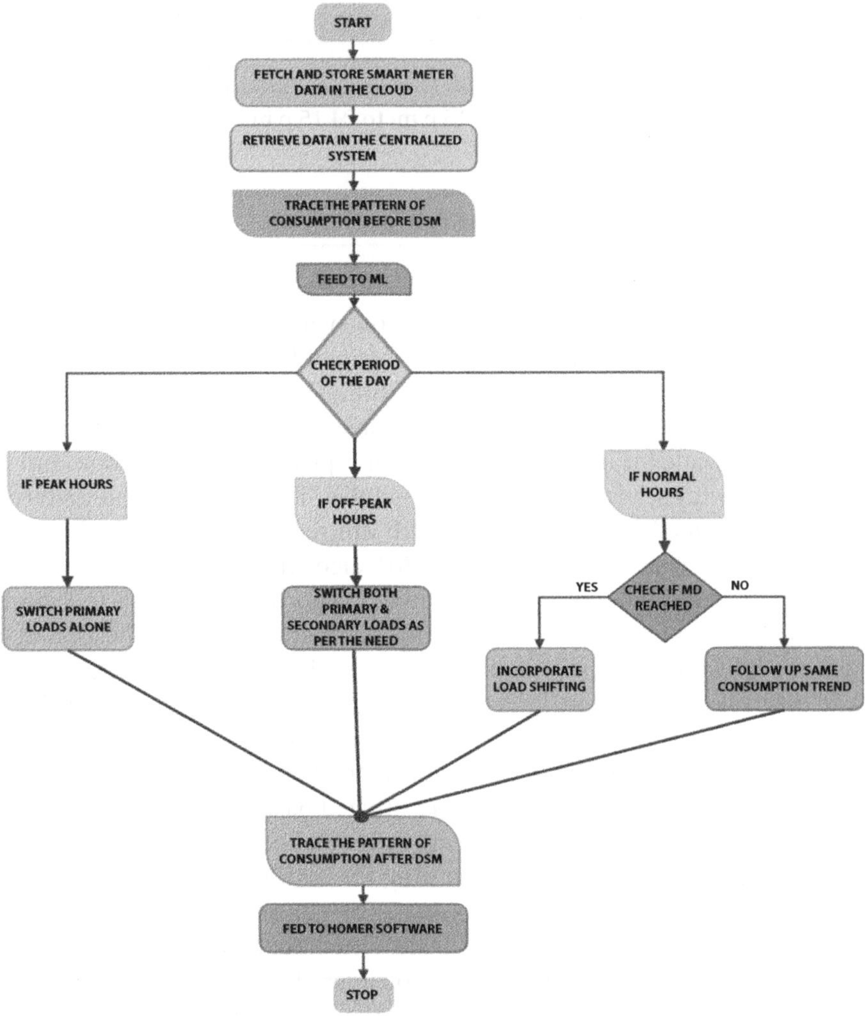

Figure 1.5 Working of ML algorithm.

1.6 Background for HOMER Simulation

The HOMER software is initialized with the exact location of the EEE block in the TCE building, as shown in Figure 1.7. It falls at 9°53′ N latitude and 78°4.5′ E longitude and located at Thiruparankundram, Madurai, Tamil Nadu, 625015, India.

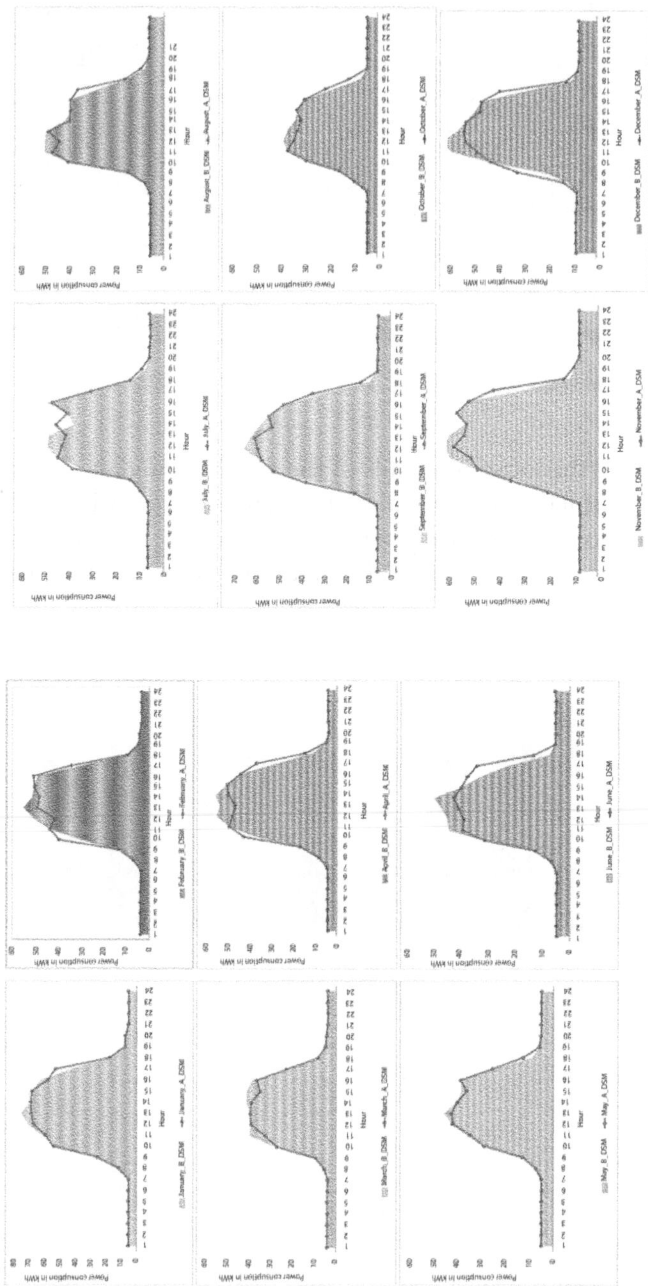

Figure 1.6 Monthly patterns before and after DSM.

Figure 1.7 Location of the TCE-EEE department.

After affixing the location details of the test system, data on renewable resources is obtained from the National Renewable Energy Laboratory and NASA's prediction of worldwide energy resources. The data on solar global horizontal irradiance and average temperature persisting in the particular

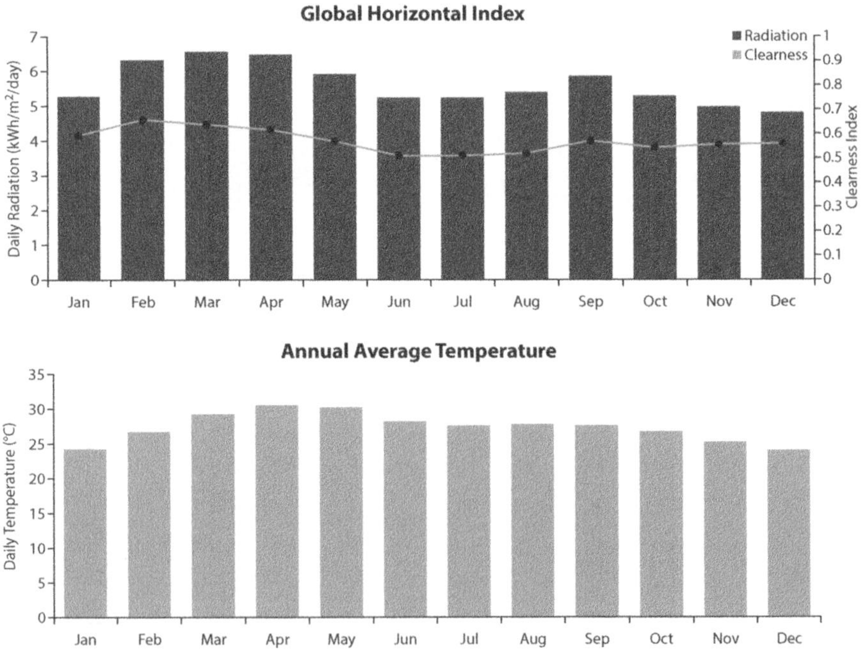

Figure 1.8 Availability of resources.

location is collected for estimating the effective utilization of available renewable energy resources. The yearly average solar irradiance is about 5.61 kWh/m^2/day, and it is represented in Figure 1.8.

1.6.1 Economical Input Data for Simulation

The load data before DSM is incorporated initially to map out appropriate energy configuration without DSM, and then load data after DSM is fed for observation. The data on all three sources, excluding the grid which needed to be imputed for simulating the real-time system, is tabulated in Table 1.3.

The load categorizing and optimization with the DSM technique promotes the management of load consumption for achieving the desired objective. As per the DSM strategy, loads are prioritized and shifted with hourly data using the load shifting strategy. The comparative seasonal profile in Figure 1.9 shows the impact of DSM in the institution's annual load profile.

1.6.2 Simulation—Energy Configurations

The HOMER software is accounted for comparing four energy configurations under two scenarios (before and after DSM) as depicted in Table 1.4. The configurations are depicted by tabulation.

Table 1.3 Economical inputs.

S. no.	Resource component	Manufacturer	Size (kW)	Initial cost ($)	Replacement cost ($)	O&M cost ($)	Lifetime
1	PV array	TATA Solar Power system	1–18	27,000	24,300	0	20 years
2	Converter	Schneider Electric	1–7	2,625	2,345	10.00	15 years
3	Diesel generator	Cummins	10	1,600	1,440	0.100	15,000 hours
4	Battery storage system	EnerSys	1 kWh	247.67	227	10.00	10 years

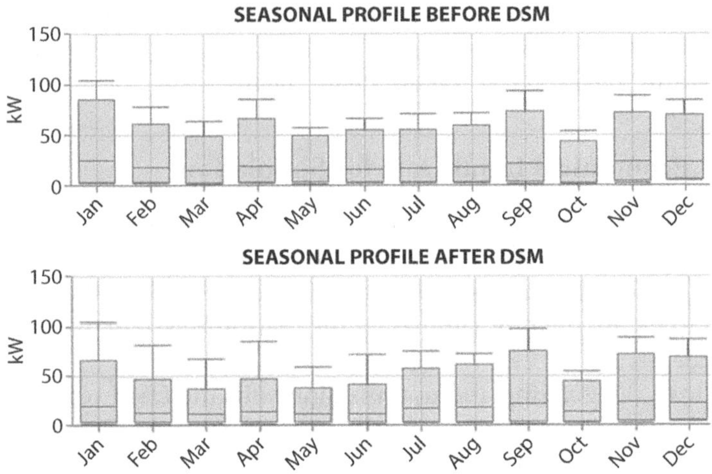

Figure 1.9 Monthly average energy demand of the EEE department (before and after DSM).

The results witnessed a reduction in the peak load and load factor from 153.90 kW and 0.16 to 133.63 kW and 0.19. Following it, the seasonal profile with a box plot comparatively depicts the variation in the range of consumption over each month. The average peak of every month is reduced to a greater extent and is notified clearly.

1.6.3 Comparative Analysis

The configurations are compared based on economical, technical, and environmental aspects considering the factors—components involved, net generation, consumption range, emissions, and, finally, cost—which are tabulated in Table 1.5. The ratio of components involved varies with system configuration, and the tabulation shows the base system to meet the load.

On the whole, the configurations after DSM are more reliable compared with the configurations before DSM. Based on production, total range of consumption, and rate of emission, DSM results in bounteous outcomes. Accounting for each combination of supplies, configurations 1 and 5, configurations 2 and 6, configurations 3 and 7, and configurations 4 and 8 are compared.

Configurations 1 and 2 and configurations 5 and 6 possess a high renewable fraction of 9.60 and 9.70, respectively. Hence, the optimal configuration from the aspect of the ratio of integrating renewable resources is among these pairs of configurations. Finally, concerning the economic

Table 1.4 HOMER simulation configurations before and after DSM.

<table>
<tr><th>Configuration</th><th>Before DSM</th><th>After DSM</th></tr>
<tr><td>PV + grid</td><td>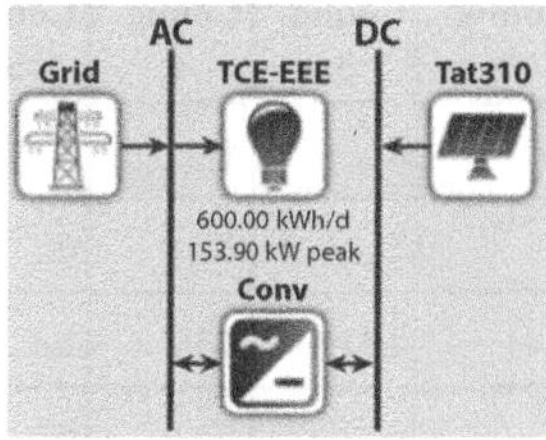

Configuration 1</td><td>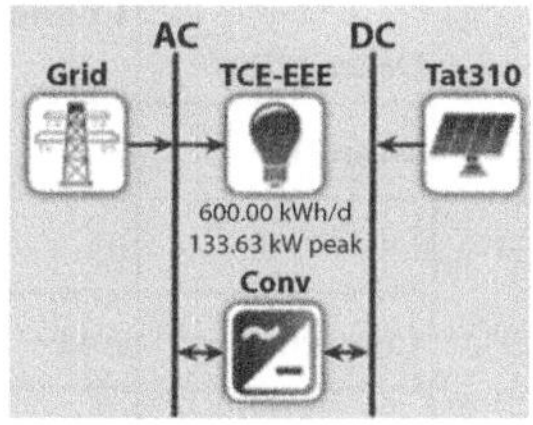

Configuration 5</td></tr>
<tr><td>PV + battery + grid</td><td>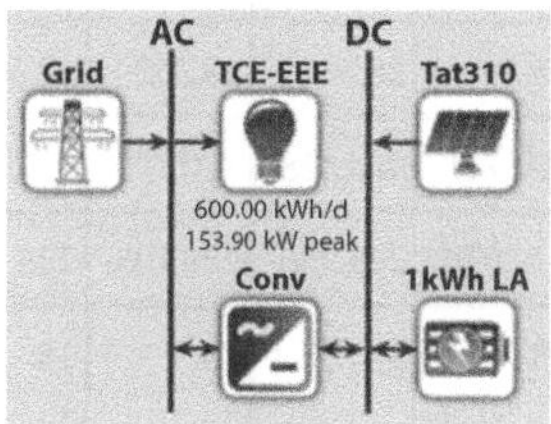

Configuration 2</td><td>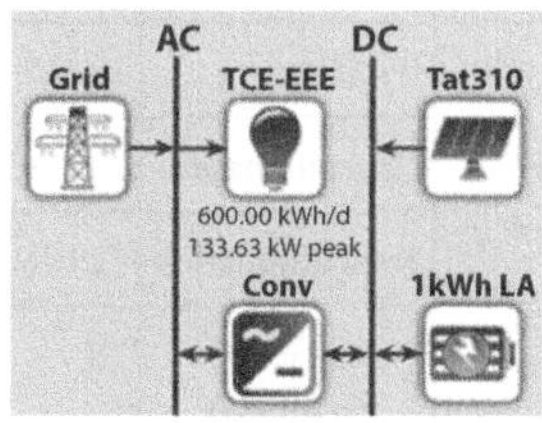

Configuration 6</td></tr>
<tr><td>PV + DG + grid</td><td>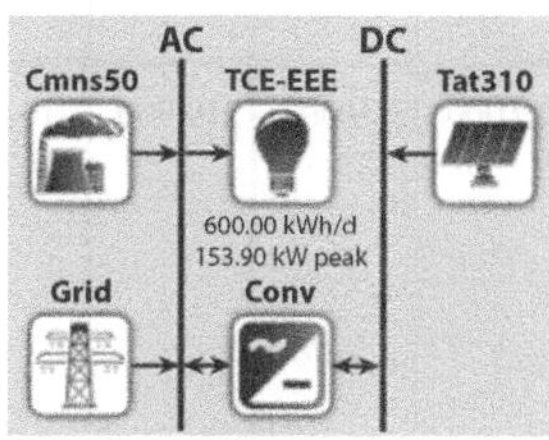

Configuration 3</td><td>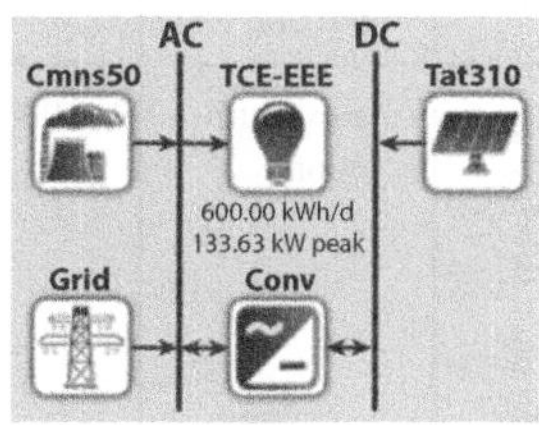

Configuration 7</td></tr>
<tr><td>PV + battery + DG + grid</td><td>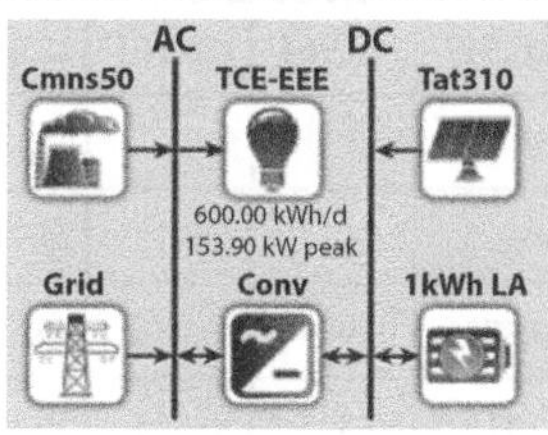

Configuration 4</td><td>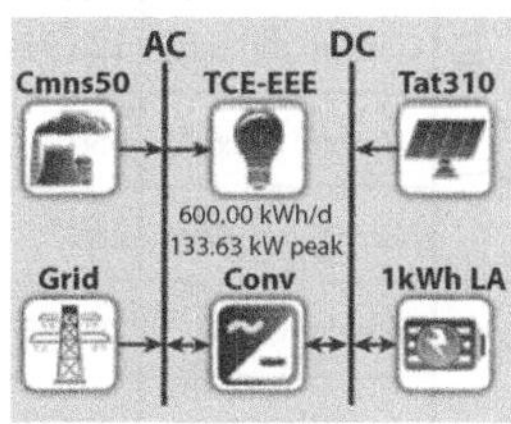

Configuration 8</td></tr>
</table>

cost involved in each configuration, the initial capital increases with the addition of new energy sources. The operating cost involved varies with configuration as it varies with accounted energy sources for that particular configuration, while levelized COE is found almost similar to each configuration, except configuration 4 and configuration 8. Next to that, NPC

Table 1.5 Comparative result analysis.

Description		Without DSM				With DSM			
		Config 1	**Config 2**	**Config 3**	**Config 4**	**Config 5**	**Config 6**	**Config 7**	**Config 8**
Component (kW)	Solar PV	3	3	4	4	4	4	3	3
	DG	0	0	2	2	0	0	2	2
	Converter	2.42	2.44	2.42	2.56	2.42	2.42	2.42	2.42
	Battery	0	1	0	1	0	1	0	1
	Grid	148	148	148	148	128	128	128	128
Production (kWh/year)	PV array	28,367	28,367	28,367	28,367	28,367	28,367	28,367	28,367
	Grid purchases	200,893	200,711	200,828	200,702	197,800	197,799	197,800	197,797
	Total production	229,260	229,078	229,260	229,105	226,167	226,167	226,167	226,187
	PV capacity factor (%)	18	18	18	18	18	18	18	18
	Renewable fraction (%)	9.60	9.60	9.50	9.50	9.70	9.70	9.68	9.68
Consumption (kWh/year)	Total consum-ption	219,000	219,000	219,000	219,900	216,000	216,000	216,000	216,000
Emission (kg/year)	Carbon dioxide	126,964	126,849	126,964	126,964	125010	125010	125010	125010
	Sulphur dioxide	550	550	550	550	542	542	542	542
	Nitrogen oxides	269	269	269	269	265	265	265	265
Cost	Initial cost ($)	29,625	29,873	29,945	30,033	29,625	29,873	29,945	30,033
	Operating cost ($/year)	20,423.2	20,415.0	20,397.2	20,584	20,114	20,137	20,088	20,261
	Total NPC ($)	293,647	293,788	293,899	296,133	289,649	290,200	290,913	291,463
	COE ($/kW)	0.1023	0.1023	0.1024	0.1031	0.102	0.103	0.103	0.1029

Table 1.6 Comparative cost analysis.

Configurations		Year-0	Year-1	Year-15		Year-20		Year-25	
		Cap. cost	Op. cost	Op. cost	Rep. cost	Op. cost	Rep. cost	Op. cost	Salvage cost
Before DSM	Config 1	$29,625	$18,982	$8,525.9	$994.92	$6,407.79	$7,747.0	$4,814.9	$4,553.20
	Config 2	$29,872	$18,974	$8,524.1	$994.92	$6,405.17	$7,747.0	$4,812.7	$4,553.20
	Config 3	$29,785	$18,982	$8,527.6	$994.92	$6,407.79	$7,747.0	$4,814.9	$4,586.83
	Config 4	$31,472	$19,124	$8,591.2	$994.92	$6,455.61	$7,819.4	$4,850.9	$4,916.73
After DSM	Config 5	$29,625	$18,690	$8,396.4	$994.92	$6,309.20	$7,747.0	$4,740.9	$4,553.20
	Config 6	$29,872	$18,681	$8,400.6	$994.92	$6,312.39	$7,747.0	$4,743.3	$4,580.39
	Config 7	$29,785	$18,690	$8,396.4	$994.92	$6,309.20	$7,747.0	$4,740.9	$4,889.54
	Config 8	$31,472	$18,899	$8,412.6	$994.92	$6,328.39	$7,819.4	$4,792.2	$4,916.73

emerges as a determining factor to trace the optimal configuration among combos of four configurations.

Focusing on the point of the Payback period or lifetime of the system under each configuration is another essential determining factor of optimal source selection depicted in Table 1.6. The HOMER simulation software estimates the system life over 25 years concerning capital investment cost, operating cost, replacement cost, and salvage cost as in Figure 1.10. As mentioned earlier, incorporating DSM upgrades the system's effectiveness by reducing the operating cost to a notifiable extent. It is also evident that configurations operating all the available resources (configuration 4 and configuration 8) are considered to be the worst configuration under both DSM scenarios. Next to that configurations 3 and 7 involve diesel generation, operating which demands non-renewable fuel of higher economic cost and also results in an adverse impact on environmental aspects i.e., emission. Finally, the combo for optimal source configuration for the test system is configurations 1&5 and configurations 2&6 which slightly vary in economic aspects.

Despite all conflicts in choices, configuration 1 and 5 is found to be more reliable, efficient, favourable, economical, and environmental. Thus, energy configuration with solar PV panels and grid-tied systems with

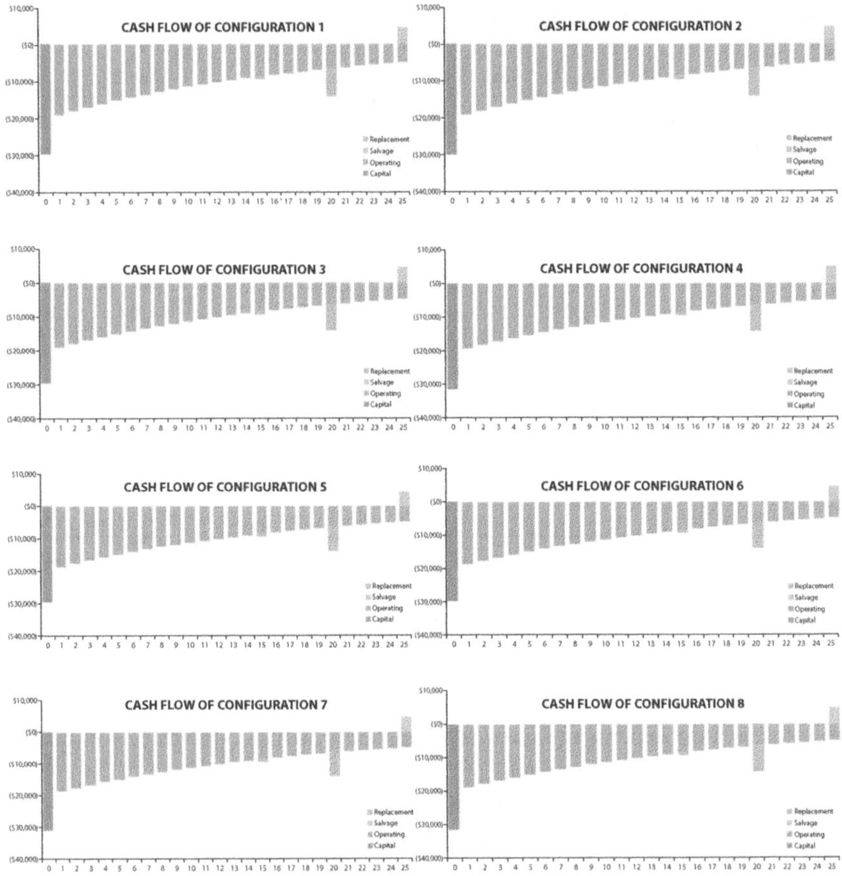

Figure 1.10 Configuration—cost analysis.

appreciable DSM strategy is found to be the best choice for promoting supply and demand side management in a building.

1.6.4 Highlights of the Proposed Framework

- Simulated real-time hybrid renewable energy system with HOMER software.
- Selected of optimal supply-side strategy along with existing Grid connection.
- Traced the seasonality trend in energy consumption.
- Compared energy configuration under both with and without DSM.
- Done economic analysis with NPC and COE metrics.

1.7 Conclusion

This study presented hybrid supply and demand side management for collected smart energy meter data on consumption in the Electrical and Electronics Engineering department of Thiagarajar College of Engineering, Madurai. Four different source configurations under two scenarios in terms of the Demand side management approach is concerned. The test system adopted consists of different sources – solar PV, Diesel, and Battery systems along with a grid connection. The strategic DSM using a machine learning algorithm renders reduced peak demand compared with the system without DSM. The following suitable results are incorporated and compared using the HOMER software tool. It compares the four combinations of resources before and after DSM. The methodology adopted highlights that configuration 5 with PV solar generation along with grid after the application of DSM is the best configuration for energy requirements in the considered test location. On the whole, the chapter emphasizes that extracting in-depth knowledge on local generations and supply sources in a smart environment facilitates customer end in planning building energy strategies concerning economic and environmental aspects. The approach of Demand cum supply side management creates a possibility of increasing the usage of Renewable energy sources among the rest and also ensures effective energy utilization at minimal cost expenditure. Thus, adopting this study will definitely aid in the sustainable operation of the Microgrid.

References

1. Yang, Y., Ren, J., Solgaard, H.S., Xu, D., Nguyen, T.T., Using multi-criteria analysis to prioritize renewable energy home heating technologies. *Sustain. Energy Technol. Assess.*, 29, 36–43, 2018.
2. Hadi Mousavi-Nasab, S. and Sotoudeh-Anvari, A., An extension of the best-worst method with D numbers: Application in the evaluation of renewable energy resources. *Sustain. Energy Technol. Assess.*, 40, 1–19, 2020.
3. Schlachtberger, D.P., Becker, S., Schramm, S., Greiner, M., Backup flexibility classes in emerging large-scale renewable electricity systems. *Energy Convers. Manag.*, 125, 336–46, 2016.
4. Elavarasan, R.M., The motivation for renewable energy and its comparison with other energy sources: A review. *Eur. J. Sustain. Dev. Res.*, 3, em0076, 2019.

5. Lampropoulos, I., Kling, W.L., Ribeiro, P.F., Van Den Berg, J., History of demand side management and classification of demand response control schemes. *IEEE Power Energy Soc. Gen. Meet.*, pp. 1–5, 2013.
6. Gellings, C.W., The concept of demand-side management for electric utilities. *Proc. IEEE*, 73, 1468–70, 1985.
7. Warren, P., A review of demand-side management policy in the UK. *Renew. Sustain. Energy Rev.*, 29, 941–51, 2014.
8. Ebubekir, S.S., Safak, B.I., Muammer, K., Demand side management opportunities, framework, and implications for sustainable development in resource-rich countries: Case study Qatar. *J. Clean. Prod.*, 241, 118332, 2019.
9. Bourhnane, S., Abid, M. R., Lghoul, R., Zine-Dine, K., Elkamoun, N., Benhaddou, D., Machine learning for energy consumption prediction and scheduling in smart buildings. *SN Appl. Sci.*, 2, 1–10, 2020.
10. Maryam, K., Gou, Z., Dupre, K., Energy use characteristics and benchmarking for higher education buildings. *Energy Build.*, 164, 61–76, 2018.
11. Cai, H., Shen, S., Lin, Q., Li, X., Xiao, H., predicting the energy consumption of residential buildings for regional electricity supply-side and demand-side management. *IEEE Access*, 7, 30386–30397, 2019, https://doi.org/10.1109/ACCESS.2019.2901257.
12. Mokhtara, C., Negrou, B., Bouferrouk, A., Yao, Y., Settou, N., Ramadan, M., Integrated supply-demand energy management for the optimal design of off-grid hybrid renewable energy systems for residential electrification in arid climates. *Energy Convers. Manag.*, 221, July, 113192, 2020, https://doi.org/10.1016/j.enconman.2020.113192.
13. Aghajani, G.R., Shayanfar, H.A., Shayeghi, H., Demand side management in a smart micro-grid in the presence of renewable generation and demand response. *Energy*, 126, 622e637, 2017.
14. Vishnupriyan, J. and Manoharan, P.S., Prospects of hybrid photovoltaic diesel standalone system for six different climate locations in the Indian state of Tamil Nadu. *J. Clean. Prod.*, 185, 309e321, 2018a.
15. Mohammad, A.V.R., Roghaye, G., Parisa, R., Soroush, M., Mehrdad, A., Techno-economic analysis of a hybrid power system based on the cost-effective hydrogen production method for rural electrification, a case study in Iran. *Energy*, 190, 116421, 2020.
16. Rebecca, N., Heejin, C., Jian, Z., Techno-economic analysis of the geothermal system in a residential building in Memphis, Tennessee. *J. Build. Eng.*, 27, 100993, 2020.

17. Chambon, C. L., Karia, T., Sandwell, P., Hallett, J. P., Techno-economic assessment of biomass gasification-based mini-grids for productive energy applications: The case of rural India. *Renew. Energ.*, 154, 432–444, 2020.
18. Petrillo, A., De Felice, F., Jannelli, E., Autorino, C., Minutillo, M., Lavadera., A.L., Life cycle assessment (LCA) and life cycle cost (LCC) analysis model for a stand-alone hybrid renewable energy system. *Renew. Energ.*, 95, 37–355, 2016.

2

Renewable Power Generation Price Prediction and Forecasting Using Machine Learning

Challa Krishna Rao[1,2*], Sarat Kumar Sahoo[2] and Franco Fernando Yanine[3]

[1]Department of Electrical and Electronics Engineering, Aditya Institute of Technology and Management Tekkali, A.P., India
[2]Dept. of Electrical Engineering, Parala Maharaja Engineering College, Berhampur, Affiliated to Biju Patnaik University of Technology, Rourkela, Odisha, India
[3]School of Engineering of Universidad, Finis Terrae, Providencia, Santiago, Chile

Abstract

In the power market, electricity price forecasting (EPF) is a crucial consideration for decision-making. The need to optimize earnings by adjusting bids in day-ahead power markets is becoming more and more important to different market participants. Prior information is required for marketers to have an advantage over the competition while controlling the risk of pricing fluctuation. However, not all marketers must accurately predict the worth of future pricing when making decisions. To make a choice, it is necessary to determine whether the cost will be prohibitive. Thus, in order to determine the prices that have an impact on marketers, electricity price classification is first performed. Based on a threshold value, prices are categorized as low class prices and high class pricing. In order to determine the precise value of prices for utility maximization, the EPF has been explored next. In order to maximize advantage or utility, buying and selling bidding techniques rely on the accurate projections of prices for the following day. Effective forecasting models can improve the performance of producers and consumers, who play important roles in the electrical markets. The EPF approaches now in use produce complicated models and are not generalizable.The best method for forecasting prices with superior generalization performance, kernel functions, and distributive prediction is to use machine learning (ML)-based models. In order to forecast prices, machine learning-based electricity price forecasting has been studied.

**Corresponding author*: Krishnarao.challa@gmail.com

P. Sivaraman, C. Sharmeela and P. Sanjeevikumar (eds.) Microgrids for Commercial Systems, (21–48) © 2024 Scrivener Publishing LLC

On the markets of Ontario, Austria, and India, the ML models have been put to the test. Neural networks, support vector machines, core vector machines, information vector machines, and relevance vector machines (RVM) have all been used with EPF, and their performances have been assessed in order to determine the optimum model. When all of these models are examined, it becomes clear that RVM is the most effective and trustworthy ML approach for various markets. The price of power fluctuates a lot, with both typical costs and spikes that might be tens to hundreds of times higher than the average price range. The prediction for the overall price of energy is created by adding together average prices and price peaks. A good and precise price spike forecast is required because price spikes have a big influence on the electrical market. As a result, a threshold value has also been used to forecast the spike value. The potential for more research in this field has also been emphasized.

Keywords: Forecasting, renewable power, electricity price forecasting, feature selection, machine learning, relevance vector machine

2.1 Introduction

Since 1980, changes have been implemented in several areas of the global economy, including banking, ports, railroads, food services, and communication. The applications listed above are required to reduce government control and replace it with free market forces. Similarly to this, global restructuring in the electrical sector is occurring [1]. The three somewhat distinct parts of the electrical sector are generation, transmission, and distribution. All three of these vertically related sectors have historically been integrated into one utility in the traditional power system [2]. Numerous power producers were compelled to switch from a single utility structure to a free market system during the 1990s. Since 1980, changes have been implemented in several areas of the global economy, including banking, ports, railroads, food services, and communication. The applications listed above are required to reduce government control and replace it with free market forces. Similarly to this, global restructuring in the electrical sector is occurring. The three somewhat distinct parts of the electrical sector are generation, transmission, and distribution. All three of these vertically related sectors have historically been integrated into one utility in the traditional power system [3]. Numerous power producers were compelled to switch from a single utility structure to a free market system during the 1990s. The first concept encourages separating the marketing, transmission, and generation of electricity. These duties are not monopolistic and can be carried out by businesses on a competitive basis in place of the government. The second tenet holds that those businesses might be privatized. Privatized businesses are more productive and have superior management techniques [4]. The third principle encourages the creation of institutions to regulate market participants and safeguard the general welfare. Chile implemented the market reforms for electricity first in 1982, after which Norway, Sweden, the United Kingdom, and Finland began the process. More respondents now think that consumer benefits from deregulation are significant [5]. A lot of work has been put into developing power markets in recent years. The physical features of the electrical market and the fluctuating cost of power contribute to market risk. Forecasting various market indicators, such as the hourly energy price of the spot market, has become vital to handle such risks in the electrical market [6]. A generating company's ability to make informed decisions about the production of energy depends on accurate market price forecasts.

2.1.1 Electricity Price Forecasting

A subset of energy forecasting known as electricity price forecasting (EPF) aims to anticipate spot and forward electricity market prices. The price projections have developed into a key component of the decision-making process for energy corporations [7]. In a market for deregulated energy, pricing decisions carry a risk. Prior knowledge is necessary for marketers to have a competitive advantage while managing the risk.

Electricity, on the other hand, is a particularly unique commodity since it cannot be economically stored and requires an equal balance between supply and demand. The generated power must be utilized because it cannot be stored. As a tradable commodity, the prices may be purchased and sold [8]. Due to their unstorability, volatility, seasonality, transportability, and unreasonable tariff regulations, the prices are crucial. As a result, the price of electricity is considered similar to other commodity markets. At times, the cost of electricity might increase by tens or even hundreds of times. It can occasionally fall to zero or even negative levels at other times [9]. Unlike demand series, price series exhibit features including non-constant mean, variation, and strong outliers, making electricity price forecasting challenging. Prices can be predicted more accurately due to their volatility. There are three different forms of forecast planning: short term, mid-term, and long term [10]. While long-term forecasting considers years, mid-term forecasting of power prices concentrates on a period of 1 to 6 months. First of all, unlike short-term forecasting, mid-term forecasting, by definition, cannot use the trend from the recent past. Furthermore, it is quite challenging to identify and estimate the power price peaks for both the short- and long-term future. Finding the peak pricing is typically highly accurate since short-term forecasting may make use of recent trend data [11]. The duration of the historical data required to train the forecasting model is the primary distinction. Typically, the data from the previous few days are all that is required to train the power price model for short-term forecasting, but to train the forecasting model, mid-term and long-term forecasting models need 1 year of historical data [12]. Short-term planning is thus taken into account in this context since it is more significant and essential in the open electricity market. The time frame for the immediate future is a few hours to a few weeks. EPF is a significantly more difficult technique compared with load forecasting [13]. It is because energy has unique qualities that set it apart from other commodities, such as its elasticity of demand, which keeps the balance between supply and demand for power constant. In Figure 2.1, a straightforward method for price forecasting is presented.

It is obvious from Figure 2.1 that the historical prices should be gathered first in the market. There are no restrictions on the data; all the information needed for the procedure is given. The input is then given for training and testing after the data has been examined to determine the threshold value. An appropriate model is chosen, and it is put into practice [14]. All machine learning models are fine-tuned to produce the desired performance. Out of all the models, the finest one is chosen and tweaked once again. However, it is challenging, if not impossible, to estimate power costs with the kind of accuracy now feasible for loads. First off, while there is seasonality in the energy price process, it is not as evident as it is for loads on a daily, weekly, monthly, and yearly basis [15]. Second, many external factors can be measured. Third, psychological and social reasons might potentially result in an unexpected and unreasonable buyout of some contracts, which will cause price increases [16].

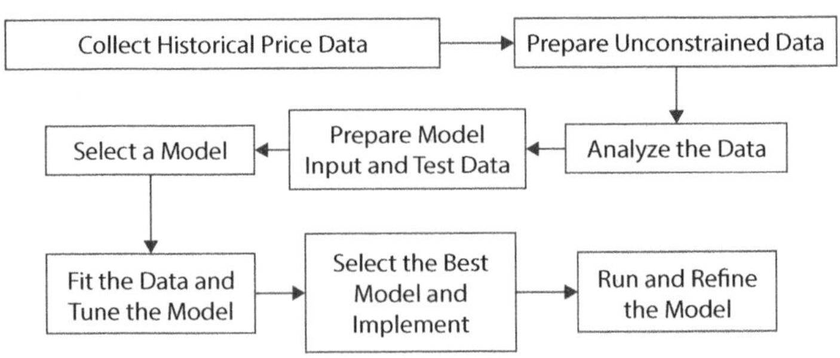

Figure 2.1 Price forecasting model [47].

Depending on the nature of energy markets and related price volatility, the price forecasting inaccuracy might range from 5% to 36%. However, compared with load forecasting, where errors typically range from 1% to 3%, this error range is rather significant [17]. Instead of predicting, price classification is done first since not all market players need to know the precise value of future prices to make decisions [18]. It is anticipated that demand-side interactions will enable a widespread participation in energy markets in the household, commercial, and industrial sectors with the advent of "smart grid" technology and new market activities. Given the nature of the majority of electric loads, particularly those at the residential level, demand-side market players could respond largely when prices go over certain criteria [19]. Beyond these limits, the precise cost of power would be immaterial since it is "too costly." Many demand response products, like the hour-ahead dispatchable load program in the markets, are created with certain energy price thresholds in mind [20]. Electricity customers with on-site generation facilities use threshold-based decision-making. Only when grid costs are less than the marginal cost of running the on-site electricity-producing equipment can these institutions buy electricity from the grid. The point price forecasting issue may be simplified to price categorization subproblems in these sorts of applications where the precise value of prices is not the primary need [21].

2.1.2 Electricity Price Classification

The demand-side actors are increasingly involved when the market structure develops and becomes more sophisticated, such as with smart grids, since they have the freedom and opportunity to react to changes in the price above and below specific thresholds [22]. Price categorization based on thresholds is preferable to identifying the exact numerical values of future prices if marketers opt to manufacture their power from the market based on market price surpassing a particular price threshold. Therefore, for market players, electricity price classification (EPC) is more significant than predicting [23]. To classify pricing, the entire dataset is taken into account. Based on a threshold value, the prices are categorized. The yearly price average is used to compute the threshold amount. To dynamically classify future, unknowable energy market prices about user-defined criteria, price categorization and thresholding are used [25]. Because not every market participant must be aware of the precise value of future prices, the price categories are crucial. In electrical markets, it is not necessary to have exact prices to reduce the price forecasting issue to a subproblem of price classification, where the class of future prices is of interest [24]. Typically, market prices for electricity are categorized following pre-established price thresholds.

Pattern recognition algorithms are typically used to classify prices. These methods go through the historical data for certain patterns and identify them [26]. The identified patterns are then used to categorize and classify upcoming occurrences into the proper "classes". The costs are often divided into low- and high-class costs. Prices fall into one of two categories: low-class prices and high-class prices, depending on whether they are above or below the threshold amount [27]. The daily price average is compared with the threshold value's yearly price average to categorize up to 24 h in advance. Since price evaluation varies among markets, there are significant differences in the price categorization accuracy attained by various models across various electrical markets [28]. Low prices are referred to as "normal prices," whereas high prices are considered "abnormal prices." Some demand response products are made with a threshold value for energy prices in mind; these devices buy power from the grid if the market prices are lower than their marginal costs and do not need an exact price value [29]. However, the majority of market players require point forecasting or numerical prediction to design bidding strategies and reduce the risk. This reduces the pricing categorization problem to a forecasting one. Only prices in the typical range are used for price forecasting [30].

Only when demand-side players' marketers are present is the categorization necessary. The future price value is not at all required for price classification. However, since their success depends on precise pricing, many market participants need to be aware of them [31]. To evaluate their potential and maintain competitiveness in a competitive market, market participants must analyze price projections and be aware of impending price increases. The prognosis for the price of electricity is based on prices in the usual range, but it is highly challenging to anticipate the anomalous value of the coming power price [32]. Extremely high prices are sometimes referred to as spikes in the electrical market.

2.1.3 Price Spike Prediction

Electricity prices are extremely variable and frequently see spikes that may be tens or even hundreds of times higher than the average price due to the complexity and unpredictability of the power infrastructure. Price spikes are extremely stochastic and may linger for many time units [33]. Predicting such electrical surges is quite challenging. Most current research on forecasting power costs is based on average electricity prices. Two categories are used to classify the anomalous price:

- A jump price is a price that is much higher than the average observed for the prices that came before it.
- A price drop, which is caused by an abrupt decrease in the prior pricing.

Since it might affect the estimate of the deterministic and stochastic components for the models of price dynamics, the identification of spikes is a crucial issue [34]. Therefore, a thorough analysis of the categorization, prediction, and price spike prediction of energy market prices has been carried out in this research work. Many techniques have been employed in recent years to predict power prices [35]. The prices have been predicted using conventional techniques including wavelet transform (WT), multivariate transfer function (TF), dynamic regression (DR), autoregressive integrated moving average (ARIMA), generalized autoregressive conditional heteroskedasticity (GARCH), and so on. However, although the price of electricity is a nonlinear function of its input variables, these models are linear predictors. Therefore, these strategies are unable to capture the behavior of the power price signal [36]. These strategies include both conventional methods and artificial intelligence (AI) technologies including neural networks, feed-forward neural networks, radial basis functions, genetic algorithms, particle swap optimization, and recurrent neural networks (RNN). Numerous studies have been conducted using these AI models for classification and prediction; however, none of the studies have produced a model that is sufficiently accurate with a mean absolute percentage error (MAPE) of less than 5%. This suggests that further work has to be done. The newest machine learning (ML)-based soft computing approaches are attempted to produce improved classifiers and predictors for energy pricing in consideration of the shortcomings of the existing methodologies [37].

2.2 Literature Review

The categorization, forecasting, and price increase of power prices are all thoroughly reviewed. Numerous strategies have been used in power markets during the past few decades. Taxonomies of forecasting techniques have been proposed in some studies [38].

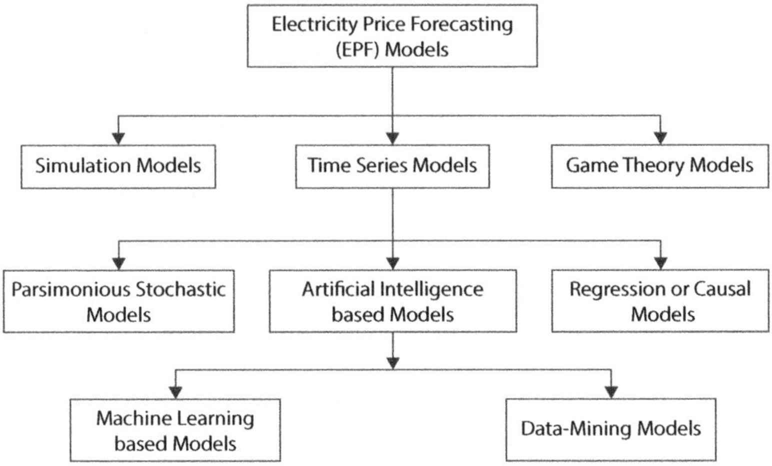

Figure 2.2 Electricity price forecasting [24].

A survey demonstrates that several forecasting techniques have been created. A crude categorization tree is depicted in Figure 2.2, although it is not all-inclusive and other approaches or techniques may be used. Price forecasting may be done using these techniques [39]. Parametric approaches and methods based on artificial intelligence are two kinds of forecasting techniques. The forecasting models are split into two categories: classic approaches and new methodologies. The methods primarily used for price forecasting fall into two groups, time series, and simulation. The time series method is based on past market price information [40]. Because there is a vast quantity of data involved, the simulation methodology necessitates accurate modeling of the electricity markets.

Time series may also be divided into non-linear heuristic models and models based on linear regression. Auto-regressive moving average (ARMA) models, their extensions, ARIMA models, and their variations are examples of regression-based models [41]. Generalized auto-regressive conditional, although these models are intended to model and predict the changing price itself, is aimed at modeling the volatility of power prices. Artificial neural networks (ANN) and other AI approaches are used in nonlinear heuristic-based models to simulate the relationship between input–output data without knowing all of the connections [42]. In recent years, machine learning and other soft computing techniques, such as data mining techniques, have been employed for price prediction [43]. There includes a thorough discussion of data mining techniques. Broadly speaking, there are six groups of methods that have been developed to assess and forecast the behavior of the power markets and the ensuing energy pricing, namely:

i. Production cost models: These models have the ability to mimic how electricity-producing plants operate as a whole. The main goal is to meet power demand at the lowest possible cost. This approach's primary flaw is that other power market participants' strategic bidding strategies are not taken into account.
ii. Equilibrium models: These models resemble production cost models, with the exception that they also take other market participants' strategic bidding prices into account. It has been noted that equilibrium models' predicting abilities in unregulated markets have not proven sufficient. The disadvantage of this strategy is that, in addition to being computationally demanding, it becomes exceedingly difficult to derive quantitative conclusions.
iii. Fundamental models: These models represent the influence of all physical and economic elements on electricity price dynamics. Compared with short-term electricity price modeling and forecasting, these models are thought to be more appropriate for medium-term energy price forecasting.
iv. Quantitative models: These models explain the statistical characteristics of energy pricing and are useful for valuing derivatives as well as risk management.
v. Statistical Analysis: These approaches are straightforward implementations of time series model-inspired techniques. Compared with those of "extremely randomly" fluctuating financial assets, it makes power costs more predictable.
vi. Models based on artificial intelligence: This method uses non-parametric techniques like neural networks and fuzzy logic, among others. AI-based models frequently have a flexible design and are capable of handling complexity and non-linearity. Their main flaw has been being counterintuitive and frequently performing below par. The best techniques from the list above for classifying and forecasting power prices are statistical and AI-based models.

2.2.1 Types of Analyses

2.2.1.1 Game Theory Models

Game theory is the foundation for the first set of models. Participants in the market are quite interested in finding solutions to such games. Because they are trying to maximize their earnings, players in oligopolistic power markets deviate from their true marginal costs while setting their bidding curves [44]. Price development may be thought of as the result of a power transaction game since these models involve the mathematical solution of these games. Numerous equilibrium models exist, including jump diffusion models, reduced form models, Nash–Cournot framework models, and supply function equilibrium models.

2.2.1.2 Model Simulations

These models belong to the second class of price-forecasting methodologies, where a precise model of the system is constructed and the answer is obtained using algorithms that take the underlying physical phenomena into account. In simulation models, price forecasting simulates the real dispatch under the limits and needs of the operating system [45]. There have been two different simulation model types examined: One is GE Power Systems Energy Consulting Market Assessment and Portfolio Strategies algorithm, while the other is LCG Consulting UPLAN software. Different future possibilities are created using simulation models, and the likelihood of such scenarios is calculated depending on the assumption of risk. These approaches, however, have two shortcomings: First, they need specific information about how the system operates, and second, simulation techniques are challenging to construct and expensive to run [46].

2.2.1.3 Models for Time Series

Time series analysis is a forecasting technique that concentrates on the dependent variable's historical history. Exogenous variables may occasionally be included in a time series structure. There are three other sorts of models that are based on time series.

2.2.1.4 Parsimonious Stochastic Models

Some of the well-known and often used methodologies, such as GARCH, moving average, ARMA, and ARIMA, have been taken into consideration [47].

To predict power market prices in California and Spain, the created univariate ARIMA models employed GARCH models to explain the fluctuation of power prices in the Spanish energy market [48]. A wavelet-ARIMA technique has been put forth. It entails three steps: (i) discrete wavelet transform decomposition of the price series, (ii) modeling of the approximation series using the ARIMA process to obtain 24-hourly predicted values, and (iii) application of the inverse transform to produce the predicted prices for the following 24 h. A wavelet-ARIMA approach often performs better than a regular ARIMA process.

A WT-based neural network (NN) model was used to estimate the price profile in a deregulated power market. To create the set of input variables for the suggested forecasting model, the historical price data was first decomposed into wavelet domain constitutive subseries using WT [49]. An innovative hybrid strategy for estimating short-term power prices in Spain has been put out. The combined WT, particle swarm optimization (PSO), and adaptive network-based fuzzy inference system (ANFIS) constitute the basis of the suggested methodology. Presented are the findings of a case study that used the power market in mainland Spain [50]. A wavelet transform was used to produce a collection of time series that behaved better in a hybrid technique for forecasting day-ahead power prices. A radial basis function (RBF) network is used to rectify the estimated error of the wavelet-ARIMA prediction after an ARIMA model has generated a linear forecast seasonal process [51]. ARIMA models have been proposed. In each of these works, a stationary price time series was created by removing the periodic and non-periodic trend components, and an ARIMA model was used to forecast the price profile. Additionally, accuracy enhancement has been made possible using a system of sequential error correction [52].

These models are evaluated using a time series of loads and prices from the California power market system from the years leading up to and including the market meltdown. Using wavelet transform, the price series has been divided into detailed and approximation sections. By using the proper model in the wavelet domain to forecast the future behavior of decomposed series, inverse WT was then utilized to provide price predictions in the time domain [53]. Two models have been constructed by Nogales *et al.* The first model, called a DR model, connected the spot price to both its own lagged data and the real demand values. In the TF model, a TF term and a disturbance term that came after an ARMA process have created the link between price and demand [54]. The second class of models with this property, the seasonal process model, is utilized since the price of electricity is a non-stationary process that shows daily, weekly, annual, and other periodicities.

2.2.1.5 Regression or Causal Models

A regression or causal model is based on the hypothesized link between a known set of independent factors and a dependent variable. Based on a correlation study of each of these independent variables and the price (dependent) variable, the explanatory variables for this model are determined [55]. Multiple regression using gas prices, available nuclear capacity, temperatures, and rainfall as regressors and a mean-reverting stochastic process for the residuals have been proposed. For each of the 48 half-hourly load periods in the British market, they developed and analyzed a basic regression model, and they contrasted its day-ahead forecasting ability with that of regime-switching regression models [56]. Similar linear regression models were used to assess various forecast-combining strategies. An algorithm for long-term EPF alternates between the predictions of many models based on a set of predetermined rules [57]. A two-step technique for EPF has been presented, with an emphasis on the effects of the anticipated system load and wind power output.

2.2.1.5.1 Artificial Intelligence Models

These models, which map the connection between input and output, are nonparametric. AI models may learn complicated, nonlinear interactions that are challenging to model using traditional methods [58]. There are two categories for these models: data-mining models and models based on machine learning. Even if the time series is unknown or extremely complicated, it may still capture the autocorrelation structure in them. RBF network, multilayer feed-forward neural network (FFNN), committee machine of NNs, and RNN are the available NN models. To predict the market values in New England, a committee machine combining cascaded neural networks and NNs was used.

With a non-cascaded FFNN and the predicted input stated as the measured input plus an extra error term representing the related uncertainties, a cascaded NN structure was built. The mapping function that an AI-based technology extracts from input and output data was learned. The price of electricity is a highly time-variant signal, and the functional relationship that underlies it changes quickly over time. Thus, the value of the retrieved NN feature depreciates quickly [59]. A preconditioned artificial neural network model for the power market in England and Wales was proposed by Wang and Ramsay in 1998. They used multi-layer perceptron (MLP) as their ANN and a front-end processor to categorize the input variables [60]. Following that, ANN and fuzzy logic-based NN approaches were created.

To estimate the short-term load, RBFNN was combined with the ANFIS. The model initially forecasts the load on the prediction day using the non-linear approaching capacity of the RBF network, without taking into consideration the factor of electricity price, and then it bases its predictions on recent changes in real-time pricing [61]. It modifies the load forecasting outcomes produced by the RBF network using the ANFIS. Extreme learning machine, a unique neural network-based approach, has been developed to address the issue of classifying future power prices concerning predetermined price thresholds. Various models have been created and tested using information from the Ontario and PJM markets. An adaptive wavelet neural network has been proposed for short-term price forecasting in the power markets [62]. The activation functions for feed-forward neural networks' hidden layer neurons were determined to be a Mexican hat wavelet (FFNN).

A fuzzy neural network divides the input space into a fuzzified categorization. The results from the Spanish power market demonstrate the potential of the suggested strategy to provide accurate projections [63]. A three-layer back-propagation neural network with modest daily mean errors was employed, using data from the Victorian power market. Two learning algorithms using neuro-fuzzy systems have been utilized for predicting the hourly power reference price in the Ontario energy market. A multilayer perceptron neural network for the prediction of electricity reference prices has been trained using a decoupled extended Kalman filter, which ignores the interdependencies of weights from various neurons [64]. A two-step training method based on the inverses of standard deviations to identify and eliminate input factors was developed since Gaussian RBF networks approximated input–output relationships by creating localized clusters. Unimportant input factors may cause the local learning of RBF networks to be inaccurate and lead to poor generalization.

Elman networks were employed to produce short-term price projections in the market of mainland Spain. They concluded that their network outperforms a variety of existing methods for predicting power prices, such as ARIMA, wavelet-ARIMA, MLP, fuzzy ANN, and wavelet-ARIMA-RBF networks [65]. A hybrid PSO-ANFIS technique was used for short-term forecasting. Different weights are assigned to the samples inside one class using the intra-class KNN approach to improve the structural information. To expedite coaching, the inter-class KNN technique deletes the expendable restrictions for the other class. A rapid clip method is used to boost the rate for large-scale tasks. Each SNPSVM model looks at the reparability across classes as well as the concentration in both classes via structural information, allowing it to immediately recover the generalization ability of the method with full prior knowledge [66].

The MLP neural network has been suggested for use in training. The Levenberg–Marquardt algorithm is the most often used training method in EPF. It trains 10–100 times quicker than back-propagation. They demonstrated that the technique, a wavelet-ARIMA model, may deliver findings that are more accurate for the Spanish market [43]. It has been shown that multilayer neural networks are used in composite topologies. Mean absolute percentage error and mean square error are used to assess performance. Data from the Australian market is used. A review of the techniques for predicting electricity prices and the contributing elements was performed [25] by using the feed-forward NN back-propagation algorithm. The result is compared with the intended output using the gradient descent method. The error function is reduced to a minimum.

MLPNN classifier was used by utilizing an electroencephalogram. The determined signals are trained under close monitoring by utilizing effective memory. For the highest classification accuracy, the hidden layer employs the tan-sigmoid transfer function. According to previously established thresholds, a categorization of power market pricing has been proposed [20]. This method uses DCT to convert all-time series from the time domain to the frequency domain. The collection of input characteristics is made up of these discriminative spectral coefficients, which are then classed using NN. The self-adaptive differential evolution-based feed-forward neural network was trained using differential evolution and genetic algorithms. The crossover and mutation rates are modified for the various stages of evolution. However, it has shortcomings due to low convergence and lack of generality [25]. By combining ARIMA-ANN models, we have presented a method to forecast hourly electricity prices for the next-day electricity markets. It is investigated with the national power market in Australia. ARIMA and ANN are used to compare the forecasting performance.

To determine the initial set, an appropriate number of clusters, and ultimately the number of rules in the proposed model, a clustering approach is also used as a pre-processing strategy. For the California day-ahead energy markets, a three-layered feed-forward back-propagation artificial neural network technique was used to estimate market clearing pricing and market is clearing quantity [35]. The authors have talked about over-fitting, one of the key problems in training neural networks. In other words, it provided an excellent fit to the training set but poor generalization to fresh data outside the training set. For price prediction in the deregulated energy market, a hybrid intelligent model was based on excitable dynamics and RNN. The Fitz–Hugh Nagumo (FHN) model was used to simulate the spiking activity, the RNN unit to regulate FHN, and feed-forward neural network to represent the residue of RNN-FHN, making up the created model's three parts [25]. In their thorough research, extreme distributions and feed-forward back-propagation neural networks were used to estimate the power pool prices for the Australian national energy market from 1998 to 2013.

2.3 Data Mining Models

For data interpretation and inference, data mining approaches including the Bayesian categorization method, genetic algorithm, and reasoning-based classification have become increasingly prominent. Using wavelet and NN-based models, the usual price and a data mining framework were used to forecast the price spike. Databases have been mined using Bayesian classification and similarity searching techniques [18]. The two price signals are separated by the pre-processing module. The NN-wavelet module forecasts both the average price and the likelihood of price increases on particular days.

2.3.1 Machine Learning Techniques

Even though NN is the most widely used method, it fails because of its extensive learning path and complex design procedure. Furthermore, NN performs well in interpolation but less well in extrapolation, with less generalization capacity [38]. Therefore, a new setting known as ML has been taken into consideration for EPC and EPF challenges to resolve these problems.

2.3.1.1 Supervised Learning

Algorithms that require outside help are those that fall under the category of supervised machine learning. Train and test datasets are created from the input dataset. The output variable in the training dataset has to be predicted or categorized. Figure 2.3 provides the workflow for supervised machine learning methods. Here we explore three of the most well-known supervised machine learning techniques [18].

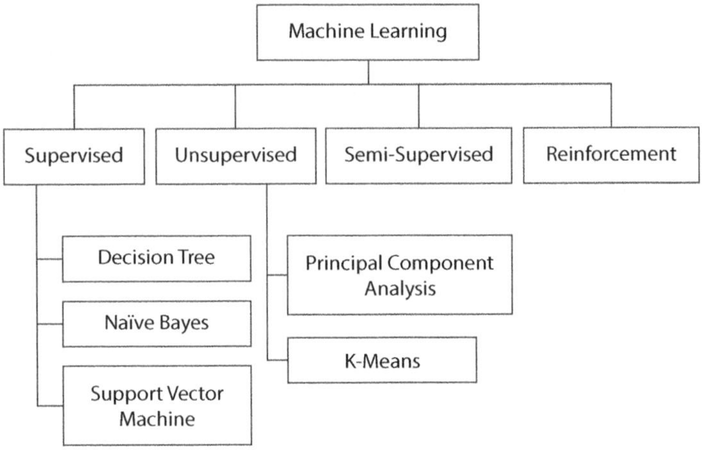

Figure 2.3 Machine learning methods [43].

2.3.2 Decision Trees

Decision trees are a subset of trees that organize qualities into groups by ranking them according to their values. It serves primarily as a categorization tool. For beginners and experts in the field of machine learning, the work in decision tree and a sample of its use are available to first time users and experts in the field of machine learning. A novel paradigm for composing decision trees utilizing interval-valued fuzzy membership values has been put out [22]. However, accurate values of fuzzy membership values are not always achievable. The majority of fuzzy decision trees now in use do not take the concerns connected with their membership values into account. By combining the various example-reliant costs into a new cost-based impurity measure and new cost-based pruning criteria, an example-reliant cost-sensitive decision tree algorithm was developed [29]. The suggested technique was then assessed using three separate databases from three real-world applications, including credit card fraud detection, credit scoring, and direct marketing. It has been suggested to predict learning styles using fuzzy decision trees in conversational intelligent tutoring systems. During the tutoring observation, the independent behavior of variables with the highest value variable is used to predict the learning style. Their method has the drawback of ignoring the interconnections between behavior variables, and as learning style modeling is inherently imprecise, even little variations in behavior might result in inaccurate predictions.

To determine the class of an attribute with an ID3 and apply fuzzification for better results, the classification approach of data mining was employed.

Concept drift is frequently too much for online decision trees based on current facts to handle [51]. In response to an approach drift, Blanco *et al.* developed an incremental algorithm that is error-margin-driven and primarily executes two operations. The energy price is the dependent variable in a prediction model, while the independent variables are primarily time series datasets from the Brazilian system. To group qualities with a similar behavior, the data samples are first clustered. After that, further if-then rules from the database are applied using a decision tree method. The decision tree was based on a short-term energy price categorization model. The decision tree's properties are found using a feature selection approach [46].

2.4 Objectives

The purpose of this thesis is to create a forecasting tool that is more precise and efficient for classifying and forecasting power prices in the electrical markets of Ontario, Austria, and India. Therefore, by utilizing machine learning algorithms together with feature selection techniques, an effort has been made in this research to create an intelligent model for various marketplaces. The following objectives are listed step by step for a machine learning algorithm:

1. To create a machine learning-based price classifier that is effective and trustworthy for the large-scale electricity market.
2. To create a viable feature selection methodology for price forecasting and categorization.
3. Utilizing the anticipated pricing to spot the price increase.
4. In order to demonstrate the suggested ML-based approach's capacity for generalization.
5. To demonstrate that the suggested price classifier is adoptable since just a few characteristics are employed across all electricity markets.
6. To demonstrate that the implementation of the chosen feature outperforms those without taking the feature selection procedure into account.

Because energy prices are stochastic, this research solely focuses on the deregulated power market. The findings of past research projects conducted to accomplish the aforementioned goals have been succinctly described in this chapter.

2.4.1 Forecasting Results for the Seasons of Indian Market

In this study, the training period is based on the prices from the preceding 42 days, producing 1,008 training instances. The information for energy pricing was gathered from hourly-presented daily trade reports of an Indian market that were tailored to annual reports. Winter: October 26–November 1, spring: February 5–11, summer: June 22–28, and rainy: September 21–27 are the test dates for the seasons. Given that the test period is 1 week and a day has a length of 24 h, a week has a total of 168 h. The SFS algorithm is used to decrease the total number of variables from 168 to 83. The results demonstrate the efficacy of the suggested RVM technique. The proposed work is assessed on a 2019 Indian market. The

Table 2.1 Comparative analysis for the four seasons in the Indian market.

Market	Forecast week	MAPE (%)	
		RVM	RVM-SFS
Indian	Winter	6.93	6.69
	Spring	5.96	4.24
	Summer	5.68	4.27
	Rainy	5.66	4.66
	Average	6.05	4.96

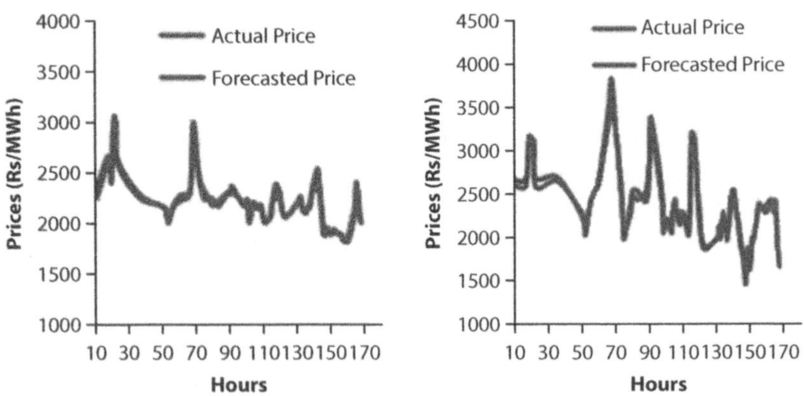

Figure 2.4 Forecasting results for the month of October to November and February.

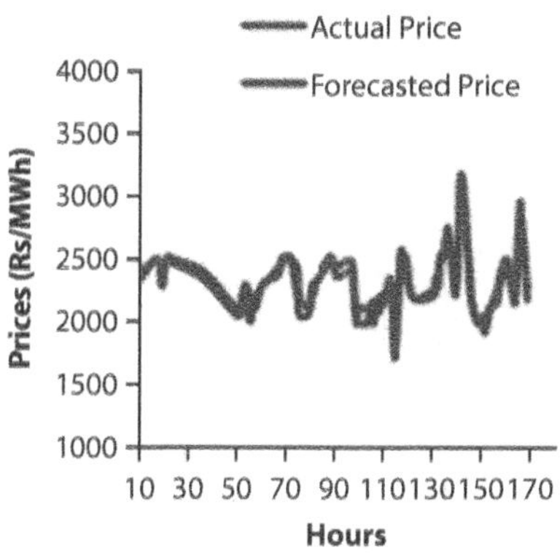

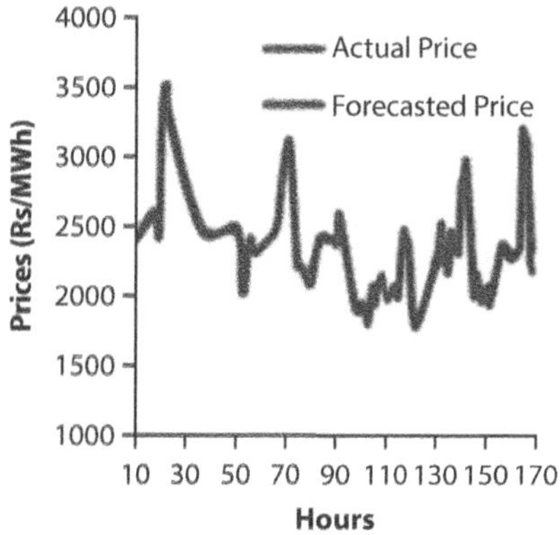

Figure 2.5 Forecasting results for the months of June and December.

dataset's dimension is down to 50.4%. The significant aspects are chosen based on the hour. The results of the evaluation of the suggested model's performance on the Indian market are shown in Table 2.1.

Table 2.1 demonstrates how, for each of the investigated seasons, the MAPE varies depending on whether all attributes are taken into account or just a few selected ones. The suggested RVM-SFS model's average value of MAPE in this market is lower than that of the RVM model. Figures 2.4 to 2.6 depict the actual and predicted pricing values for the four seasons in the Indian market using the RVM-SFS model. Each graph displays real pricing in Rs/MWh together with the anticipated prices.

Figures 2.4 and 2.5 display the predicted price determined by the RVM-SFS for the winter, spring, summer, and rainy seasons together with the actual price as well as the mean absolute value of predicted errors. Figures 2.4 to 2.6 show how closely the anticipated prices for each seasonal week match the actual price.

2.4.2 Day-Ahead Forecasting of Prices for the Indian Market

Since the emergence of increasingly competitive power markets, short-term projections have grown in significance. The producers and consumers must also develop their bid strategy for the power market. The high prices are taken into account for projecting daily mistakes, as shown in Figure 2.6. Considered are the ranges of 150 to 200 days and 310 to 350 days, which correspond to June and December, respectively.

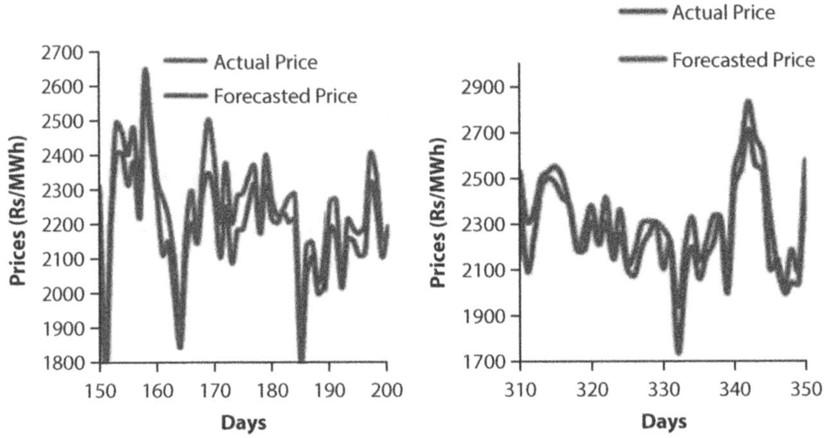

Figure 2.6 Forecasting results for January and October.

2.4.2.1 Forecasts of Cases A and B

Electricity prices are changed hourly in Indian power markets. The number of training points and testing points is chosen for case study A from June 1 through 15 and June 16 through 30, respectively. In the case of study B, the exam period is December 1–15, and there is a total of 15 training points, which are divided into 24-h time chunks from November 16 to 30.

The MAPE findings for daily forecasting errors are displayed in Tables 8.14 and 8.15. These tables compare the ML-based price classifiers with FS and all variables. It is clear from the simulation results that the RVM with FS performs well when compared with other classifiers.

The findings of the daily MAPE from June 16 to June 30, 2016 and from December 1 to December 15, 2016 are examined in Tables 2.1 and 2.2, respectively. Through this comparison, it can be seen that the RVM-SFS model outperforms the other models in terms of predicting the accuracy. Table 2.2 shows the machine learning price classifiers with SFS feature selection. It is determined that RVM with SFS is the most effective strategy for the Indian market. For June and December, the daily MAPE results are computed as percentages. A low error rate is provided by the RVM-SFS model.

Table 2.2 Daily MAPE results in percentage for the period June 16–30, 2019.

Days	NN		SVM	CVM		IVM		RVM		
June	With all features	With FS	With all features	With FS	With all features	With FS	With all features	With FS	With all features	With FS
16	8.69	7.17	8.41	6.38	8.08		5.89	6.76	5.46	3.39
17	10.40	10.25	10.19	10.17	9.63		9.49	9.15	8.83	2.01
18	10.70	9.72	8.94	8.48	5.67		5.45	4.69	3.90	1.53
19	7.79	7.16	6.37	6.03	6.12		5.03	6.89	3.99	3.10
20	14.25	10.29	12.32	10.10	11.28		9.82	10.02	6.70	5.01
21	11.97	10.23	11.02	9.16	9.07		8.48	4.50	3.96	1.53
22	12.72	9.15	10.35	8.73	8.03		7.88	4.27	3.99	3.61
23	11.24	7.16	7.12	6.41	5.14		3.78	3.5	2.78	2.78
24	12.64	12.26	8.86	7.92	7.54		6.91	4.31	4.02	1.10
25	11.29	10.31	10.24	9.23	8.20		6.30	7.64	3.55	1.32
26	10.05	6.14	9.54	4.23	8.64		4.23	5.42	4.15	1.11
27	13.25	11.47	11.72	11.27	10.90		10.05	9.83	7.67	1.01
28	10.33	8.05	9.07	7.95	7.78		7.31	5.15	4.21	3.75
29	8.07	8.05	7.56	7.55	6.95		6.90	4.72	3.80	2.10
30	10.47	4.61	9.57	4.25	7.33		4.16	6.75	3.84	2.70

2.5 Conclusions

The electricity supply has been transformed into a private, unregulated product as a result of the power market reform. Profitability following privatization has evolved into the primary goal for all market participants in such a market environment. Therefore, an initial classification of electricity pricing was done to determine the costs that have an impact on marketers. Based on a threshold value, the prices have been divided into low- and high-class pricing. Thresholds have been developed by numerous demand-side market actors to categorize power prices. However, as individuals make decisions based on their decision-making, marketers typically need to know the exact worth of prices. Thus, to compete for market share and to either bid or hedge against volatility in the current upheaval of the deregulated electricity market, electricity price forecasting has been conducted. The price prediction was made in conjunction with the impending price increase. After then, the increase was discovered to lessen the impact of prices on the markets. As a result, a novel method of machine learning-based classifiers for classifying and forecasting electricity prices has been developed and is described in this study. It has given timely assessments of electricity market pricing. Forecasting has been assessed, and spike values have been determined based on the classification.

References

1. An, J.Y., You, Z.H., Meng, F.R., Xu, S.J., Wang, Y., RVMAB: Using the relevance vector machine model combined with average blocks to predict the interactions of proteins from protein sequences. *Int. J. Mol. Sci.*, 17, 5, 757–762, 2016, May.
2. Anamika, and Kumar, N., Market clearing price forecasting for Indian electricity markets. *Proceedings of International Conference on Intelligent Communication Control and Devices*, India, pp. 633–642, 2017, Sep.
3. Anbazhagan, S. and Kumarappan, N., Binary classification of dayahead deregulated electricity market prices using neural network input featured by DCT. *ICTACT J. Soft Comput.*, 2, 4, 384–390, 2012, Jul.
4. Anbazhagan, S. and Kumarappan, N., Day-ahead deregulated electricity market price forecasting using recurrent neural network. *IEEE Syst. J.*, 7, 4, 866–872, 2013, Dec.
5. Anbazhagan, S., Pravin, K., Kumarappan, N., Classification of day-ahead prices in Asia's first liberalized electricity market using PNN. *Proceedings of IEEE Conference Symposium Series on Computational Intelligence for Engineering Solutions*, Singapore, pp. 176–179, 2013, April.

6. Areekul, P., Senjyu, T., Urasaki, N., Yona, A., Next-day price forecasting in deregulated market by combination of artificial neural network and ARIMA time series models. *Proceedings of Fifth IEEE Conference on Industrial Electronics and Applications (ICIEA)*, Taichung, Taiwan, pp. 1451–1456, June, 2010.
7. Asharaf, S., Narasimha Murty, M., Shevade, S.K., Multiclass core vector machine. *Proceedings of the 24th International Conference on Machine Learning*, New York, USA, pp. 41–48, 2007, June.
8. Azadeh, A., Moghaddam, M., Mahdi, M., Seyedmahmoudi, S.H., Optimum long-term electricity price forecasting in noisy and complex environments. *Energy Sources B: Econ. Plan. Policy*, 8, 3, 235–244, 2013, Feb.
9. Badoui, M. and Clarkson, K.L., Optimal core-sets for balls. *Comput. Geom.*, 40, 1, 14–22, 2008, May.
10. Bahnsen, A.C., Aouada, D., Ottersten, D., Example-dependent cost-sensitive decision trees. *Expert Syst. Appl.*, 42, 19, 6609–6619, 2015, Nov.
11. Bajpai, P. and Singh, S.N., Bidding and gaming in electricity market: An overview and key issues. *Proceedings of National Power System Conference (NPSC)*, Chennai, India, pp. 338–346, 2004, December.
12. Cerjan, M., Krzelj, I., Vidak, M., Delimar, M., A literature review with statistical analysis of electricity price forecasting methods. *Proceedings of IEEE European Conference*, Zagreb, Croatia, pp. 756–763, 2013, July.
13. Cestnik, B., Estimating probabilities: A crucial task in machine learning. *Proceedings of the European Conference on Artificial Intelligence*, Jamova, Yugoslavia, pp. 147–149, 1990, August.
14. Cestnik, B., Kononenko, I., Bratko, I., Assistant 86: A knowledge elicitation tool for sophisticated users. *Proceedings of the 2nd European Conference on European Working Session on Learning*, Bled, Yugoslavia, pp. 31–45, 1987 May.
15. Chaabane, N., A novel auto-regressive fractionally integrated moving average-least-squares support vector machine model for electricity spot prices prediction. *J. Appl. Stat.*, 41, 3, 635–651, 2014, Sep.
16. Che, J. and Wang, J., Short-term electricity prices forecasting based on support vector regression and auto-regressive integrated moving average modelling. *Energy Conver. Manag.*, 51, 10, 1911–1917, 2010, Oct.
17. Chen, D., Tian, Y., Liu, X., Structural nonparallel support vector machine for pattern recognition. *Pattern Recognit.*, 60, 1, 296–305, 2016, Dec.
18. Conejo, A.J., Plazas, M.A., Espinola, R., Molina, A.B., Day-ahead electricity price forecasting using the wavelet transform and ARIMA models. *IEEE Trans. Power Syst.*, 20, 2, 1035–1042, 2005, May.
19. Contreras, J., Espinola, R., Nogales, F.J., Conejo, A.J., ARIMA models to predict next-day electricity prices. *IEEE Trans. Power Syst.*, 18, 3, 1014–1020, 2003, Jul.
20. Costanza, M.C. and Afifi, A.A., Comparison of stopping rules in forward stepwise discriminant analysis. *J. Am. Stat. Assoc.*, 74, 8, 777–785, 1979, Dec.

21. Crockett, K., Latham, A., Whitton, N., On predicting learning styles in conversational intelligent tutoring systems using fuzzy decision trees. *Int. J. Hum. Comput. Stud.*, 97, 1, 98–115, 2017, Jan.
22. Rao, C.K., Sahoo, S.K., Balamurugan, M., Yanine, F.F., Design of smart socket for monitoring of IoT-based intelligent smart energy management system, in: *Lecture Notes in Electrical Engineering*, pp. 503–518, Springer, Singapore, 2021.
23. Girish, G.P., Spot electricity price forecasting in indian electricity market using autoregressive-GARCH models. *Energy Strategy Rev.*, 11, 52–57, 2016, Jun.
24. Girish, G.P. and Vijayalakshmi, S., Determinants of electricity price in competitive power market. *Int. J. Bus. Manag.*, 8, 21, 70–75, 2013, Sep.
25. Girish, G.P., Vijayalakshmi, S., Panda, A.K., Rath, B.N., Forecasting electricity prices in deregulated wholesale spot electricity market: A review. *Int. J. Energy Econ. Policy*, 4, 1, 32–42, 2014, Oct.
26. Rao, C. K., Sahoo, S. K., Yanine, F. F. Demand side energy management algorithms integrated with the IoT framework in the PV smart grid system, in: *Advanced Frequency Regulation Strategies in Renewable-Dominated Power Systems*, pp. 255–277, 2024.
27. Guo, J.J. and Luh, P.B., Selecting input factors for clusters of gaussian radial basis function networks to improve market clearing price prediction. *IEEE Trans. Power Syst.*, 18, 2, 665–672, 2003, May.
28. Guo, J.J. and Luh, P.B., Improving market clearing price prediction by using a committee machine of neural networks. *IEEE Trans. Power Syst.*, 19, 4, 1867–1876, 2004, Nov.
29. Haiteng, X.H. and Niimura, T., Short-term electricity price modelling and forecasting using wavelets and multivariate time series. *Proceedings of IEEE Power Systems Conference and Exposition*, New York, USA, pp. 208–212, 2004, October.
30. Hall, M., Frank, E., Holmes, G., Pfahringer, B., Reutemann, P., Witten, I.H., The WEKA data mining software: An update. *ACM SIGKDD Explor. Newslett.*, 11, 1, 10–18, 2009, Jun.
31. Hebrich, R., *Learning kernel classifiers: Theory and algorithms*, First Edition, MIT Press, Cambridge, USA, 2001.
32. Hu, L. and Taylor, G., A novel hybrid technique for short-term electricity price forecasting in UK electricity markets, in: *J. Int. Counc. Electr. Eng. (JICEE)*, 4, 2, 114–120, Informa UK Limited, 2014.
33. Rao, C.K., Sahoo, S.K., Balamurugan, M., Satapathy, S.R., Patnaik, A., Yanine, F.F., Applications of sensors in solar energy systems, in: *2020 International Conference on Renewable Energy Integration into Smart Grids: A Multidisciplinary Approach to Technology Modelling and Simulation (ICREISG)*, IEEE, 2020.
34. Hu, Z., Yu, Y., Wang, Z., Sun, W., Gan, D., Han, Z., Price forecasting using an integrated approach. *Proceedings of IEEE International Conference on Electric*

Utility Deregulation, Restructuring and Power Technologies, Hong Kong, China, pp. 28–31, 2004, April.
35. Kim, K., Velas, J.P., Lee, J., Park, J., Shin, J., Lee, K.W., Short-term system marginal price forecasting using system-type neural network architecture. *Proceedings of Power Systems Conference and Exposition*, Atlanta, USA, pp. 1753–1758, 2006, October.
36. Kim, C.I., Yu, I.K., Song, Y.H., Prediction of system marginal price of electricity using wavelet transform analysis. *Energy Convers. Manag.*, 43, 14, 1839–1851, 2002, Sep.
37. Kononenko, I., Estimating attributes: Analysis and extensions of relief. *Proceedings of European Conference on Machine Learning, (ECML)*, Catania, Italy, pp. 171–182, 1994, May.
38. Kumar, V. and Minz, S., Feature selection: A literature review. *Smart Comput. Rev.*, 4, 3, 211–229, 2014, Jun.
39. Kurata, E. and Mori, H., Short-term load forecasting using informative vector machine. *Electr. Eng. Jpn.*, 166, 2, 23–31, 2009, Apr.
40. Ladha, L. and Deepa, T., Feature selection methods and algorithms. *Int. J. Comput. Sci. Eng. (IJCSE)*, 3, 5, 1787–1797, 2011, May.
41. Lawrence, N.D. and Jordon, M.I., Semi-supervised learning via gaussian processes. *Proceedings of Advances in Neural Information Processing Systems 17 (NIPS 2004)*, Vancouver, British Columbia, Canada, pp. 753–760, 2005.
42. Lawrence, N.D., Platt, J.C., Jordan, M.I., Extensions of the informative vector machine. *Proceedings of the 1st International Conference on Deterministic and Statistical methods in Machine Learning*, Sheffield, UK, August, pp. 56–87, 2005, June.
43. Lertworaprachaya, Y., Yang, Y., John, R., Interval-valued fuzzy decision trees with optimal neighbourhood perimeter. *Appl. Soft Comput.*, 24, 1, 851–866, 2014, Nov.
44. Rao, C.K., Sahoo, S.K., Yanine, F.F., Demand response for renewable generation in an IoT based intelligent smart energy management system. *2021 Innovations in Power and Advanced Computing Technologies (i-PACT)*, IEEE, 2021.
45. Min, J. and Lee, Y., Bankruptcy prediction using support vector machine with optimal choice of kernel function parameter. *Expert Syst. Appl.*, 28, 4, 603–614, 2005, May.
46. Moghram, I. and Rahman, S., Analysis and evaluation of five short-term load forecasting techniques. *IEEE Trans. Power Syst.*, 4, 4, 1484–1491, 1989, Nov.
47. Pandey, N. and Upadhyay, K.G., Different price forecasting techniques and their application in deregulated electricity market: A comprehensive study. *Proceedings of IEEE International Conference on Emerging Trends in Electrical, Electronics and Sustainable Energy Systems (ICETEESES)*, Sultanpur, India, pp. 1–4, 2016, March.

48. Peng, X., Rafferty, K., Ferguson, S., Building support vector machines in the context of regularized least squares. *Neurocomputing*, 211, 26, 129–142, 2016, Oct.
49. Peter, S.E. and Jacob Raglend, I., Sequential wavelet-ANN with embedded ANN-PSO Hybrid electricity price forecasting model for Indian energy exchange. *Neural Comput. Appl.*, 28, 8, 2277–2292, 2016, Aug.
50. Peter, S.E., Jacob Raglend, I., Sishaj, P.S., An architectural frame work of ANN based electricity price forecasting engine for indian energy exchange using similar day approach. *Int. J. Res. Eng. Technol.*, 2, 4, 111–122, 2014, Apr.
51. Phyu, T.Z. and Oo, N.N., Performance comparison of feature selection methods. *Proceedings of Third International Conference on Control, Mechatronics and Automation (ICCMA 2015)*, Yangon, Myanmar, pp. 1–4, 2016, September.
52. Pindoriya, N.M., Singh, S.N., Singh, S.K., An adaptive wavelet neural network-based energy price forecasting in electricity markets. *IEEE Trans. Power Syst.*, 23, 3, 1423–1432, 2008, Aug.
53. Pousinho, H.M.I., Mendes, V.M.F., Catalao, J.P.S., Short-term electricity price forecasting in a competitive market by a hybrid PSO-ANFIS approach. *Int. J. Electr. Power*, 39, 1, 29–35, 2012, Feb.
54. Prasanta Kumar, P. and Ghoshal, S.P., Dynamic electricity price forecasting using local linear wavelet neural network. *Neural Comput. Appl.*, 26, 8, 2039–2047, 2015, Feb.
55. Rakotomalala., R., Tanagra: A free for research and academic purposes. *Proceedings of European Grid Conference*, Amsterdam, Netherland, pp. 697–702, 2005, January.
56. Reston Filho, J.C., Affonso, C.M., De Oliveira, R.C.L., Pricing analysis in the Brazilian energy market: A decision tree approach. *Proceedings of IEEE Conference on Bucharest Power Tech*, Bucharest, Romania, pp. 1–6, 2009, June.
57. Reston Filho, J.C., Affonso, C.M., De Oliveira, R.C.L., Energy price classification in North Brazilian market using decision tree. *Proceedings of the 12th International Conference on the European Energy Market*, Lisbon, Portugal, pp. 1–5, 2015, May.
58. Rodriguez, C.P. and Anders, G.J., Energy price forecasting in the Ontario competitive power system market. *IEEE Trans. Power Syst.*, 19, 1, 366–374, 2004, Feb.
59. Saini, L., Aggarwal, S., Kumar, A., Parameter optimisation using genetic algorithm for support vector machine-based price forecasting model in national electricity market. *IET Gener. Transm. Distrib.*, 4, 1, 36–49, 2010, Jan.
60. Sajal, G. and Kakali, K., Modelling and forecasting of day-ahead electricity price in indian energy exchange – evidence from SARIMAEGARCH model. *Int. J. Indian Cult. Manag.*, 8, 3, 413–422, 2014, Feb.
61. Samuel, A., Some studies in machine learning using the game of checkers. *J. Int. Bus. Manag. (IBM)*, 3, 3, 210–229, 1959, Jul.

62. Sansom, D.C., Downs, T., Saha, T.K., Evaluation of support vector machine based forecasting tool in electricity price forecasting for Australian national electricity market participants. *J. Electr. Electron. Eng.*, 22, 3, 227–233, 2003, May.
63. Rao, C.K., Sahoo, S.K., Yanine, F.F., Forecasting electric power generation in a photovoltaic power systems for smart energy management. *2022 International Conference on Intelligent Controller and Computing for Smart Power (ICICCSP)*, pp. 1–6, 2022.
64. Schmutz, A. and Elkuch, P., Electricity price forecasting: Application and experience in the European power markets. *Proceedings of the 6th IEEE Transactions on Power and Experience in the European Power Markets*, Zurich, Switzerland, pp. 1859–1867, 2004, July.
65. Shafie-Khah, M., Moghaddam, M.P., Sheikh-El-Elsami, M.K., Price forecasting of day-ahead electricity markets using a hybrid forecast method. *Energy Convers. Manag.*, 52, 5, 2165–2169, 2011, May.
66. Sharma, V. and Shrinivasan, D., A hybrid intelligent model based on recurrent neural network and excitable dynamics for price prediction in deregulated electricity markets. *Eng. Appl. Artif. Intell.*, 26, 5, 1562–1574, 2013, May.

3

Energy Storage System for Microgrid for Commercial Systems

Anushree Ramanath

Form Energy, California, USA

Abstract

The intermittent nature of renewable energy sources poses technical and economic challenges especially when integrated on a large-scale for commercial applications. This is mostly due to the introduction of significant uncertainties in the planning and operation phases. In an effort to ease this concern, standalone distribution systems with a network of distributed energy resources, called microgrids, are utilized to cater to the loads. They can function independently in an islanded mode or act as a smaller power system by connecting to the main grid while supporting several renewable energy sources. Energy storage plays an important role in microgrids by serving as a secondary-generation voltage control or regulation element. They help cope with the variability resulting due to renewables, capacity flexibility, and power supply reliability. They also enhance the system efficiency while reducing the overall cost. The economic benefits are realized as they store energy when renewables are producing or when electricity prices are low and discharge when the prices are high. There are different types of energy storage systems that can cater to different use cases based on application, sizing, and other requirements. Batteries are some of the commonly used storage devices in a commercial microgrid, and they can be implemented based on a wide variety of chemistries. Thus, modeling the same with the rest of the microgrid can help with finalizing the design aspects, validation, control, and optimization. A critical subsystem, called battery management system, helps with maximizing the system performance while minimizing the power consumption in order to extend the battery life. This chapter provides insights regarding state-of-the-art energy storage systems, storage options used for a commercial microgrid, classification of energy storage, a deep-dive of battery and its chemistries, modeling, sizing, and control of storage systems, and also a glimpse of possible technological advancements in this area.

Email: anushree.ramanath@gmail.com; raman074@umn.edu

P. Sivaraman, C. Sharmeela and P. Sanjeevikumar (eds.) Microgrids for Commercial Systems, (49–72) © 2024 Scrivener Publishing LLC

Keywords: Microgrid, distributed energy resources (DERs), energy storage systems (ESS), renewable energy sources, batteries, battery management system (BMS)

3.1 Introduction

Due to increased environmental damage leading to the climate crisis, the push for choosing renewable energy sources over conventional ones is increasing rapidly. Microgrids are known to integrate several renewable energy resources along with energy storage [1]. Energy storage is essential to reap the maximum benefits out of the renewable sources which are otherwise less reliable. Energy storage can also come in handy to ensure the seamless integration of renewable energy sources in the grid. There have been different classifications of energy storage based on physical form, mechanism in which the energy is stored, and the duration of storage it offers. However, for commercial applications, electro-chemical batteries have been the most popular and widely used option—be it microgrids and large-scale residential or industrial needs. The chemistries and the advantages they offer also vary widely from Li-ion-based batteries, Ni-based batteries, and the more recent Fe-air batteries.

The shift towards microgrids is well justified, as this kind of integration is proven to enhance the controllability, stability, and reliability. Microgrids comprise of interconnected distributed energy resources (DERs), battery energy storage systems, and loads [2]. A single line diagram of a microgrid with some DERs, loads, and energy storage systems is shown in Figure 3.1. Along with the improvements in the field of reaping the maximum benefits out of the renewable energy resources, it is critical to focus on advances in the field of energy storage as they both go hand in hand. It is critical to focus on enhancing the parameters like state of charge of the batteries and battery life while optimizing the degradation rate so that the overall system is cost-effective and efficient. The optimal system would rely on performance factors including, but not limited to, reliability, power quality, and ease of integration.

In this chapter, we will learn more about the nuances of design and implementation of electrical energy storage in microgrids from theoretical aspects to real-world case studies. Starting from the state-of-the-art technologies in Section 3.2, we will dive deep into the significance and recent developments in the field of energy storage systems. Section 3.3 delves into the types of energy storage systems, sizing, and control strategies. It deals with the different types of electrical energy storage, their pros and cons,

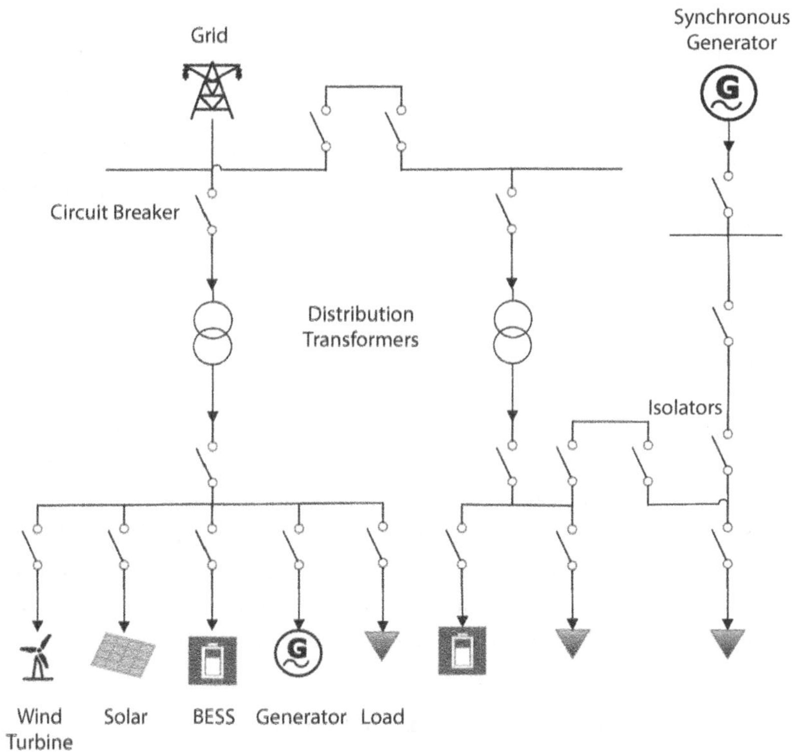

Figure 3.1 Single-line diagram of a microgrid with renewable energy sources, energy storage, and loads.

applications to which they are best suited, and also the design challenges. Furthermore, we will understand the modeling and simulation of such systems which serve as a design verification stage and also a workbench for algorithm development in Section 3.4. Different battery chemistries are also compared and contrasted. We look at some examples pertinent to the design and implementation of energy storage systems for microgrids as well as current trends and future scope as relevant to commercial systems in Section 3.5. Section 3.6 captures the summary of the importance of energy storage systems in a microgrid.

3.2 State of the Art

The electric power and transportation industries form a major portion of the primary energy consumption with a consumption of a few trillion kilowatt hours of energy and emission of a thousand million metric tons of carbon dioxide in the US alone [3]. This is in a span of a few months, which translates to the challenges we are facing for reducing the carbon footprint while pushing for cleaner energy technologies, especially in the commercial sector. The transition to cleaner energy sources is only possible with significant advances in the design and implementation of semiconductors, power electronic components, electric machines, and battery technologies [4].

Energy storage systems are crucial for the operation of microgrids, as they help in compensating for the fluctuations in the energy generation caused due to reliance on renewable energy sources. This also helps in meeting the demand fluctuations as the peak power demand will not always coincide with the peak generation [5]. Over the years, there has been lot of research around enhancing the chemistries of the battery components, advances in power electronic components used for energy conversion, and optimization of battery models to determine the right kind of energy storage for the application in hand. In the process, energy storage systems have become more cost-effective and technologically viable as an integral part of microgrids of various scale and applications.

3.2.1 History of Energy Storage Systems

Since the beginning of the nineteenth century, a number of energy storage devices have been used for several applications. The use of power electronic components in conjunction with traditional energy storage devices is relatively new but dates back to 1960s. The key point to observe is that, since the power electronics-based energy storage devices have been introduced, they have gained significant momentum in replacing the conventional energy storage devices based on mechanical or electrochemical technologies. The reason for the shift is the focus on determining the best possible energy storage type based on high efficiency, reliability, and less environmental impact.

The main disadvantages observed in the conventional solutions were high cost, low reliability, low conversion efficiency, slow response time, bad power quality, bulky size or low power density, and less tolerance to extreme environmental conditions [6]. Power electronics-based storage solutions helped overcome some of these while reducing the issues on other fronts, thus becoming the choice of system for both low- and high-voltage applications in microgrids. Based on the charge and discharge requirements of the microgrids, storage devices can be chosen. They also offer flexibility in terms of operation as they can be connected in series or parallel and can be controlled to serve as voltage or current generators based on the application in hand.

3.2.2 Significance of Power Electronics-Based Systems in Energy Storage

Power electronics-based energy storage devices are one of the fastest growing technologies that help in providing ancillary services and solving power problems while enabling access to clean energy-based technologies for multiple segments. They have proven to be advantageous for use in a wide variety of segments spanning from renewable energy sources, electric vehicles to heavy machinery and a range of dynamic loads. They are flexible, compact, modular, and cost effective and offer superior performance compared to its counterparts. They can be used for critical applications as they can respond instantaneously, have good stability, are robust, are reliable, and have less losses comparatively. They also comply with most of the power quality standards and grid codes, thus making it very sustainable for use.

The introduction of energy storage devices with power electronics-based switches revolutionized storage as they started serving as replacement technology for conventional solutions like mechanically switched energy storage media and standby generators that have traditionally faced issues with power quality and adherence to codes and standards. Not to forget the fact that the conventional solutions have other disadvantages like limited power quality, less efficient, slow response time, and extreme bulkiness. Moreover, with the cost factor coming into play, the cost of power electronics-based switches started becoming more effective, thus facilitating the transition to these devices in low- and high-voltage applications.

3.2.3 Recent Developments in Storage Systems for Microgrids

There has been significant research related to battery energy storage systems and their evolution mostly over the last decade. There has been significant progress made both in hard core research and in terms of advanced technological implementation. Based on a detailed study over the last 7 years [7], it is observed that significant research had evidence related to energy storage optimization methods in 2016. In 2017, the hottest area of research was related to lithium-ion-based batteries, which is one of the main types used in microgrids and electric vehicles due to its high energy density.

The year 2018 saw a great deal of research around the topics related to battery technology, hybrid energy storage systems, and electrochemical cell models. Electrochemical cell models were studied extensively due to the fact that lead acid-based batteries are one of the most commonly used storage technologies for grid-based applications.

More recently, 2019 saw a huge drift of articles focusing on energy storage systems catering to microgrids and their optimization. Technological advancements related to battery management systems that facilitate the monitoring, control, and performance optimization of battery modules inside the energy storage system also gained traction. The timing was perfect as that is when microgrids attracted high interest from several application areas due to their ability to integrate and serve as a network of several renewable energy resources, grid, loads, and storage. In 2020, hybrid energy storage systems gained attention as they were able to utilize multiple energy sources at the generation end and service several types of loads at the consumption side. Lithium-ion-based batteries still held the center stage and remained to be the key energy storage option especially for increasing energy resilience in solar installations.

In 2021, the focus on lead acid batteries was noted to be significant again along with storage options to cater to renewable energy resources, mostly wind energy. With the research around integrating solar and storage in the previous year followed by harnessing of wind energy with storage in 2021, it was apparent that energy storage systems need to complement renewable energy sources in order to make them viable solutions, both technologically and economically. It also helps in mitigating the risk of energy fluctuations due to increased reliance on renewable energy resources and the possible power shortage during peak demand intervals. The major areas of application for battery energy storage systems were deemed to be microgrids, electric vehicles, and small-scale residential installations. There was also interest in determining the quality of batteries by means of state of charge (SoC) and state of health (SoH) values, implementing analytical tools to help decide the type of chemistry based on application and perform optimizations and developing models and simulations to better understand the performance parameters of energy storage devices.

3.3 Energy Storage Systems

The classification of energy storage systems (ESS) can be accomplished in multiple ways. Most commonly, ESS are classified as electrochemical, electrical, mechanical, thermal, and hybrid energy storage systems. They can be further classified based on the process of implementation, materials used, or end application. They are also classified as compressed air energy storage, batteries, flywheel energy storage, hydrogen fuel cell, superconducting magnetic energy storage, and combinations of these for hybrid storage systems. These are the most commonly employed forms for microgrid applications. The focus of further discussion in this chapter is mostly electrical energy-based storage systems which include batteries, ultracapacitors, and fuel cells. While batteries are the most commonly used storage devices, others serve as supplementary storage options in case of hybrid energy storage systems. Batteries are most commonly used as the energy can be stored by transforming electrical energy into any other form like chemical or mechanical energy.

In recent years, the battery energy storage systems for microgrids have mostly lithium-ion batteries, lead acid batteries, and redox flow batteries. Lithium-ion-based batteries possess high energy and power densities, making them an ideal candidate for microgrid-like stationary and automotive applications [8]. On the other hand, lead acid batteries that have been used traditionally have a decent efficiency range but has an advantage of lower cost, making it suitable for large-scale stationary back-up applications [9]. Redox flow batteries are more recent and are geared towards enhancing renewable energy source integration with the grid [10]. These have features such as low self-discharge, fast response characteristics, and longer lifetime, making them ideal for stationary applications like a commercial microgrid.

3.3.1 Definition and Classification

The modern power electronics-based energy storage devices have several advantages compared with their traditional counterparts. They can also be controlled to act as current or voltage generators, having an energy storage media that is capable of providing active power for a stipulated amount of time as required. It can also be connected either in series or in parallel with the power system based on the application. Power electronics-based energy storage systems are broadly classified into electrical energy storage systems, electrochemical energy storage systems, mechanical energy storage systems, and hybrid energy storage systems as shown in Figure 3.2. Hybrid energy storage systems typically combine two or more other types of storage systems in order to achieve better performance, added advantages, and ideal characteristics to cater to specific applications—for example, in the case of a microgrid, there is a need for high power and high

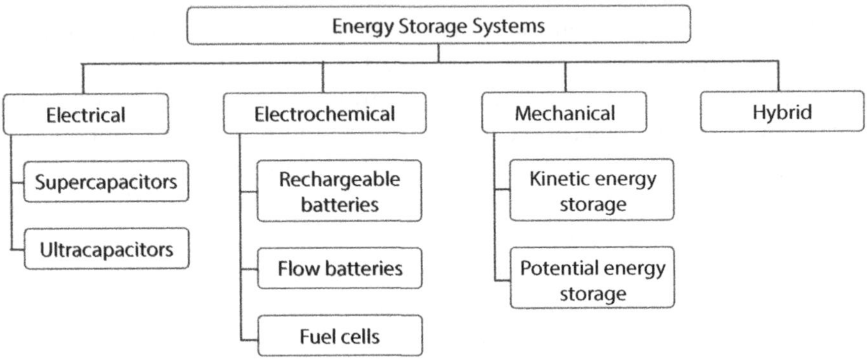

Figure 3.2 Classification of energy storage devices.

energy density to enhance the overall stability and reliability, which can be achieved by a combination of different energy storage systems.

Energy storage technologies have been constantly evolving. They also have diverse classifications based on their storage media, based on the functions or services they provide, and so on. Based on the storage media, they can be classified into electrical—including capacitors, supercapacitors, and magnetic energy storage, electrochemical—including rechargeable batteries, flow batteries, and fuel cells, and mechanical—including potential energy storage and kinetic energy storage. Based on the functionalities that they support, they can be categorized based on their energy density, their power density, their cycle time, their efficiency, or their response time. They can be configured as aggregated or distributed systems based on the type of microgrid application. For cases where the control of large energy storage systems is difficult and also when they are hard to manufacture, small-scale and distributed storage systems come in handy to help attain reliable and effective power regulation. In a distributed configuration, there are multiple interfaces that are connected to numerous resources, which makes the control of power flow extremely challenging.

Electrochemical storage devices convert chemical energy into electrical energy by means of a chemical reaction. The stored energy is in the form of electric current stored at a particular voltage for a defined time duration, which is achieved by a series or parallel combination of cells. Conventional rechargeable batteries and flow batteries are two types of devices that come under electrochemical storage and are implemented in several form factors to cater to a wide array of applications. The chemical reaction may reduce the lifetime, as it poses environmental adversities, but the maintenance required for the operation of these is minimal. They include two main categories—flow batteries and battery energy storage systems. Some of the well-known examples include lithium-ion batteries, lead acid batteries, nickel–cadmium batteries, nickel-metal hydride batteries, and sodium–sulphur batteries.

Mechanical energy storage systems can be advantageous as they can operate flexibly and convert energy from several other forms of energy. The most commonly used ones are flywheel, pumped hydro, and compressed air. Chemical storage systems are ideal for long-duration storage of high capacity. They mostly focus on hydrogen technology as it can store a large amount of electrical energy and is referred to as hydrogen fuel cell storage systems. Electrical storage systems store electrical energy and have the ability to produce electricity and supply the same to cater to the load based on the demand. This can be accomplished using electrical or magnetic means and thus can serve additionally to help support the microgrid at times of peak demand and also improve the power quality. Ultracapacitors and super magnetic energy storage systems are some examples for this type and are ideal as supplementary sources of energy at the end of the power plant to help stabilize the output or in case of peak consumption in the industrial sector [11]. Thermal energy storage systems have the capability to store energy in the form of heat or ice and then release them later based on the need. This is suitable for some industrial and residential applications like heating or cooling systems.

The design of power electronic systems has been focused towards specific functionalities. Under the grid code compliance, they have been used for power quality improvement and to cater to ancillary services. This translates to the ability of providing protection against power supply interruptions, compensating voltage-based distortions like voltage harmonics, voltage fluctuation, and voltage imbalances, and current-based distortions like current harmonics and current imbalances. In terms of the ancillary services provided, it includes voltage support, frequency support by means of containment, reactive power control, black start capability, system restoration after islanding, fault ride through capability, and others. The other major aspect is the ability to support the development of clean energy by means of backup power supply, peak shaving, and load shifting.

3.3.2 Sizing of Primary Storage

The optimal sizing and operation of storage devices is more important than the choice of right type. If the storage is undersized, it cannot cater to all the necessary operations of the microgrid and, on the other hand, if it is oversized, it will add on to the capital costs, overall system size, and effectiveness of system operation [12]. However, it is also true that adding energy storage itself helps in cost reduction even though it is not optimally sized. The sizing of an energy storage system is optimizing on size or capacity such that minimum cost is achieved. It is crucial to understand that the investment costs increase linearly with the size of the energy storage systems, while operation costs decrease exponentially, making the objective to find the right size of the system at a minimum cost.

Several optimization techniques are used to size energy storage systems appropriately. Mixed integer linear programming (MILP), dynamic programming, and particle swarm optimization are some of the most commonly used ones. The solution of the optimization problem, sometimes with added reliability constraints, depends solely on the storage system application. However, MILP is the most commonly used technique, in general, as it is known to provide the most accurate results. Dynamic programming, on the other hand, is found to take a much longer time and has constraints on the number of buses in the system under study. Sizing of energy storage systems has been complicated due to the independent power and energy ratings inherent in battery storage systems (BSS). Sizing a BSS for both required power output and energy storage capacity requires in-depth analysis to cater to optimal scheduling capabilities while keeping the investment costs at a minimum [13].

3.3.3 Supplementary Storage

The wide use of sophisticated microgrids with energy storage systems in commercial applications is limited mostly because of high form factor, increased refueling or recharging time, and high cost. All three of these concepts are closely tied to the energy storage system. As we have been discussing, the main considerations for any energy storage device for commercial microgrid applications are power density, energy density, weight, compactness, cycle life, and cost.

The other considerations include safety, recycling capability of materials used, maintenance requirements, and operating temperature. The main energy storage device is battery for most of the microgrids. The other supplementary storage devices used in conjunction with batteries are ultracapacitors and fuel cells in small-scale hybrid energy storage systems. Figure 3.3 shows the cross-section of an individual ultra-capacitor (UC) and a hydrogen fuel cell (FC).

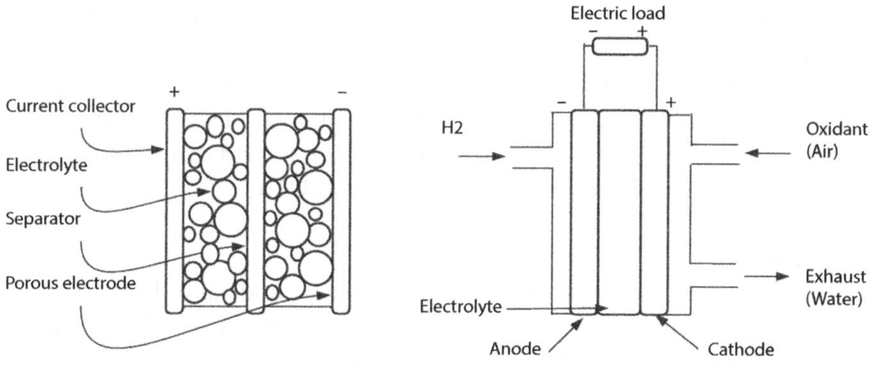

Figure 3.3 Cross-section of an individual UC and a hydrogen FC [14].

Ultra-Capacitors

Ultra-capacitors store energy by physically separating positive and negative charges. The setup is of two parallel plates that store charges and are separated by an insulator. The positive plate functions as a current collector, and the plates are porous in nature. The insulator serves as a separator and is surrounded by electrolyte. The applied potential on the positive electrode attracts the negative ions in the electrolyte, whereas the potential built-up on the negative electrode attracts positive ions. UCs typically have a long life cycle due to the absence of chemical variations on the electrodes but suffer from low energy density. The power density of the UC is considerably higher than that of the battery as the charges are physically stored on the electrodes.

The terminal voltage of the UC is directly proportional to the SOC. This makes it a good backup energy storage device especially for small-scale hybrid energy storage systems. When the required total power is relatively low in the case of short power outages or in between transitioning to other sources of energy seamlessly, UCs are the best choice of energy storage. This is feasible due to its very fast charge and discharge rates. The development of power electronics interface also allows the UC to operate well throughout its variable voltage range. There are several UC technologies that are being actively researched and developed, including carbon or metal-based fiber composites, foamed carbon, a carbon particulate with a binder, doped conducting polymer films on a carbon cloth, and mixed metal oxide coatings on a metal foil [15].

Higher energy density can be achieved by the use of carbon composite electrode by using an organic electrolyte rather than using a carbon or metal fiber composite electrode with an aqueous electrolyte. All these technologies still point toward UCs having higher power densities than batteries which have higher energy densities. The long lifetime and low maintenance requirement of UCs also make it cost-effective and an ideal candidate for the combined use with batteries in hybrid storage systems to maximize the benefits of both the batteries and UCs.

Fuel Cell

The FC generates electricity from the fuel on the anode and the oxidant on the cathode while reacting in the electrolyte. During the electricity generation process, the reactants flow into the fuel cell, whereas the products of the reaction flow out. The advantages of using the FC include long life, reliability, high efficiency, less emission, quiet operation, and ability to recover waste heat. One of the most popular types of FC is a hydrogen FC that has hydrogen as the fuel. It is advantageous as hydrogen is nonpolluting, has highest energy density compared to other fuels, and the end product of cell reaction is water. Other combinations of fuels and oxidants are also possible for FCs like hydrocarbons and alcohols as fuels with chlorine and its oxide form as oxidants.

The efficiency of a FC is directly dependent on the amount of power drawn from it, which means that the higher the power drawn, the lower the efficiency of the FC. The reactants must also be refilled before they are emptied, making it necessary to carry a fuel tank in house, which is not very convenient. The response time of FCs is relatively longer compared to that of batteries and UCs, while it costs a fortune more—about five times the cost of conventional systems, making it a difficult choice for use in hybrid energy storage systems even though it is technologically feasible at this point in time [16].

3.3.4 Control Strategies

The energy storage systems can be categorized into three stages, including central storage stage, power transformation stage, and control stage [17]. Energy that is obtained after a suitable form of conversion is stored in the central storage. The power transformation stage functions as an interface between the central stage and the power grid with bidirectional power transfer capability. The control stage helps with the modulation by means of issuing commands for the operation of relevant components in a timely manner, determination of charge or discharge levels using sensors, and providing support in monitoring the health of the system by means of measuring critical electric performance parameters.

The battery-based energy storage systems typically require bidirectional AC–DC converters and DC–AC converters for its operation. This demands highly efficient converters in place for the EES along with sophisticated controls to cater to the efficient operation of the system [18]. However, the electric power sources for storage systems in a commercial microgrid setting are now including batteries, UCs, FCs, and other energy storage systems along with it. With multiple sources of energy, the design of intelligent control strategies is crucial, inevitable, and also a critical issue.

The control strategies employed depends on the type of storage system chemistry, application, and size of microgrid and accompanying storage system, along with several other factors. However, most of the strategies of the past have focused on improving the battery life and overall system efficiency. It is important to also consider peak power demand, duration of critical operation at the time of an outage, and SoC of all the power sources employed along with batteries. This makes the design of optimal controls a very important part of the design and implementation of any commercial microgrid. Moreover, the control strategy in case of hybrid energy storage systems is way more complicated than individual systems [19].

3.4 Batteries for Microgrids in Commercial Applications

Battery energy storage system is the primary choice for energy storage in commercial microgrids. There have been several studies related to hybridizing batteries with other supplemental sources like supercapacitors or fuel cells. This is to ensure that the batteries take care of the power demand while the supercapacitors offer high energy density. This ensures that large capacity is obtained along with fast charging and discharging characteristics [20]. The result is also enhanced overall system efficiency and increased life expectancy.

3.4.1 Battery Chemistry

Batteries typically have higher energy densities and help store majority of the electric energy, whereas ultracapacitors can be used along as a hybrid source due to its high power density, reliability, and faster response time. The table below highlights the characteristics of most commonly used batteries in microgrids. The most commonly used battery types are discussed in the following subsections.

Characteristics of commonly used batteries in microgrids [21].

Battery type	Lead acid	NiMH	Lithium-ion	NaS	Vanadium redox
Capacity (MWh)	0.25–50	0.25–25	0.25–25	≤300	≤250
Power (MW)	≤100	≤100	≤100	≤50	≤50
Discharge time (h)	≤4	≤2	≤1	≤6	≤8
Lifetime (y)	≤20	≤15	≤15	≤15	≤10
Efficiency (%)	≤85	≤80	≤90	≤80	≤80

Lead-Acid Batteries

Lead-acid (PbA) batteries are available commercially and quite mature in terms of their technology, thus making them one of the simplest choices. It is known to be the most widely used rechargeable storage option with a wide range of form factors and designs to cater to different applications. Lead oxide serves as the positive active material, while spongy lead works as the negative active material with diluted sulfuric acid as the electrolyte [22]. They are less expensive than their Ni-based counterparts and deemed to be fit for large-scale microgrid applications. They also possess excellent charge retention capability along with high energy density, long life cycle, and fast response time. However, it is found that at a deep rate of state of charge, the battery's life cycle reduces drastically, making it less fit for higher discharges of its rated capacity [23].

Nickel-Metal Hydride Batteries

The nickel-metal hydride battery is known to have double the energy density as compared to the lead-acid battery [24]. It is also safe to operate at a high voltage, has wide operating temperature range, and has a long cycle life while facilitating a safer operation as the batteries can be recycled without harming the environment. It is composed of nickel hydroxide on the positive electrode and engineered alloy of vanadium, titanium, nickel, and other metals as the negative electrode, with alkaline solution as the electrolyte [14]. It has been the most commonly used energy storage system for automotive applications but lesser for large-scale microgrids. However, it is a candidate for hybrid energy storage systems of a smaller scale.

Other Nickel-Based Batteries

Nickel–zinc batteries have high energy and power densities, deep cycling capability, are recyclable, and have the ability to operate under severe temperatures. However, they suffer from poor life cycles. Nickel–cadmium batteries have a good range of operation, are recyclable, and have a long lifetime, but it is expensive to be used in large-scale commercial microgrid applications, and the cost does not seem to be outweighed by any other performance factors.

Lithium-Ion Batteries

Li-ion batteries have high energy density, are recyclable, have a long battery life, and have the ability to sustain high external temperatures. It was commercialized in the 1990s and has become one of the fastest growing technologies in the recent years. It is capable of storing energy at the megawatt scale along with high efficiency, high energy density, and rapid response time [25]. It has been proven to have great performance in other critical fields like medical devices, safety equipment, and consumer electronics. The positive electrode is made of an oxidized cobalt material, and the negative electrode is made of a carbon material with lithium salt in an organic solvent used as an electrolyte. This has huge potential and serves as the best option in terms of safety and compliance with regulatory requirements, making it an ideal candidate for residential, automotive, and medium-scale microgrid applications [26].

Sodium–Sulphur Batteries

Sodium–sulphur batteries are made of molten electrodes, with the sulphur-treated ones as the positive electrode and the sodium-treated ones as the negative electrode, and non-aqueous beta alumina as the electrolyte. This technology is known to be suitable for voltage sag minimization and stabilizing renewable energy power generation in commercial microgrid applications [27]. It seems to operate at a high efficiency and offers a long cycle period and fast response time.

Iron-Air Batteries

There are several recent developments with respect to iron-air batteries that can fill the gap of missing long-term energy storage solution. Iron is submerged in an electrolyte solution, and the process of natural rusting is the phenomenon behind the charge and discharge cycles. The active material in this case is the oxygen in the air. This kind of battery seems to be more reliable, cost effective, and environment-friendly. If commercialized at a large scale, it is ideal for large-scale microgrids and data center applications.

3.4.2 Modeling and Simulation of Batteries

Due to the ever-growing applications of battery storage in the commercial sector, there is a constant need for innovating and experimenting with different battery chemistries. Simulation is one of the most commonly used techniques to evaluate the design cost effectively and reduce the development cycle time [28]. This enables efficient parallel operation as the same model can be deployed on multiple setups to evaluate different use cases without the need for expensive test and measuring equipment. There are several types of modeling work that has been done in the past, including the mathematical ones, electrochemistry-based models, physics-based models which mostly use equations, and electrical equivalent models that use circuits for analysis. The goal is to build models that are simple, accurate, and as close to reality as possible.

An example for the use of battery models can be that of a battery management system (BMS) implementation in a commercial microgrid. BMS typically has a limited amount of memory and computational ability, thus demanding a simple yet accurate model representation of the same [29]. This model could come in handy for designing algorithms that predict the state of the battery in terms of SoC and also helps with the implementation of controls. Thus, there is a possibility of building different types of models that cater to different end uses based on the fidelity level. However, it is vital to focus on the trade-offs between complexity, performance, and parameterization efforts.

The most commonly used electrical battery models are shown in Figure 3.4. The models include the open circuit voltage, resistance of the contacts, electrodes, and the electrolyte, and the resistances and capacitances that characterize the transient response of the battery cell electrodes. The parameterization efforts for these models can be carried out either in a linear manner or by use of look-up tables. These result in differences in terms of accuracy and complexity which is driven by the use-case in hand. The linear models typically require significantly less parameterization time but does suffer from a small loss of accuracy in the output. The battery model presented with one resistor and capacitor combination, which uses linear parameterization of the internal impedance, seems to offer the best trade-off between the parameterization efforts and the accuracy of the results.

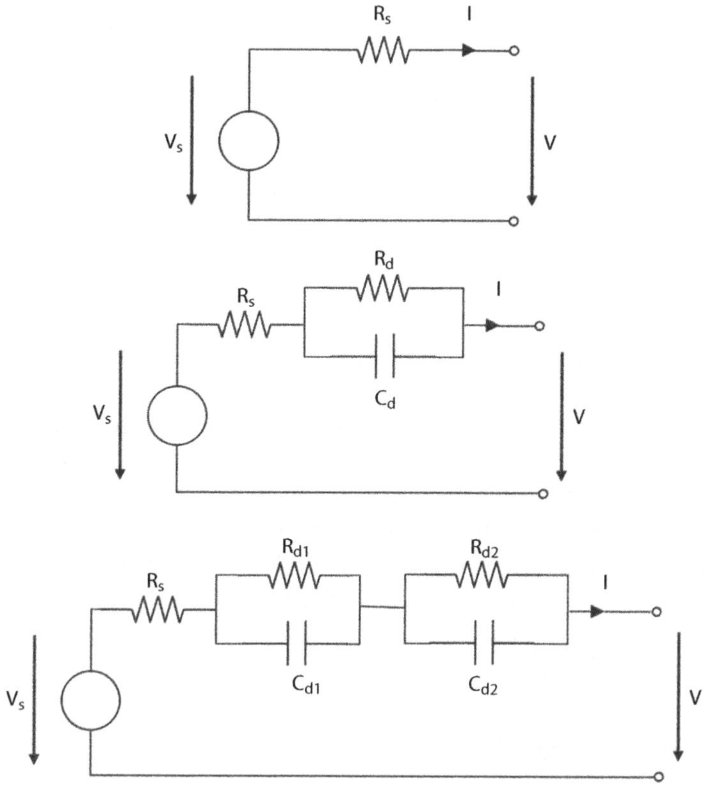

Figure 3.4 Battery models [30].

There are more complicated models based on hybrid energy storage systems and also based on the microgrid configuration that houses the energy storage system.

3.4.3 Battery Management System

The goal of BMS is to achieve maximum system performance while minimizing the power consumption in order to extend the battery life. In order to achieve these objectives, it is crucial for BMS to accurately communicate the state of the battery to the system controller. The key functions performed by BMS include undervoltage protection, overvoltage protection, short-circuit protection, or maximum current limit and thermal protection [31]. The additional functions performed by BMS may include monitoring of SoC, SoH, and state of function and cell balancing or equalization on the cell or module level [15].

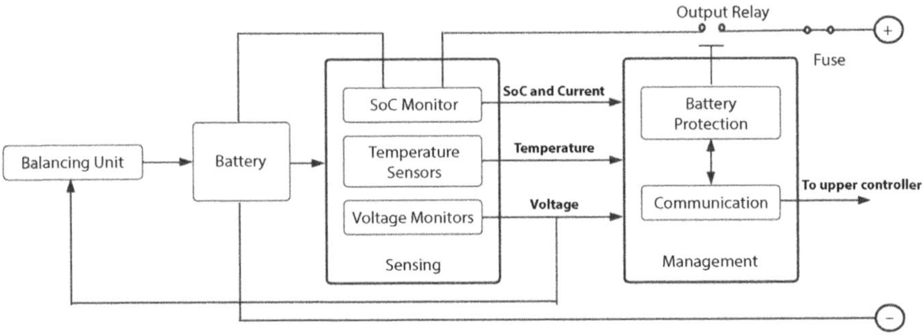

Figure 3.5 Block diagram of a generic BMS.

The block diagram of a generic BMS is shown in Figure 3.5 [15]. The voltage information gathered by the sensing subsystem is sent to the balancing circuit, which uses the information to ensure that all cell voltages are within the allowed limit. The balancing is typically done on the module level rather than on the cell-to-cell level. The sensing block also records temperature and current values. Voltage, current, and temperature information along with the battery history are then used to determine the state of charge of the battery.

3.5 Future Trends

The future of commercial microgrid battery could be based on lithium air technology. These batteries have an anode made of lithium and an air cathode made of a porous material that draws in oxygen from the surrounding air. When lithium combines with oxygen, it forms lithium oxide and releases energy. As oxygen does not have to be stored in the battery, the cathode is much lighter than that of a lithium-ion battery, which gives lithium-air batteries their high energy density. These batteries seem to have the capability to significantly increase the longevity and reliability due to their inherent high energy density. For long duration storage, there is a lot of research ongoing related to iron–air chemistry for the batteries. These aimed to serve at the generation end for long duration storage when integrated with renewable energy sources and also as back-up for data center kind of applications.

3.5.1 Energy Storage System Challenges

One of the major bottlenecks for increasing the implementation of large-scale commercial microgrids is the effective scaling of energy storage systems.

The most common issues related to rechargeable lithium cells is that of cell equalization where the capacity and internal resistance of all the cells stacked up are not equal, leading to numerous balancing and performance issues. Power electronics-based battery cell voltage equalizers help in battery cycle life improvement and support increase in power, reliability, and safety. It is a controller that helps equalize the voltage in each cell by continuously measuring the actual capacity, internal resistance, and SoC. This helps ensure that the SoC of each cell is the same during charging and discharging, thus ensuring that they will degrade equally over time. The equalizers can be resistive, inductive, or capacitive in nature.

Hybrid energy storage systems have proven to be better than some of the individual storage systems in terms of enhanced efficiency, storage capacity, and energy density. However, there is fear of complexity and calendric and cyclic aging. Materials used to build the batteries directly affect the performance as they determine the life cycle of the storage device [32]. The material cost and ease of obtaining raw materials also influence the rate of production and thus the manufacturability of necessary components or subsystems. The power electronic interface also directly influences the power quality, durability, and efficiency of the system [33]. Advanced research is essential to overcome the prevalent shortcomings like ripples, high cost, and size disadvantage [34]. Size and cost are mostly proportional in the case of many energy storage systems. It is essential to consider a sweet spot where power is balanced such that cost is reduced and reliability is increased for microgrid applications.

Safety is of utmost importance and is one of the critical issues in an electrical energy storage system. It is inevitable to adhere to standards and comply with regulatory requirements while ensuring that the materials used are of the highest quality and that safety is carefully monitored at every stage of development and installation. All the critical battery parameters like cell level voltages, currents, SoC, overall temperature, and others need to be monitored along with taking necessary corrective measures when the values are out of range. It has also been found that the cell equalizers not only help with maintaining the SoC and cycle life of the cells but also improve the overall calendar life while ensuring system safety.

3.5.2 Technological Advancements

Extensive research is focused on sizing or the form factor of the batteries, cost effectiveness, and efficient energy management when integrated with renewable energy sources as in a commercial microgrid. The key challenges remain to be around material selection, power electronic interfacing, and energy balancing. Other significant issues are related to manufacturing capabilities, reliability, robustness of the technology, and so on. Since the development of commercial microgrids is driven by environmental concerns, the recyclability of the materials used in the energy storage system is also an important consideration along other factors that have an environmental impact. Moreover, there are concerns around charge equalizing, safety, and reliability. It is important to holistically address these concerns while designing, developing, and implementing the next generation of energy storage systems.

Several battery companies, utilities, and other manufacturers are investing heavily to be able to build cheaper, denser, and compact batteries. The innovation in terms of possible battery chemistries is still ongoing. Research includes building up on the existing battery chemistries, looking at newer materials, or modifying the form factor or battery assembly aspects to realize performance gains or cost advantages, even if incremental. There is also a scope for taking a deeper look at other energy storage technologies—batteries, electrochemical storage solutions, flywheels, advanced chemistry batteries, flow batteries, and different grades of capacitors to act as primary energy storage systems in varied scales of commercial microgrid installations.

3.6 Summary

Energy storage system is a critical component of a commercial microgrid and is one of the weakest points as well. Even though technological advances in power electronic design, cost reduction in semiconductors, and enhanced reliability have made microgrids and renewable energy integration with storage more promising, the detection of accurate battery health and its SoC have remained as issues of concern.

The choice of battery or energy storage system is thus a crucial aspect of the microgrid due to increased focus on electrical performance, reliability, and safety of operation. The ideal characteristics for a battery used in microgrid applications are high power capacity, high energy density, high charge–discharge efficiency, long life, ability to sustain a wide range of temperatures, good load characteristics, fast charging, minimal self-discharge, low internal resistance, good recycle ability, low cost, high reliability, and safety.

Section 3.2 provides an overview of the state-of-the-art techniques used for energy storage while touching upon the history behind its development over the years and the power electronics involved in building energy storage devices. It highlights the disadvantages in conventional solutions, thus emphasizing the path of research and development in this area thus far. It also highlights the recent developments in storage systems for microgrids over the past decade while elaborating on research focuses and progress in implementation, manufacturability, and adoption of the same.

Section 3.3 describes the energy storage systems by elaborating on the advantages that they offer, the applications in which they are essential, and the control strategies essential for their effective operation. It also delves deeper into the classification, sizing, and application of these systems in a microgrid setting. Energy storage systems, in general, are electrical, mechanical, electromechanical, or hybrid in nature. Each of these have their own advantages and disadvantages which have been contrasted. Commonly employed supplemental storage devices are also explained in detail. This gives an overview of the kind of batteries and other storage systems being used for commercial microgrids individually or in a hybrid configuration.

Section 3.4 delves into the details of the electrical energy storage by contrasting the different battery chemistries used for microgrid applications followed by the design and creation of battery models for microgrids. The most commonly employed batteries as primary energy storage devices in microgrids include lead-acid, lithium-ion, nickel-based, and vanadium redox flow batteries. It also discusses the battery management system that is used to control the necessary performance parameters. The goal of battery management system is to achieve maximum system performance while minimizing the power consumption in order to extend the battery life.

The future trends of energy storage systems for commercial microgrids are highlighted in Section 3.5. Current challenges for energy storage system development include safety concerns, monitoring of critical battery parameters, and cell balancing. Technical solutions for these concerns are emphasized and prioritized for analysis and implementation. Looking forward, research and development focused towards overcoming challenges around material selection, interfacing of power electronics, energy balancing, and other research endeavors aimed to address safety and environmental issues are proposed.

References

1. Xie, C. *et al.*, Optimal sizing of battery energy storage system in smart microgrid considering virtual energy storage system and high photovoltaic penetration. *J. Clean. Prod.*, 281, 2021. https://www.sciencedirect.com/science/article/abs/pii/S0959652620353531

2. Delfino, F. *et al.*, *Microgrid design and operation toward smart energy in cities*, Artech House, Norwood, MA, 2018.
3. USA US Dept., *U.S. climate action report*, Washington DC, U.S. Department of State, 2014.
4. Boukettaya, G. and Krichen, L., A dynamic power management strategy of a grid connected hybrid generation system using wind, photovoltaic and fly-wheel energy storage system in residential applications. *Energy*, 71, 148–159, 2014.
5. Baldinelli, A. *et al.*, Economics of innovative high capacity-to-power energy storage technologies pointing at 100% renewable micro-grid. *J. Energy Storage*, 28, 2020. https://www.sciencedirect.com/science/article/abs/pii/S2352152X19308977
6. Switzerland Int. Electrotech., *Electrical energy storage: White paper*, pp. 1–78, Commission Geneva, Geneva, Switzerland, 2011.
7. Zarate-Perez, E. *et al.*, Battery energy storage performance in micro-grids: A scientific mapping perspective. *Energy Rep.*, 8, 9, 259–268, 2022.
8. Van Mierlo, J., Kebede, A.A., Kalogiannis, T., Berecibar, M., A comprehensive review of stationary energy storage devices for large scale renewable energy sources grid integration. *Renew Sustain Energy*, 159, 2022. https://www.sciencedirect.com/science/article/pii/S1364032122001368
9. Sanders, M., The rechargeable battery market and main trends 2016-2025, in: *The Battery Show North America 2017*, 2017.
10. Kang, J. *et al.*, Comparison of comprehensive properties of Ni-MH (nickel-metal hydride) and li-ion (lithium-ion) batteries in terms of energy efficiency. *Energy*, 70, 618–625, 2014.
11. Guney, M.S. and Tepe, Y., Classification and assessment of energy storage systems. *Renew. Sustain. Energy Rev.*, 75, 1187–1197, Aug. 2017.
12. Abdulgalil, M.A. *et al.*, Sizing of energy storage systems to enhance microgrid reliability, in: *2018 IEEE International Conference on Industrial Technology (ICIT)*, pp. 1302–1307, 2018.
13. Masaud, T.M., Eluyemi, F., Challoo, R., Optimal sizing of battery storage systems for microgrid expansion applications, in: *2018 IEEE Power & Energy Society Innovative Smart Grid Technologies Conference (ISGT)*, pp. 1–5, 2018.
14. Khaligh, A. and Li, Z., Battery, ultracapacitor, fuel cell, and hybrid energy storage systems for electric, hybrid electric, fuel cell, and plug-in hybrid electric vehicles state of the art. *IEEE Trans. Veh. Technol.*, 59, 6, 2806–2814, 2010.
15. Lukic, S.M. *et al.*, Energy storage systems for automotive applications. *IEEE Trans. Ind. Electron.*, 55, 6, 2258–2267, 2008.
16. Zakeri, B. and Syri, S., Electrical energy storage systems: A comparative life cycle cost analysis. *Renew. Sustain. Energy Rev.*, 42, 569–596, 2015.
17. Faisal, M. *et al.*, Review of energy storage system technologies in microgrid applications: Issues and challenges. *IEEE Access*, 6, 1–1, May 2018.

18. Wang, P., Xiao, J., Setyawan, L., Hierarchical control of hybrid energy storage system in DC microgrids. *IEEE Trans. Ind. Electron.*, 72, 4915–4924, 2015.
19. Hemmati, R. and Saboori, H., Emergence of hybrid energy storage systems in renewable energy and transport applications - A review. *Renew. Sustain. Energy Rev.*, 16, 11–23, Nov. 2016.
20. MacKay, D. and Winser, N., *The future role for energy storage in the UK main report*, Energy Res. Partnership, London, U.K., June 2011.
21. Palizban, O. and Kauhaniemi, K., Energy storage systems in modern grids - Matrix of technologies and applications. *J. Energy Storage*, 6, 248–259, May 2016.
22. Spanos, C., Turney, D.E., Fthenakis, V., Life-cycle analysis of flow assisted nickel-zinc, manganese-dioxide, and valve-regulated lead-acid batteries designed for demand-charge reduction. *Renew. Sustain. Energy Rev.*, 43, 478–494, 2015.
23. Cooper, A. and Moseley, P., Progress in the development of lead-acid batteries for hybrid electric vehicles, in: *2006 IEEE Vehicle Power and Propulsion Conference*, pp. 1–6, 2006.
24. Wehrey, M.C., What is new with hybrid electric vehicles. *IEEE Power Energy Mag.*, 2, 6, 34–39, 2004.
25. Kousksou, T. *et al.*, Energy storage: Applications and challenges. *Sol. Energy Mater. Sol. Cells*, 120, 59–80, Jan. 2014.
26. Lippert, M., Li-ion energy storage takes microgrids to the next level. *Renew. Energy Focus*, 17, 4, 159–161, 2016.
27. Kawakami, N. *et al.*, Development and field experiences of stabilization system using 34 MW NAS batteries for a 51 MW wind farm, in: *Proc. IEEE Int. Symp. Ind. Electron*, pp. 2371–2376, July 2010.
28. Chan, H.L., A new battery model for use with battery energy storage systems and electric vehicles power systems, in: *2000 IEEE Power Engineering Society Winter Meeting. Conference Proceedings (Cat. No.00CH37077)*, vol. 1, pp. 470–475, 2000.
29. Linden, D. and Reddy, T.B., *Handbook of batteries*, vol. 3, McGraw-Hill, New York, NY, 2001.
30. Tremblay, O., Dessain, L.-A., Dekkiche, A.-I., A generic battery model for the dynamic simulation of hybrid electric vehicles, in: *2007 IEEE Vehicle Power and Propulsion Conference*, pp. 284–289, 2007.
31. Affanni, A. *et al.*, Battery choice and management for new-generation electric vehicles. *IEEE Trans. Ind. Electron.*, 52, 5, 1343–1349, 2005.
32. Guerrero, J.M. *et al.*, Advanced control architectures for intelligent microgrids—Part I decentralized and hierarchical control. *IEEE Trans. Ind. Electron*, 60, 1254–1262, Apr. 2013.
33. Rashid, M.H., *Power electronics handbook*, vol. 3, Butterworth-Heinemann, Portsmouth, NH, 2011.
34. Lin, P. *et al.*, An integral droop for transient power allocation and output impedance shaping of hybrid energy storage system in DC microgrid. *IEEE Trans. Power. Electron.*, 33, 6262–6277, July 2017.

4

Emerging Topologies of DC–DC Converters for Microgrid Applications

Stutee Patra, Aditi Chatterjee* and K.C. Patra

Department of Electrical Engineering, C V Raman Global University, Bhubaneswar, Odisha, India

Abstract

This chapter presents a review of the DC–DC converter topologies implemented for microgrid applications. The adverse environmental effect of modernization urged for the wide-scale utilization of renewable energy resources, bringing forth the concept of hybrid renewable energy resources. The microgrid has emerged as a definite solution for environmental issues, autonomous power facilities, definite power allocation during grid failure, and long-distance transmission and distribution losses. Power electronics converters are the major interfacing devices in microgrids, which feature efficient power conversion, power factor correction, voltage regulation, power quality improvement, etc. There has been a detailed study on the DC–DC converter topology with a major emphasis on isolated DC–DC converter topology and non-isolated DC–DC converter topology. A concise discussion on the different switching techniques and control schemes implemented in operating a DC–DC converter is presented. Furthermore, a comparative analysis is presented to find out the most suitable converter configuration among the discussed categories.

***Keywords*:** Microgrid, isolated DC–DC converters, non-isolated DC–DC converters, switching techniques

4.1 Introduction

Technological advancement and improvement have driven humans towards modernizing industrialization, transportation, communication,

**Corresponding author*: aditichatterjee.ee@gmail.com

P. Sivaraman, C. Sharmeela and P. Sanjeevikumar (eds.) Microgrids for Commercial Systems, (73–132) © 2024 Scrivener Publishing LLC

navigation, business, healthcare, and infrastructure. This has increased the global energy demand from 116,214 to 136,761 TWh in the last 10 years [1]. Climate change has become an agenda for the UN, leading to the formation of the United Nations Climate Change Conference (UNCC), also known as COP. Recently, data published by the World Resource Institute, as shown in Figure 4.1, depicts the energy sector's scenario as the major contributor to climatic change [2]. Nearly 77% of the energy is generated worldwide by fossil fuels, i.e., coal, gas, and oil, which contributes to the rising carbon footprint. Hence, our concern is to reduce the dependence on fossil fuels and increase the reliance on renewable energy resources. The inclination towards clean and green technology of energy production has paved the way for solar, wind, biomass, tidal, geothermal, and other replenished sources. Currently, 18% of the total energy production is from renewable energy resources and is expected to reach 30% by 2030. As per the IEA, solar energy, as shown in Figure 4.2 [3], has emerged as the leader in providing efficient and reliable energy solutions for a customer-centric global demand.

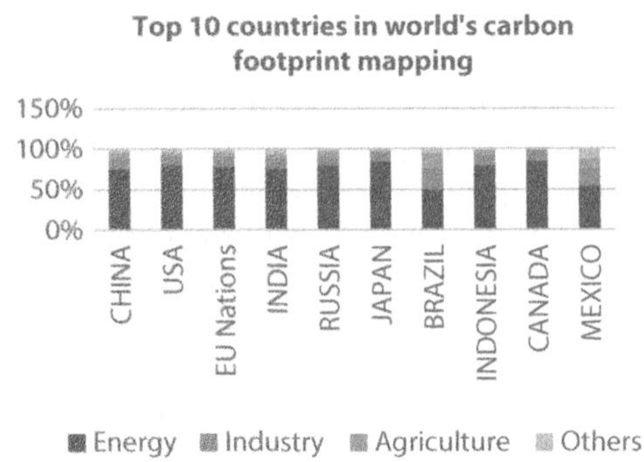

Figure 4.1 Data from the World Resources Institute.

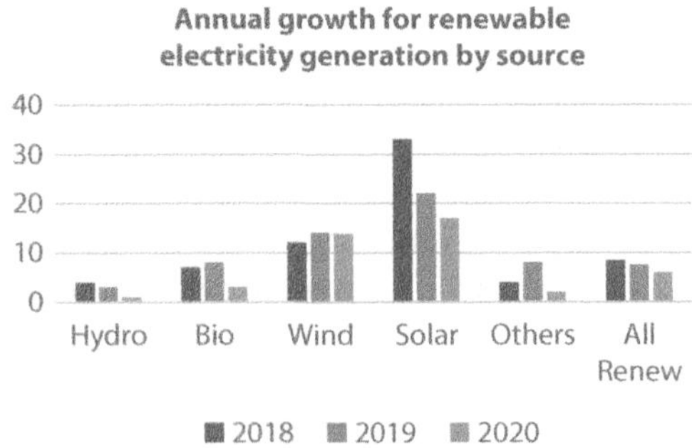

Figure 4.2 Data from the International EnergyAgency.

4.2 Microgrid

Hybrid renewable energy resource (HRER) systems are integrated into the main grid with power converters that may be either DC–DC or AC–DC or DC–AC or AC–AC type depending on the connected resources or the connected bus (i.e., AC bus or DC bus) [4]. The power converters have proven to be efficient due to their compact size, lesser number of components, negligible maintenance, less power loss, and faster dynamic response [5]. Despite the number of advantages that the power converters have, they fail to address certain problems such as power quality, power flow control, power sharing and management, load side voltage deviation, synchronization, economic despatching and optimization, load sharing, stable operation with different types of loads, etc. Hence, various control techniques have been implemented in the form of centralized, decentralized, and distributed control schemes.

A HRER system supported by energy storage systems is integrated into the utility grid with efficiently suitable power electronic converters whose functioning is monitored with basic control and/or multilevel control strategies to supply critical and non-critical load demands, collectively known as a microgrid. There are three types of microgrids: DC microgrid, AC microgrid, and hybrid microgrid. The schematic diagram of a hybrid microgrid arrangement is shown in Figure 4.3. Microgrids can operate in two modes, i.e., autonomous mode and grid-connected mode [6]. The majority of renewable energy resources provide DC power that requires DC–AC power converters in AC microgrids—for example, wind turbines require an AC–DC converter (for power transfer) and DC–AC converter (for grid synchronization); solar PV requires a DC–DC converter (for high output and better efficiency) and DC–AC converter (for grid integration), whereas DC microgrids can directly interface many renewable energy resources (solar PV, fuel cells, etc.) and energy storage systems (batteries, supercapacitors, etc.) using DC–DC power converters. This reduces the complexity of system architecture, reduces the cost of interface, operation, and distribution, improves the power flow control, and improves the power quality of the system due to the inherent features of a DC system.

DC microgrids are the most reliable and efficient ones, but a hybrid microgrid has emerged as a definite solution that resolves the problems of power mismatch, synchronization, power harmonics, etc., (concerned with AC microgrid) and immature power system protection, lower compatibility with AC loads, higher initial cost, complex system, etc. (concerned with DC microgrid) [7, 8].

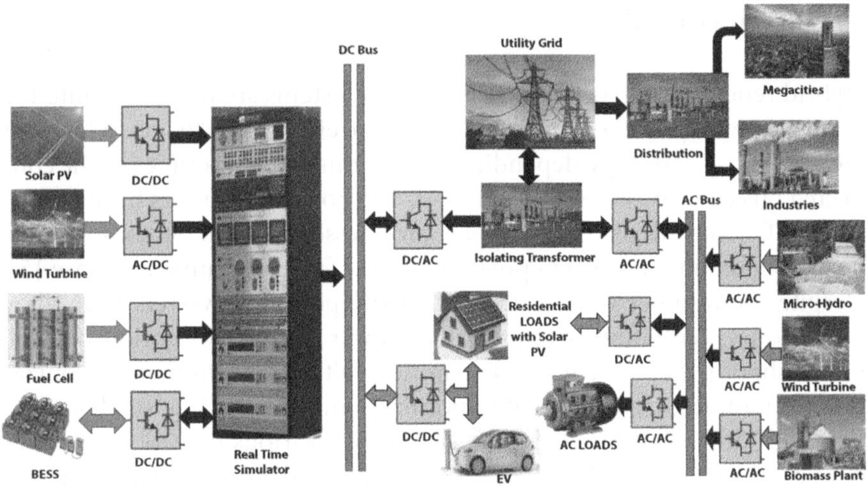

Figure 4.3 Schematic diagram of a hybrid microgrid.

4.3 DC–DC Converter Topologies

This chapter envisions various DC–DC converter topologies that are continuously evolving with advancements in power electronics technology in the field of microgrid application. In Figure 4.4; the DC-DC converter topology presents two types of converters unidirectional type & bidirectional type; and further bidirectional is divided into isolated type & non-isolated type.

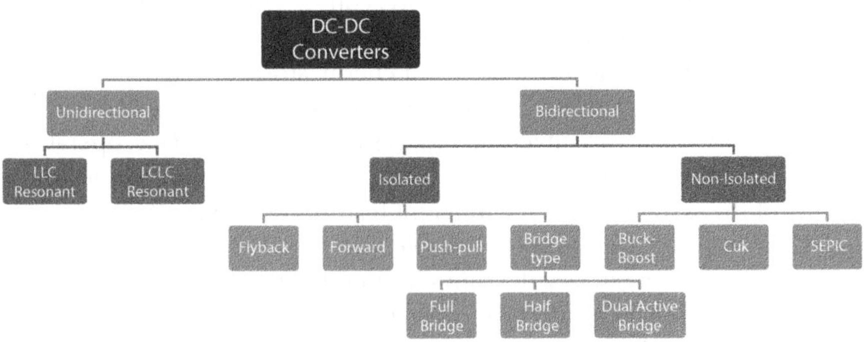

Figure 4.4 DC-DC converter topology.

A. Unidirectional DC–DC Converters

There are various types of unidirectional converters such as isolated DC–DC converters, non-isolated DC–DC converters, and resonant DC–DC converters. A resonant DC–DC converter has a tank circuit with inductors and capacitors, the resonant element in conjunction with a switch; it operates at a resonant frequency [10]. Resonant converters are superior to PWM converters due to their high efficiency, negligible switching losses, lossless snubber circuit operation, soft switching, and no electromagnetic interference (EMI) [11]. A US-patented unidirectional DC–DC converter design based on a buck-boost configuration has been presented by Uruno *et al.* that enables both step-up and step-down operations [12]. Figure 4.5 presents a variety of unidirectional DC-DC converters as discussed below in separate sub-headings.

A.1 LLC Resonant Converters

The resonant power converters are mainly classified as series resonant converters, parallel resonant converters, and series–parallel resonant converters. According to the arrangement of resonating elements, resonant converters are classified as LLC, LCC, or LCLC types. It is a modified form of a series resonant converter [10] that consists of the series combination of capacitor C_r and inductor L_r in addition to a parallel inductor L_{rp}. This circuit exhibits two resonant frequencies f_{r1} and f_{r2} that can be expressed as follows:

$$f_{r1} = \frac{1}{2\pi\sqrt{(L_r + L_{rp})C_r}}; f_{r2} = \frac{1}{2\pi\sqrt{L_r C_r}} \tag{4.1}$$

A.1.1 Half-Bridge LLC Resonant Converters

This circuit configuration is mostly used in bipolar DC microgrid systems. The LLC resonant converter is preferred due to simple control, soft-switching, high voltage gain, and high-efficiency features [10]. This type of configuration uses two resonant frequencies to operate depending on the inductance that participates in resonance.

Zubeita and Lehn discussed a conventional LLC resonant converter design that operated at a series resonant frequency with ZVS switching to turn on power MOSFETS and ZCS switching to turn on power diodes [13]. However, certain negative features were pointed out such as IGBT devices with 1,200 V rating that were subjected to serious switching losses during turn-off when operated below 400 kHz. Hence, a new type of series–parallel LLC structure was proposed in [16].

A.1.2 Full-Bridge LLC Resonant Converters

A commercial converter board was developed by ARDA Power, Canada, for integrating renewable energy sources with a bipolar DC bus [13]. During unbalanced load conditions, currents to the positive bus increase and decrease at the negative bus, thus maintaining a bipolar DC bus criterion with the proper switching operation of the converter. Bhajana *et al.* designed an 8-kW full-bridge LLC resonant converter that was used for auxiliary drives in traction applications [14]. The simulation results showed that the efficiency of the full-bridge LLC resonant converter was higher than the hard-switched full-bridge converter.

A.1.3 Series–Parallel LLC Resonant Converters

An LLC resonant converter whose input is connected in series with a balanced input voltage and output in parallel with load configures a series–parallel LLC resonant converter. This circuit configuration has been implemented by Bor-Ren Lin [16] in DC microgrid applications (where the voltage is 750–800 V) such as DC light rail vehicles, DC microgrids with bipolar voltage systems, etc. There are two resonant structures in a series–parallel combination as shown in Figure 4.5. A flying capacitor is implemented to equalize the voltage between two input capacitors. The circuit was simulated for varying input, varying loads as well as rated power conditions to fulfil the ZVS and ZCS criteria satisfactorily.

A.2 LCLC Resonant Converters

This is a form of parallel resonant converter that combines the features of both LLC and LCC type resonant converters. Generally, the circuit consists of series elements L_r and C_r with a parallel resonant capacitor C_{rp} and a parallel resonant inductor L_{rp}. This topology exhibits three resonant frequencies: f_{rp1}, f_{rp2}, and series resonant frequency f_{rs} [21].

A.2.1 Half-Bridge LCLC Resonant Converters

LCLC resonant converters have high power density, high-frequency operation, and high efficiency and are suitable for high voltage applications [22]. One of the striking features of LCLC converter is that the power switches can operate in both ZVS and ZCS modes. This reduces the switching losses to a minimum value even lesser than that of LLC converters. Computer and telecommunication systems suffer from hold-up problem: when the AC power is lost, the output bus voltage will drop due to discharge of the capacitor. Hence, a DC–DC converter is required to operate at a lower bus voltage to save time for the UPS to operate. An LCLC resonant converter can solve this problem in switched-mode power supplies. It attains high

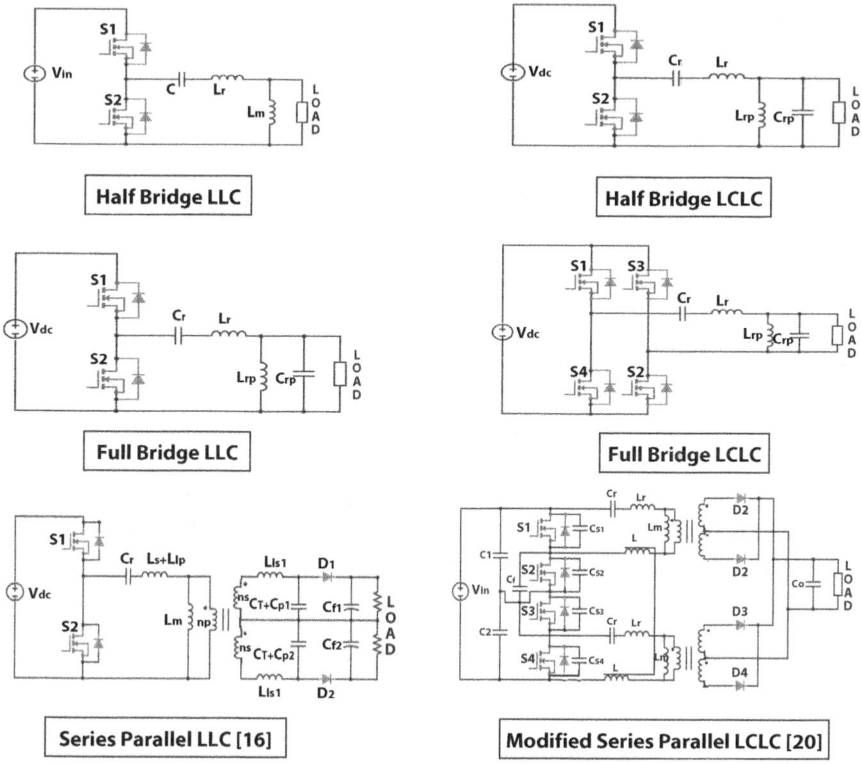

Figure 4.5 Unidirectional resonant DC–DC converters.

efficiency in normal operating conditions and provides high reliability with simple control [22].

A.2.2 Full-Bridge LCLC Resonant Converters

LCLC resonant converters were used as maximum power point trackers in microgrid applications [17] that provided constant output voltage by varying the input resistance. A 40-V input, 4.8-kV output 295-W prototype was built and tested [19].

Another type of CLLC resonant DC–DC converter was used for bidirectional electric vehicle charging in DC microgrids [18]. An algorithm was formulated to select the proper working mode of resonant LCLC bridge for attaining the minimum transformer RMS current.

A.2.3 Modified Series–Parallel LCLC Resonant Converters

This is one of a kind of LCLC resonant converter with a half-bridge configuration on the input side and two ports on the output side. This configuration was modeled and simulated in SPICE [20]. Generally, this type of configuration ensures reliable output regulation, better transient characteristics at the output under different operating conditions, and steady-state measurements for duty cycle control and frequency control methods [20].

B. Bidirectional DC–DC Converters

The bidirectional DC–DC converters can be classified into isolated DC–DC converters and non-isolated DC–DC converters, and they are further classified in the preceding sections.

B.1 Isolated DC–DC Converters

The isolated DC–DC converters are also called galvanically isolated DC–DC converters; they are classified as flyback type, forward type, push–pull type, and bridge type, and these converters are further classified as interleaved type, cascaded type, actively clamped type, and bridgeless type [9]. Figures 4.6 and 4.7 displays different types of bidirectional isolated DC-DC converter configuration implemented for DC microgrid.

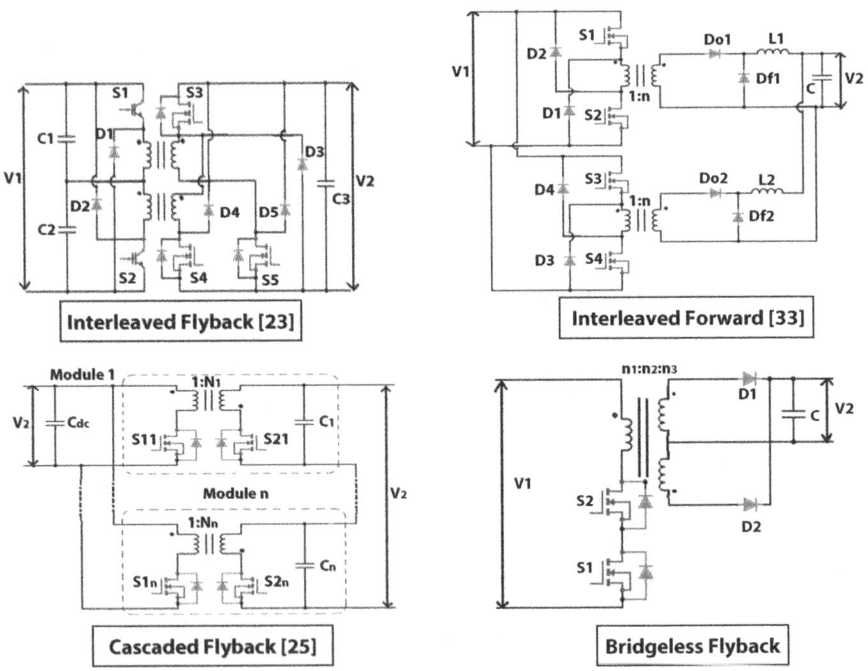

Figure 4.6 Bidirectional isolated DC–DC converter.

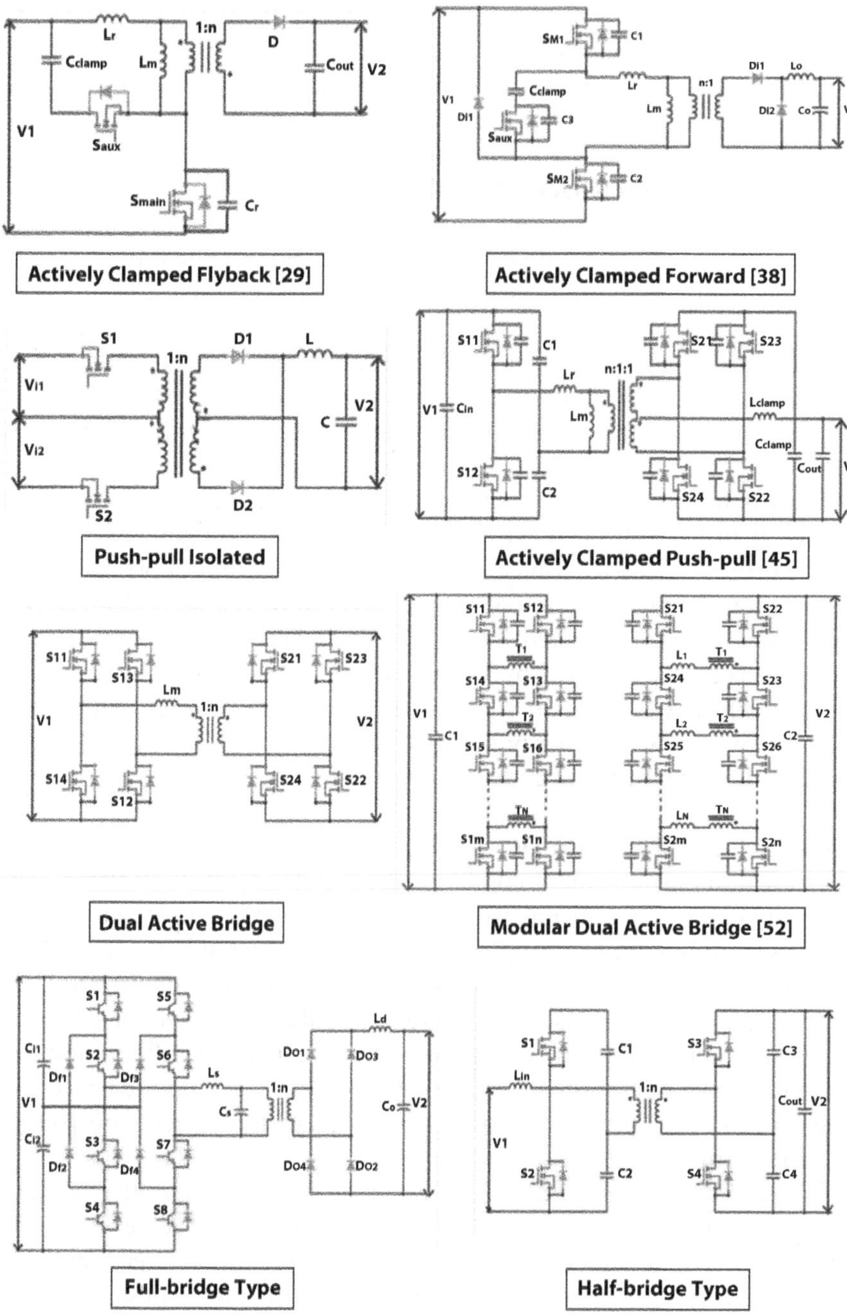

Figure 4.7 Bidirectional isolated DC–DC converter.

B.1.1 Flyback DC–DC Converters

Various types of flyback DC–DC converter topologies have been designed and implemented according to specific applications. This paper discusses four major types of flyback DC–DC converter configurations.

B.1.1.1 Interleaved Flyback DC–DC Converters

In this type of circuit configuration, the presence of galvanic isolation enables either the input side in a series connection or the output side in a parallel connection or vice versa. An interleaved configuration reduces peak currents, improves transient response, and reduces EMI [23, 24]. The low-power application in DC microgrid has been discussed here. Chen, Liang, and Huang designed a prototype of interleaved flyback DC–DC converter system for battery backup applications with 400 V/48 V, 500 W capacity. This configuration reduced the voltage stresses on primary switches, reduced the current stresses on secondary switches, and efficiently utilized the leakage inductance energy of transformers [23].

The process of interleaving could be modified as a multiport converter; the hardware implementation of a three-port interleaved flyback converter with a PV source integrated into a 230-V, 25-W bulb, and 12 V, 7 Ahr battery was built and tested by Venugopal and Kanaksabapathy [24]. This arrangement gave high gain without high duty ratio change, could integrate many renewable energy sources with different V–I characteristics, and could improve the life cycle of a connected battery.

B.1.1.2 Cascaded Flyback DC–DC Converters

The cascading arrangement combines the input side of the isolated converter in series or parallel as well as the output side in series or parallel. There is no difference in cascading and interleaving arrangement, but there is a lot of difference in the method of operation and control. Eitzen, Graf, and Maas proposed a cascaded bidirectional flyback converter with five converter modules for wave energy harvesting [25]. This configuration had three modes of operation: CCM, DCM, and BCM. The boundary condition mode (BCM) of operation was performed with a current and voltage control method.

This topology presented certain advantages, i.e., increased reliability, reduced component rating, the high output voltage was achieved due to series connection, standard semiconductors were chosen over high voltage semiconductors that showed better electrical properties, improved converter efficiency, the parasitic effects due to high voltage isolation was reduced, and capacitors with low voltage could be used [25]. Manikandan and Sunderrajan implemented closed-loop PI and fuzzy logic control for a cascaded flyback converter system for a PMDC motor [26]. The focus was mainly on the dynamic response of the converter, and finally, it was concluded that the fuzzy logic controller improved the time domain response (with reduced rise time, peak time, and settling time) and reduced the steady-state error.

B.1.1.3 Actively Clamped Flyback DC–DC Converters

The power semiconductor device manufacturers are focusing on higher power densities, limited power dissipation, minimum switching and conduction losses, efficient driving circuits for high current density, and high switching speeds during normal and overload conditions in a semiconductor module that developed the concept of active clamping [27]. Active clamping involves the conjunction of a clamping element with the auxiliary semiconductor switch to provide efficient protection. Parasitic inductances cause serious problems with voltage transients, increased switching losses, and turn-off delays. This problem was minimized by the soft turn-off technique, but it was unable to provide optimal driving under different operating conditions. Thus, an active clamping process was implemented to demagnetize all the stray inductances present in the circuit. Chub *et al.* implemented a three-port actively clamped CCM flyback DC–DC converter as a building-integrated photovoltaic solution (BIPV) in a bipolar DC microgrid [28]. Another configuration of the active clamped flyback converter was developed by Jailyn *et al.* for low voltage and low power applications. Active clamping could efficiently minimize voltage stresses, reduce switching losses, and monitor ZVS operation with proper delay time [29].

B.1.1.4 Bridgeless Flyback DC–DC Converters

The bridgeless concept evolved to improve the converter efficiency, achieve high power density, reduce conduction losses, and reduce the cost and size of the equipment. Most bridgeless flyback converters are used as power factor correctors in hybrid renewable energy systems, motor drives, and low-power lighting applications. Chen *et al.* developed a 90-W, 20/4.5-V prototype of a single-stage bridgeless flyback PFC for adapter application. The size of the transformer was reduced by using a clamping diode that produced a clamped sinusoid peak current, reducing the magnetizing inductance of the transformer. The net area of the flyback transformer was reduced from 0.81 to 0.5, which is around 40% reduction in size [30].

Tseng *et al.* proposed a 48-V power system supplying DC loads with a PFC and connecting a battery system with a DC–DC converter [31]. This configuration removed the bridge diodes to reduce the conduction losses and increase the efficiency; also, instead of two transformers, a three-winding transformer is used to make the circuit bridgeless. This typical arrangement can achieve a higher power factor by eliminating harmonics from the line source, increasing the power factor between the input voltage and input current.

B.1.2 Forward DC–DC Converters

This transformer-isolated forward converter is based on a buck converter. It is well suited for high-output current applications. Various sub-categories of forward DC–DC converters are discussed below.

B.1.2.1 Interleaved Forward DC–DC Converters

The interleaving technique involves paralleling two forward DC–DC converters whose switching instants are phase-shifted over an interval. Texas Instruments have developed a 200-W interleaved forward DC–DC converter design for voltage regulation module in CPUs [32]. The interleaving of converters reduced the input and output capacitor ripple current, reduced the electromagnetic interference, and reduced the size of input and output capacitor banks, thus reducing the cost of the design.

A circuit configuration of an interleaved dual double-ended forward converter was proposed as an LED driver in the DC microgrid lighting application [33]. The parallel connection of two double-ended forward converters was operating 180° out of phase, resulting in reduced ripple currents and making the output current like DC. This circuit topology removes the need for bulky electrolytic capacitors for filtering. Even if the number of components required increases, the cost and size of the converter also decrease because the requirement of a high-value transformer and large filter inductance is eliminated. Another interleaved forward DC–DC converter was presented by Habumugisha *et al.* to step-up voltage for DC microgrids. The advantages it possessed are reduced current and voltage ripples, reduced size of the output filter, reduced transformer size, efficient safety of DC microgrid, and ensured reliability of resources [34, 35].

B.1.2.2 Bridgeless Forward DC–DC Converters

The bridgeless concept, as already discussed in the previous sections, has numerous advantages; the only variation is in the technique of achieving bridgeless configuration in galvanic isolations. Lan *et al.* have developed a prototype of an 80-W bridgeless forward PFC converter with 90-V DC output [36]. An improved method of constant on-time control was enabled to generate a half-sine wave proportional to AC rectified input voltage that was added to the error of two comparators, thus producing more sinusoidal AC input current. The converter provided an improved power factor, increased efficiency, reduced harmonics, and reduced conduction losses. Soldano *et al.* have patented a bridgeless bidirectional forward converter for multistage applications [37].

B.1.2.3 Actively Clamped Forward DC–DC Converters

Actively clamped forward converters are quite efficient to reduce the cost and size of the transformer, implement the soft-switching technique, and provide a simple configuration. However, conventional active clamped forward converters suffer from voltage stress with an increase in duty cycle ratio. Even if the duty cycle is reduced with low voltage stress on switches, the current stress on the switches also increases, further increasing the size of the output filter circuit [40]. Park, Moon, and Youn have designed a 300-W, 200–400-V input and 48-V output converter operating at 100 kHz as the prototype of a two-switch actively clamped forward DC–DC converter [38]. They have countered this problem by using a single-clamped diode and a delayed turn-off gate signal for the main and auxiliary switches. Thus, the duty cycle was expanded, which reduces the amplitude of current across primary leakage inductance, further reducing the conduction and switch turn-off losses.

An exceptional case of an actively clamped forward DC–DC converter has been integrated with smart power IC technology [39]. They have implemented an optimal four switch active clamping H-bridge to drive the pulse transformer of the actively clamped forward DC–DC converter, thus providing a wide range of duty cycle operations and allowing integration in junction-isolated smart ICs [39].

B.1.3 Push–Pull DC–DC Converter

A bidirectional push–pull DC–DC converter has been designed by Solanki *et al.* for boosting the voltage from 48 to 400 V on the input side of the inverter tied to the grid in a microgrid application [41]. This converter isolates the DC bus from the AC bus to prevent any serial damage in the system and reduces the overall size of the system, and due to less input current ripples, the converter becomes more efficient at higher loads. However, this design increased the power-to-weight ratio, thus limiting its applications to certain areas only.

Various clamped configurations of push–pull DC–DC converters have also been discussed. Bal *et al.* have simulated and verified a special type of naturally clamped snubberless bidirectional current-fed push–pull DC–DC converter conjoining two DC buses of 48 and 380 V in a microgrid

[44]. A unique secondary side modulation technique is implemented to achieve natural clamping of primary side devices, thus eliminating the need for a snubber circuit. Soft switching is also employed, i.e., ZCS for primary-side low-voltage devices and ZVS for secondary-side high-voltage devices.

Another application of a push–pull DC–DC converter for driving a DC motor load is reported and simulated using PSIM [43]. It enhanced the system efficiency with the P and O technique, increased reliability, provided good regulation, and provided a high input power factor. Schulz and Wild have bridged an actively clamped current-fed push–pull DC–DC converter in between a 960-Wh mobile battery system and a low voltage DC (LVDC) microgrid [45]. The converter operates in both step-up and step-down modes to discharge and charges the battery utilizing a unique modulation scheme to achieve ZVS. The converter operates in three different modes—CCM, DCM, and BCM—depending on the direction of power flow, i.e., either from the grid to the battery or vice versa. The converter was highly efficient in step-up and step-down modes and reduced the EMI.

B.1.4 Isolated Bridge-Type DC–DC Converter

B.1.4.1 Dual Active Bridge DC–DC Converter

The expanse and familiarity of the microgrid technology showcased the need for storage systems, i.e., highly efficient batteries to charge during excess generation and discharge during intermittent supply. Many problems faced by converters could not be resolved, such as inflexible power flow, electrical stress on power semiconductor switches, dynamic bidirectional control, maintaining a constant voltage profile, and active switching. Hence, dual active bridge DC–DC converters employed two active bridges (either in full-bridge or half-bridge form) conjoined by a high-frequency transformer and leakage inductance to provide a better solution for the above-stated problems.

The dual active bridge (DAB) converters are efficient in maintaining grid stability with the better enhancement of the battery's SoC [46]. The authors in [47] developed a DAB converter that could interconnect a battery energy storage system (BESS) of 48 V to a DC bus of 450 V for providing better reliability and good dynamic response under transient conditions. The single-phase shift modulation technique enabled faster power transfer as a master converter during islanded operation. DAB converters are controlled as a virtual DC machine in DC microgrids to counter voltage dip and voltage fluctuation for improving power quality during load transition modes. Virtual DC machines (VDM) are analogous to virtual synchronous generators (VSG) that have been mostly used in AC microgrids for maintaining a constant frequency [48]. A virtual DC machine control strategy utilizes the DC machine's dynamic torque equation and emf equation of the armature circuit to generate the phase shift ratio for the DAB converter in islanding and interconnected modes of operation.

An active bridge active clamped configuration of DAB converter interfaces a 270-V HVDC network with a 28-V LVDC network [49]. This configuration has the advantages of compact design, best weight-to-volume ratio, and active control of output current, is highly efficient, and finds application in HVDC aerospace microgrids. A modified configuration of the DAB converter has been developed as a fault-tolerant controlled three-phase current-fed DAB DC–DC converter [50]. The converter is analyzed under open-circuit fault conditions, and blocking capacitors have been provided to block the DC component for avoiding the saturation of the transformer. During a severe fault condition, the switch of the faulty phase is turned off, and the converter is operated as a single-phase DAB [50].

B.1.4.1.1 Modular DAB DC–DC Converter

This topology is developed for sufficing the needs of grid extension in terms of high power and medium power applications. Modular connection of converters can be achieved by cascading or interleaving DAB converters. Hosseini *et al.* have proposed a modular DAB DC–DC converter configuration that has bidirectional power flow, reduced switching losses, reduced cost and size, flexible ZVS switching option, and the possibility of configuration extension in the future [52]. Steineker and Doncker have focused on an inherent feature of modular converters, i.e., redundancy, that makes the system more reliable and available during critical conditions [53]. Several other topologies have also been proposed and developed by many researchers such as triple-active bridges [54], matrix-based DAB [55], single-active bridge [56], etc.

B.1.4.2 Full-Bridge DC–DC Converter

The full-bridge type converters are the most conventional DC–DC converters used particularly for bidirectional applications. The advancement in switching technology, control strategy, modulation technique, and evolution in different converter configurations has led to an improved full-bridge converter configuration. Deng and Chen have developed a prototype of a 3-level full-bridge DC–DC converter for integrating a wind turbine in a DC microgrid [57]. A novel control strategy and modulation technique was implemented for attaining suitability in medium voltage and high power applications.

Zhang *et al.* have developed a Simulink model of a single module phase-shifted full-bridge DC–DC converter [58]. Similar modules of DC–DC converter were connected in series to implement an improved structure of the solar power generation system. This particular system provided efficient power point tracking, minimize switching losses due to resonant switching, maximized performance with minimum investment on construction cost, and also subdue the effect of line parasitic parameters. Another distinctive model of a modular full-bridge ZCS DC–DC converter was employed to integrate a PV system with a medium voltage DC grid [59]. This model utilized the features of reverse blocking switches, phase shift frequency modulation, series resonant capacitor, and galvanic isolation to provide benefits over conventional converter topologies.

B.1.4.3 Half-Bridge DC–DC Converter

The half-bridge configuration is primitively derived from the full-bridge DC–DC converter configuration and is utilized as bidirectional converters for cost-effectiveness. In certain half-bridge configurations, a hidden feature of synchronous rectification is discovered by implementing power devices with feedback diodes [61]. Jing *et al.* have developed a Simulink model of a 3-level half-bridge bidirectional DC–DC converter that operates under two modes i.e., charging mode and discharging mode. The converter's equivalent AC circuit analysis has been derived to analyse the full control range optimized by the converter voltage gain during soft-switching operation implemented in both the primary side and secondary side of the half-bridge circuit [62]. Phetphimoon and Bhumkittipich have modeled and simulated a resonant half-bridge bidirectional DC–DC converter that operates in boost and buck mode for channelizing power flow from the HV side to the LV side and vice versa [63].

B.2 Non-Isolated DC–DC Converter

The non-isolated DC–DC converters are major of buck type, boost type, buck-boost type, Cúk type and SEPIC type; further classified as multi-switch type, MIMO type, resonant type and hybrid type converters. Figures 4.8 and 4.9 presents a new variety of non-isolated DC-DC converter widely used in microgrid; as per the recent research literature.

B.2.1 Buck-Boost DC–DC Converter

B.2.1.1 Multi-Switch Buck-Boost DC–DC Converter

Multi-switch topology is mostly preferred over the conventional 2-switch buck-boost converters due to improved bidirectional power flow capability,

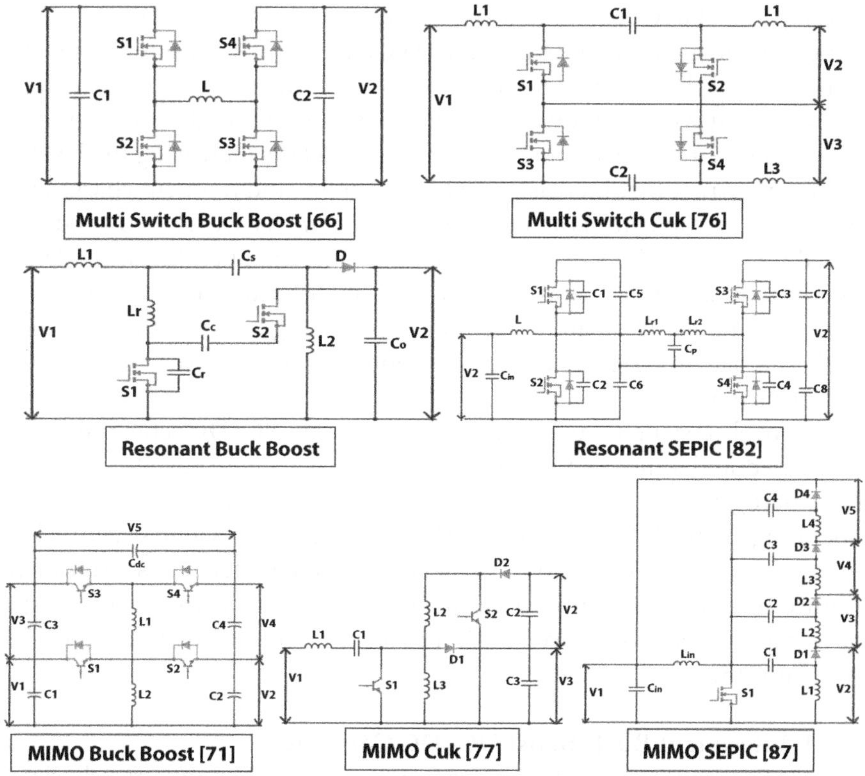

Figure 4.8 Bidirectional non-isolated DC–DC converter.

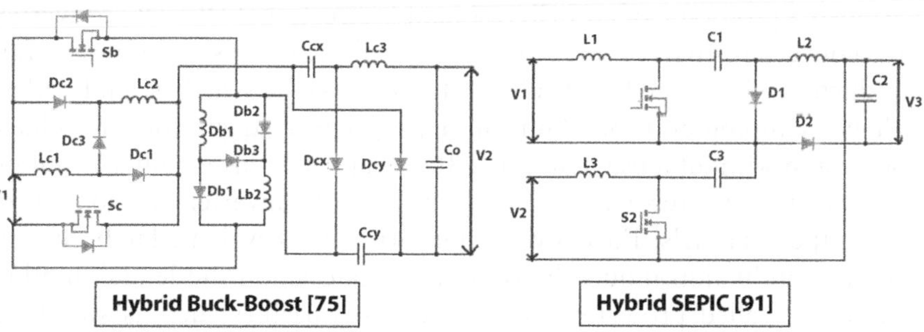

Figure 4.9 Bidirectional non-isolated DC–DC converter.

reduction of conduction losses over switches, reduction of voltage stress over individual switches, efficient switching technique implementation to cater to problems such as power harmonics, voltage transients, switching losses, etc., improved speed of response, and many more. Schulz *et al.* have developed a 300W prototype of a 4-switch buck-boost DC–DC converter to be implemented in a mobile battery system. An H-bridge topology of 4 switches buck-boost DC–DC converter was developed with soft switching and operated with a unique continuous conduction mode with a middle active short-circuit technique for limiting the THD, increasing the level of inductor current that increases total DC current, and frequent charging of the bootstrap capacitor is avoided [64]. Pires *et al.* have designed a bidirectional 6-switch DC–DC converter operating in the buck-boost mode for interlinking a unipolar and a bipolar DC microgrid. The converter possessed the inherent capability of voltage balancing between the two poles of the bipolar microgrid and allows smooth energy transfer between the two grids [65].

A novel 4-switch buck-boost bidirectional DC–DC converter was developed and simulated in PSIM to interface a supercapacitor with the DC microgrid [66]. Jiya *et al.* presented a bidirectional DC–DC converter operating in buck-boost mode with four-quadrant switches also called fully controllable bidirectional switches. The converter has a prominent feature of bi-directionality to be incorporated in hybrid microgrids that have been tested and proven with a real-time HIL device i.e., OPAL-RT simulator [67].

B.2.1.2 Resonant Buck-Boost DC–DC Converter

Mostly Resonant converters are preferred over PWM converters because of loss reduction during switching, elimination of electromagnetic interference, operation in higher switching frequency, etc. Rathore *et al.* have designed a prototype of an LCL resonant bidirectional DC–DC converter. This converter provides soft switching for all the diodes and power semiconductor switches, reduced voltage stress across the switches, snubberless voltage clamping and maintains an appropriate voltage ratio for voltage conversion without a transformer [68]. Litvani *et al.* have built a Simulink model of a 5-level resonant DC–DC buck-boost converter that consists of four output channels. The converter performance is evaluated in both continuous conduction mode, discontinuous conduction mode and another new mode called protection mode. Certain variables such as switching angle, capacitor peak voltage and per unit load resistance were considered as the control variables as they were analysed independently for each operation [69].

B.2.1.3 MIMO Buck-Boost DC–DC Converter

To explore and address various aspects of a multifunctional Microgrid different types of multi-input multi-output (MIMO) bidirectional buck-boost DC–DC converters have been reported. Akar *et al.* have developed a prototype of a 1kW power-efficient MIMO buck-boost DC–DC converter. A detailed analysis in the form of a comparative analysis of the proposed converter with existing configurations, operational analysis, small-signal modeling and controller designing is done [70]. A simulated version of a MIMO-integrated buck-boost DC–DC converter was reported in [71] sufficing the wide range of voltage and power level integration of resources, storage system and loads. This converter was designed with conjoined MPPT and PI controller action that provided higher efficiency due to reduced components, reduced number of voltage conversions and higher boosting capability [71].

Another type of MIMO bidirectional buck-boost DC–DC converter prototype was fabricated and simulated using PSCAD with dual input and dual output ports [72]. They have tested the operation of the converter under different switching conditions i.e., source 1 and source 2 are available to both loads with a variation in inductor charging and discharging; that display both buck mode and boost mode. Liang *et al.* have devised a unique control strategy i.e., improved virtual capacitor control for a MIMO bidirectional buck-boost DC–DC converter. This converter control strategy has made the system more robust to counter the changing voltage during load disturbances and switching operations [73].

B.2.1.4 Hybrid Buck-Boost DC–DC Converter

The hybrid concept has emerged from the cascaded or interleaved combination of different types of configuration and certain examples may be buck-boost-Cúk, SEPIC-Cúk, SEPIC-Zeta, etc. Bagewadi and Dambhare have implemented a novel augmented sinusoidal pulse width modulation technique to a classic buck-boost converter for creating a hybrid buck-boost DC–DC converter that can regulate a DC input voltage to DC as well as AC output voltage. The control strategy includes two separate PI controllers that can regulate the AC and DC outputs differentiating them as two single-input single-output systems [74]. Gupta *et al.* have presented a comparative analysis of their proposed hybrid buck-boost DC–DC converter i.e., formed by the integration of the Cúk and the buck-boost converter. They have shown the importance of high voltage gain, reduced stress across switches and increased number of components [75].

B.2.2 Cúk DC–DC Converter

B.2.2.1 Multi-Switch Cúk DC–DC Converter

The conventional Cúk converter implements an additional inductor and capacitor compared with a buck-boost converter configuration. The multi-switch bidirectional Cúk converter utilizes 4 power semiconductor switches to integrate a battery energy storage system in a DC microgrid as reported in [76]. The Cúk converter poses as an ideal DC source thus enabling the injection of DC currents into the microgrid and promising effective charging or discharging of the battery system. The converter configuration also forms a voltage doubler circuit that helps in reducing the voltage stress across individual switches to a factor of two. The authors have also proposed an isolated multi-switch bidirectional Cúk converter configuration that can be implemented for specific applications.

B.2.2.2 MIMO Cúk DC–DC Converter

The multi-input multi-output functionality is required in microgrid applications to integrate various energy sources, storage systems and different critical and non-critical loads. A unique dual output DC–DC converter with the conjoined configuration of SEPIC and Cúk converter has been designed using Simulink that possesses the advantages of providing a continuous input current, multilevel output voltage, reduced voltage stress across the switches, magnified voltage gains and feasible for multiple applications in the field of machine drives, power supply, microgrids, etc [77].

Asthana and Verma have presented a MIMO interleaved bidirectional DC–DC converter to provide battery backup for a telecom power system interfaced in a 72V LVDC microgrid [78]. They have analysed the steady-state operation of charging and discharging mode, analysed the performance of the converter for parameters such as battery currents, SoC, output currents and voltages, capacitor voltages and currents, etc. and picturized comparative analysis of the presented converter and conventional buck-boost converter in terms of efficiency and ripple current over the complete duty cycle range. An isolated version of a 3-port bidirectional Cúk DC–DC converter has been reported in [79] with inherent features of zero-ripple output currents, low parts count, reduced weight, and EMI due to novel design methodology, a wide range of applications also includes a plug-in hybrid vehicle. They have used the ANSYS platform for FEM (Finite Element Method) analysis to design an appropriate transformer core.

B.2.2.3 Hybrid Cúk DC–DC Converter

There are various configurations of hybrid Cúk converter available for microgrid applications, Lica *et al.* have proposed a hybrid Cúk converter with a minimum number of components that can operate in continuous conduction and discontinuous conduction mode. The hybridized version is brought about by modeling the coupled inductors as an ideal transformer with a magnetizing inductance and turns ratio. The converter is designed to operate in a wide range of duty cycle i.e., 0.1 to 0.9, reduce the stress across the switch, minimizes the voltage ripple at the input side to 10% and output side to 5% and can perform both step-up and step-down operations [80].

Sarath and Kanaksabapathy have designed a hybrid Cúk converter that is employed to supply both AC and DC loads. The converter configuration is derived from the basic Cúk converter that is hybridized by replacing the power electronic switch with a voltage source inverter to provide AC output voltage along with DC output voltage. The converter has 3 modes of operation where it can operate as a buck or boost converter for the DC load and also provides inversion operation for the AC load. The converter has inherent features of high gain voltage ratio, a unique PWM control technique that generates gate pulses for the converter regulation as well as inversion operation and provides protection against unwanted gate turn-on of switches due to spurious noise signals and EMI [81].

B.2.3 SEPIC DC–DC Converter

B.2.3.1 Resonant SEPIC DC–DC Converter

Single-ended primary inductance converter abbreviated as SEPIC has extensive features even better than the Cúk converter with minimum voltage stress across the switch, minimum input and output voltage and current ripples and efficient noise filtering. Ikeda *et al.* have proposed an active clamped resonant SEPIC DC–DC converter that provides soft switching at high frequencies and medium voltages. Active clamped resonant circuits are efficient in reducing voltage stress across the switches which can be inferred from this configuration [82].

Hu *et al.* have designed a prototype of a resonant SEPIC DC–DC Converter that can operate in high and very high frequencies. The designing process includes rectifier design, inverter design, coupling the rectifier and inverter together and re-tuning them to create a link between the resonant inverter and the resonant rectifier and finally a gate driver is designed to make the switching process faster [83]. The converter is controlled by PWM on-off control with a hysteresis override feature to avoid any load disturbance effect.

B.2.3.2 MIMO SEPIC DC–DC Converter

Hailing the performance of multiport converters or MIMO Converters, their application has been extensive in the field of microgrids. Tomar *et al.* have proposed a multi-input single-output SEPIC DC–DC converter for a solar power-operated water pumping system that provides an interface between the number of PV panels and a 3-phase voltage source inverter connected to a BLDC motor coupled to a centrifugal pump [84]. The multi-input single-output SEPIC DC–DC Converter was designed to operate with MPPT under different unsuitable practical conditions such as shading effect, partial shading effect, clouding, dust, etc. Another multi-input SEPIC DC–DC converter was reported in [85] that could perform buck converter operation and boost converter for charging and discharging an energy storage system.

A typical isolated version of a multi-output SEPIC DC–DC converter has been designed to suffice the various load demands extracting power from renewable energy resources [86]. The converter design integrated with a multilevel inverter has been tested with open-loop and closed-loop simulation under the PSIM environment. Uno *et al.* fabricated a single switch multi-stacked SEPIC DC–DC converter for equalizing the state of charge of super capacitors [87]. The converter was operated in discontinuous mode with a fixed duty cycle to limit the inductor ripple currents [87]. A prototype was developed that exhibited an efficiency of 57%, the reduced efficiency was due to the frequent operation of the converter and the energies of the supercapacitors were wasted in the power conversion.

B.2.3.3 Hybrid SEPIC DC–DC Converter

Sebaje *et al.* have simulated a 1kW, 400V hybrid SEPIC-Zeta DC–DC converter with a switched capacitor cell for integrating a battery storage system with a microgrid. This converter reduced the voltage stress across the switches, provided low ripple input current, presented current source characteristics at the input side and voltage source characteristics at the output side, was efficient to operate in the partial charge region and requires low filtering [89]. A hybrid buck-SEPIC DC–DC converter with switched inductor capacitor was designed to operate by a discrete controller based on QFT (Quantitative Feedback Theory) [90]. This converter was specific to address uncertainties during renewables operation, noise during switching devices and load disturbance.

Shastry *et al.* proposed a hybrid Cúk-SEPIC DC–DC converter controlled with MPPT for a microgrid that interfaces solar PV and wind turbine; connected to a space vector PWM controlled Quasi Z-source inverter [91]. Maurya and Patle simulated a hybrid quadratic SEPIC-derived DC–DC converter to be utilized in the microgrid application. This converter offered various advantages such as a high voltage conversion ratio, an ideal current source, applicability for AC and DC loads and reduced ripple effects [92].

4.4 Modulation of DC–DC Converters With Different Control Strategies

Various modulation techniques are available in the research literature to trigger the power semiconductor switches and operate the power diodes in DC–DC Converters. The modulation technique can be a combination of classical switching techniques and general control methods for operating a converter in the microgrid. Depending on the type of operation such as resonant switching [10–22], bidirectional power flow capability [41, 42, 46–51], dynamic load flow monitoring [47, 48], bridgeless operation [30, 31, 36, 37], power circuit isolation [23–51], cascaded operation [25, 26], high voltage gain conversion [70–76] and multiple features are enhanced based on the control scheme implemented in switching the DC–DC Converters.

A. Classical Switching Technique With Linear Control

The commonly used switching methods in DC–DC Converters are duty ratio control or pulse width modulation and frequency modulation technique. A linear control involves a PI/PID controller acting as a voltage and/or current controller generating the required duty ratio or pulses for switching a DC–DC converter. The linear control can be formed up of single loop feedback or double loop feedback path as showcased in the schematic diagram Figure 4.10. A cascaded flyback DC–DC converter was controlled with a PI controller that acts as a voltage controller and a two-state current controller for switching the converter modules; it enables bidirectional energy flow and equal voltage distribution between the modules [25].

A PID controller was implemented for switching the MOSFETs during the charging and discharging of the battery using a DAB DC–DC Converter [46]. An augmented sinusoidal PWM technique was used for switching a hybrid buck-boost converter feeding AC and DC loads [74]. A unipolar sinusoidal PWM technique is applied to operate a hybrid Cúk DC–DC converter designed to replace a voltage source inverter bridge network; the

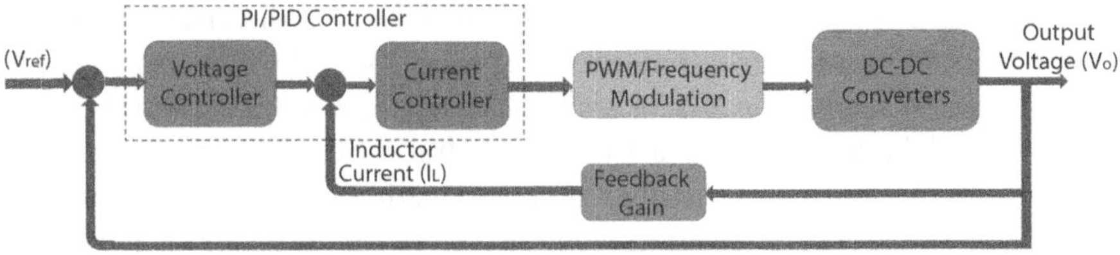

Figure 4.10 Classical switching with linear control.

converter interfaces both AC and DC loads [81]. The closed loop simulation of an isolated multi-output SEPIC DC–DC converter is carried out by a PI controller that works as a voltage controller to generate switching pulses for the input port IGBT [85].

B. Classical Switching Technique With Nonlinear Control

Various types of nonlinear control techniques are present in research literature such as sliding mode control (SMC), model predictive control (MPC), nonlinear modeling approach (Small-signal Modeling/Large-signal Modeling), passivity-based control (PBC) and backstepping control (BSC). PI controllers were designed based on a small signal modeling analysis of MIMO buck-boost DC–DC converters; to provide PWM signals to the converter switches [70]. Cascaded PI control loops with MPPT control were implemented for a MIMO buck-boost converter interfacing multiple renewables and loads altogether [71]. A virtual capacitor control was imbibed for a four-port isolated buck-boost DC–DC converter; monitoring parallel coordination control of the DC microgrid using a PI controller for voltage and current closed-loop feedforward control deduced from the converter transfer function [73]. A buck-SEPIC hybrid DC–DC converter is controlled using quantitative feedback theory-based controller; to stabilize the converter during source and load disturbances and provide proper

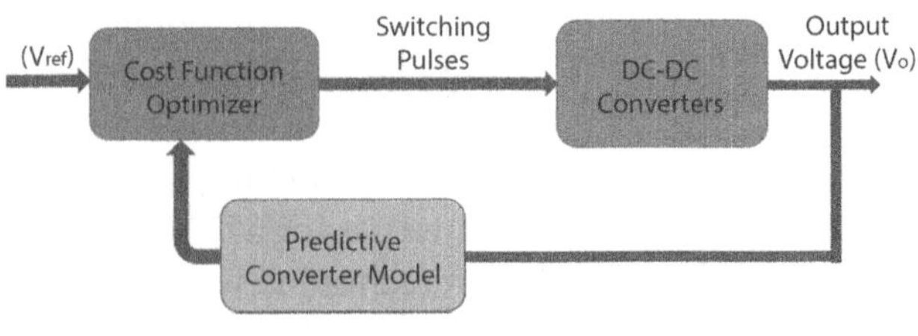

Figure 4.11 Classical switching with model predictive control.

voltage distribution and uniform load sharing [90]. The basic block diagram for model predictive control is shown in Figure 4.11.

An improved sliding mode control technique is applied to a Hybrid boost DC–DC converter for providing better transient performance during dynamic load conditions and input voltage variations [93]. Model predictive control (MPC) based MPPT algorithm was formulated for controlling a high gain DC–DC converter that provided better efficiency, faster tracking of PV power and reliable technique during irradiance variation or dust/sand deposition [94]. A distributed model predictive control with proper optimization was proposed for a multiport DC–DC Converter integrating a hybrid wind-solar system; that proved to be speeder than a centralized model predictive control [95]. A detailed discussion on various types of control techniques based on small-signal modeling and large-signal modeling along with the tuning of controllers is presented for the pulse width modulated DC–DC converters [96]. A passivity-based control technique was applied to a grid-connected bidirectional buck-boost DC–DC Converter to monitor the charging and discharging of a battery energy storage system during dynamic battery behaviour, load and supply variations [97].

C. Classical Switching Technique With Intelligent Control

Fuzzy logic, neural networks, heuristic approach and reinforcement-based control can be grouped under intelligent control methods [98]. Figure 4.12 is a block diagram representation of an intelligent controller implemented to tune the classical controller in a conventional closed loop control technique of DC-DC converter. A fuzzy logic control scheme was proposed for a smart grid integrating solar PV, fuel cell, BESS, main grid and residential loads [99]. A high-gain non-isolated DC–DC converter was controlled using a fuzzy logic control that could track the maximum power of the solar PV module [100]. A DC–DC converter is controlled by an optimal controller based on approximate dynamic programming that was trained offline to maintain the

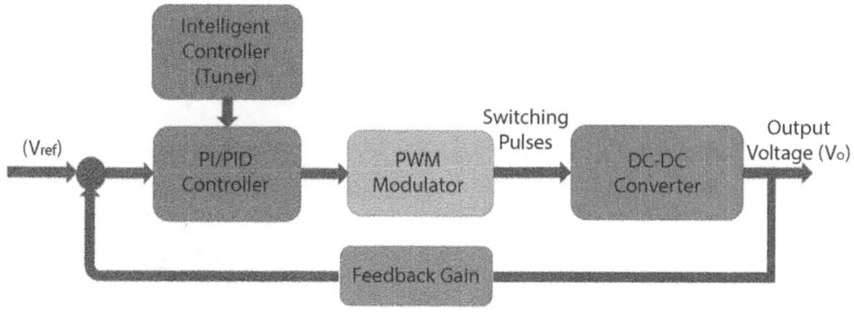

Figure 4.12 Classical switching with intelligent control.

stability of the network and could generate numerous gain values for the artificial neural network (ANN) controller [101]. The deep reinforcement learning technique was implemented in controlling a buck-boost DC–DC converter using a real-time interface i.e., dSPACE micro lab box [102]. An AI (Artificial Intelligence) controller was designed to control the switching of a buck DC–DC converter and trained using a deep learning algorithm [103].

D. Classical Switching Technique With Digital Control

Figure 4.13 shows a diagrammatic representation of the classical technique implemented with digital controller. A DSP controller was implemented as a central controller for closed-loop voltage control of an interleaved flyback DC–DC converter; that ensured simultaneous operation of solar PV and battery for a resistive load using MPPT with PI control [24]. An interleaved PWM controller was designed for an interleaved forward DC–DC converter and fabricated as a UCC28211 controller for PCs [32]. A cascaded control structure with PI controllers could vary different control parameters using a graphical user interface (GUI) [67]. A DSpace-based real-time controller was installed for a prototype of the MIMO buck-boost DC–DC converter facilitating two inputs and two outputs [72].

E. Soft Switching Technique With Intelligent Control

A fuzzy logic controller was designed for an LLC resonant DC–DC converter to control the switching frequency of its power MOSFETs; providing a satisfactory transient response of the output voltage [104]. A CLLC resonant DC–DC converter was modeled using deep learning and MPC (Model Predictive Control) was implemented to provide a stable operation under dynamic load conditions [105]. An LLC resonant DC–DC converter interfaced between the AC grid and PV module was controlled using a dual MPPT algorithm using an adaptive neuro-fuzzy interface system for producing phase shift and P and O for generating duty cycle; ZVS was realized in PV side connected switches and ZCS was realized in grid side

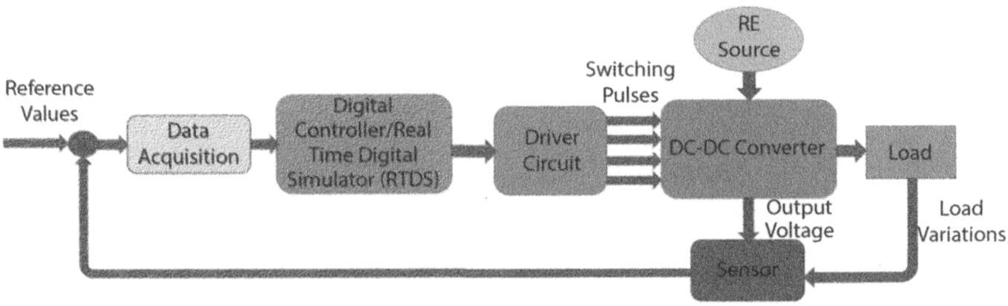

Figure 4.13 Classical switching with digital control.

connected diodes [106]. A multi-output push-pull DC–DC converter realizing a quasi-resonant condition and closed loop operation was attained by the fuzzy logic controller [107].

F. Soft Switching Technique With Digital Control

ZCS turn-off and ZVZCS turn-on for switches are realized by driver circuits and DSP processor in a full-bridge LLC resonant unidirectional DC–DC converter [14]. The converter becomes more efficient than hard-switched converters. An actively clamped push–pull converter's soft switching is achieved through the Xilinx FPGA board [44]. ZVS turn-on of switches is obtained for power factor correction of a bridgeless flyback DC–DC converter that supplies DC loads with a single chip control circuit [31]. Toshiba designed a 200-W, 24-V active clamped forward DC–DC converter to operate with Texas Instrument's active clamp controller [40]. ZCS turn-on and turn-off for switches as well as ZVS turn-on and ZCS turn-off for diodes is achieved, and digital control is implemented with TI's DSP controller that performs hybrid-phase shift frequency modulation to provide enhanced PWM operation of a full-bridge modular DC–DC converter [59]. Soft switching was implemented with a TI's DSP controller for a resonant buck-boost DC–DC converter to improve conversion ratio and efficiency and avoid magnetics in the circuit [68]. A quasi-resonant SEPIC converter is realized with ZVS operation by a resonant sinusoidal gate drive circuit for MOSFET, and a PWM ON-OFF control technique is implemented by a bi-CMOS PWM chip UCC2813 [83]. ARM processor LPC 2148 realizes a fuzzy logic controller for an LCL resonant DC–DC converter to provide efficient voltage regulation during dynamic conditions [108].

G. Bidirectional Switching Technique With Linear Control/Nonlinear Control

The bidirectional switching technique involves a unique method of providing phase shift between the operation of two or more bridges of a bidirectional DC–DC converter/bridge-type DC–DC converter. Different types of phase-shifting techniques can be termed bidirectional switching techniques such as single-phase shifting, dual-phase shifting (DPS), extended phase shifting, triple-phase shifting, hybrid-phase shifting (that includes a combination of phase shifting method and control methods such as direct power control or voltage control method or current control method), and other evolving phase shifting techniques [51, 109].

A droop control method using sigma-delta modulation, i.e., a bidirectional switching technique with Hill climbing MPPT was implemented for an actively clamped flyback DC–DC converters interfacing solar PV with a bipolar DC microgrid [28]. This method not only overcomes the stability problems linked with unbalanced bipolar DC microgrids but also maintains a perfect voltage balance between the two pole voltages [28]. A phase-shift modulation is utilized for operating a three-winding double-transformer interleaved forward DC–DC converter to provide easier operation between various micro-sources and DC microgrids [34]. A virtual DC machine control strategy was proposed for the DAB DC–DC converter operating with phase-shift technique and proven to be more efficient than the conventional droop control scheme [48]. A full-bridge three-level DC–DC converter is controlled using the permanent synchronous generator principle for integrating a variable-speed wind turbine with a DC microgrid [57].

SMC-based voltage controller and current controller were designed for a bidirectional DAB DC–DC converter to provide the necessary phase shift [110]. The double-integral sliding mode controller proved to be effective against uncertainties and disturbances. The integrated perturbation analysis and sequential quadratic programming method are incorporated to enhance the optimization of a linear model predictive controller and a nonlinear model predictive controller employed for the voltage regulation of a full-bridge DC–DC converter within certain specified constraints [111]. A model predictive control technique was proposed for a DAB DC–DC converter that placed weighting factors of a cost function based on the ANN technique to improve dynamic stability and output voltage regulation [112].

H. Bidirectional Switching Technique With Intelligent and Digital Control

A comparison of different hybrid control techniques, i.e., extended-phase shift direct power control, dual-phase shift with voltage control, and single-phase shift with voltage control, was examined with a DSP controller for a DAB DC–DC converter [51]. A multi-input SEPIC DC–DC converter integrating solar PV and a battery source connected to the bidirectional port of the converter uses an Arduino Pro mini for control purposes and S8050 as the gate driver circuit [88]. The Q-Learning algorithm, one type of reinforced learning, was applied to compute angles for triple-phase shift modulation of a DAB DC–DC converter that was implemented in the real-time domain experimentally to minimize switching and magnetic losses and reduce optimization complexity [113]. The comparison between the DPS method, fuzzy-PI, and fuzzy-particle swarm optimization (PSO)-PI controllers were presented, and the fuzzy-PSO-PI controller was more efficient in controlling a full-bridge bidirectional DC–DC converter [114]. A hybrid-phase shift technique termed as current-stress-optimized unified-phase shift modulation was implemented on a DAB DC–DC converter using a DSP28335 processor for an efficient converter operation [115]. A Cyclone IV FPGA realized a geometric sequence algorithm (based on MPC) and asymmetric half-cycle modulation of the DAB DC–DC converter that proved to be a precise, fast, and robust technique [116].

4.5 Comparative Analysis

The different types of DC–DC converters studied in the previous section are compared on their operational performance and design configuration. A comparative analysis has been done for unidirectional DC–DC converters, bidirectional isolated DC–DC converters, and bidirectional non-isolated DC–DC converters based on their efficiency and total parts count. The efficiency versus switching frequency graph in Figure 4.14 depicts that with an increase in switching frequency, the efficiency of DC–DC converters decreases. Among the different types of unidirectional resonant DC–DC converters discussed above, full-bridge LLC resonant converters are the most efficient one with maximum efficiency at the minimum resonant frequency and maximum power rating specification.

Bridge-type bidirectional isolated DC–DC converters are advantageous over the other types of bidirectional isolated DC–DC converters in terms of the ability to operate at a higher switching frequency and higher power rating. This can be observed from Figure 4.15 & Figure 4.16. These figures show efficiency versus switching frequency of bidirectional isolated converter and non-isolated converter respectively, Figure 4.17 shows efficiency versus power rating of unidirectional resonant DC-DC converter and Figure 4.18 presents efficiency versus power rating of bidirectional isolated

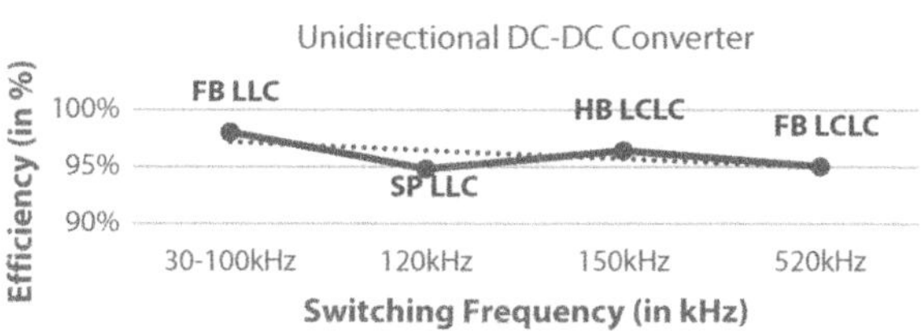

Figure 4.14 Efficiency vs switching frequency of unidirectional resonant DC–DC converter.

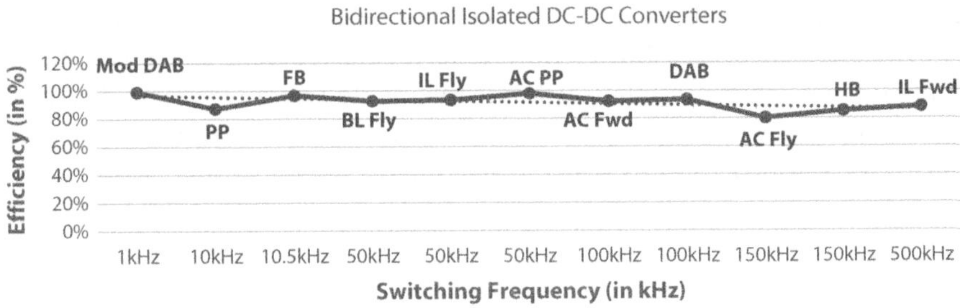

Figure 4.15 Efficiency vs switching frequency of bidirectional isolated DC–DC converter.

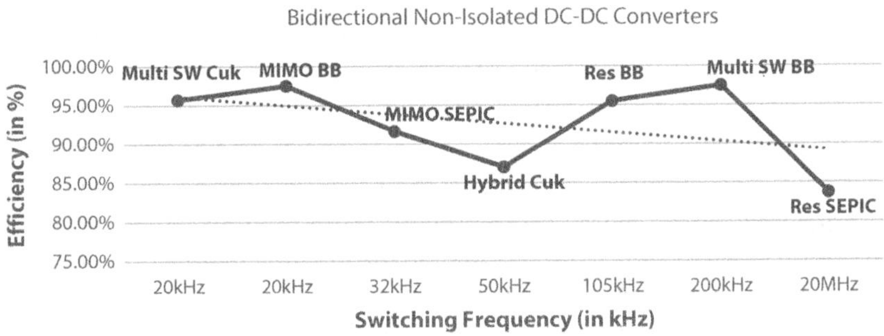

Figure 4.16 Efficiency vs switching frequency of bidirectional non-isolated DC–DC converter.

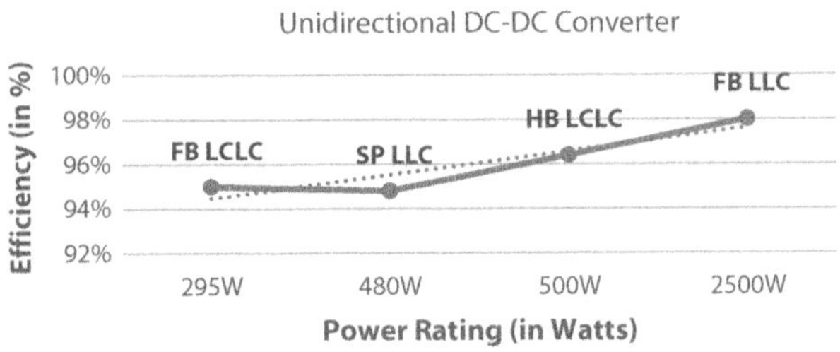

Figure 4.17 Efficiency vs power rating of unidirectional resonant DC–DC converter.

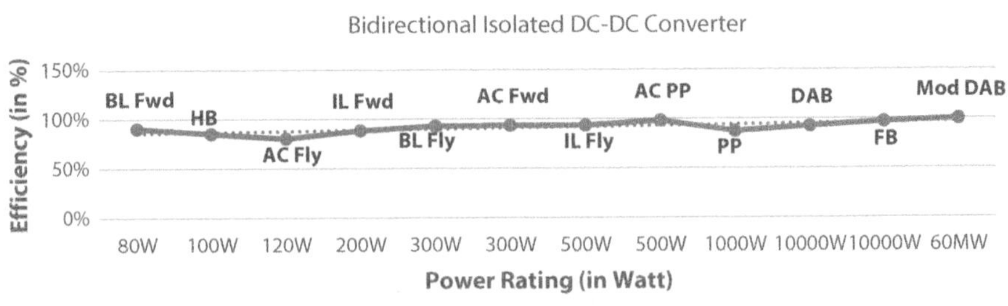

Figure 4.18 Efficiency vs power rating of bidirectional isolated DC–DC converter.

DC-DC converter. The bridge-type bidirectional isolated DC–DC converters include configurations of half-bridge type, full-bridge type, dual-active bridge type, and modular DAB-type converters. The next comparison brings the superiority of buck-boost configuration again over SEPIC and Cúk-type bidirectional non-isolated DC–DC converters in terms of efficiency, power rating, and switching frequency range. As per the different types of bidirectional buck-boost DC–DC converters discussed above (multi-switch type, resonant type multi-input multi-output type, and hybrid type), the MIMO buck-boost converter is the most efficient one among the other configurations with maximum rated power and limited range of switching frequency.

The DC–DC converters could be economical, depending on the number of components forming the circuit configuration. Most converter components include power switches that are either IGBTs or MOSFETs, power diodes, inductors, capacitors, and transformers (for isolation). A graphical representation of the total parts count has been displayed for unidirectional DC–DC converters, bidirectional isolated DC–DC converters, and bidirectional non-isolated DC–DC converters. From Figure 4.19, it can be visualized that modified configurations implement a greater number of components compared with the other configurations. From Figure 4.20, bridge-type bidirectional isolated DC–DC converters are the configurations implementing a maximum number of components, Figure 4.21 depicts bidirectional buck-boost DC–DC converters implementing a maximum number of components The data for scribing the graphs from Figure 4.20 to Figure 4.22 could be analyzed from Table 4.1 to Table 4.6 in the appendix section. Table 4.7, 4.8 & 4.9 depicts a comparative analysis of unidirectional DC-DC converters, bidirectional isolated DC-DC converters and bidirectional non-isolated DC-DC converters respectively; according to their distinct feature, control

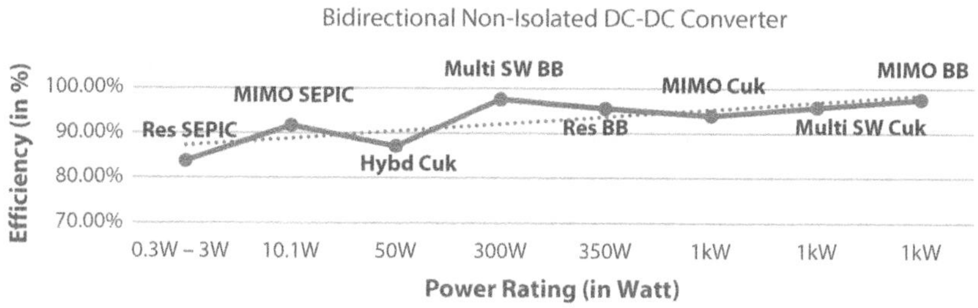

Figure 4.19 Efficiency vs power rating of bidirectional non-isolated DC–DC converter.

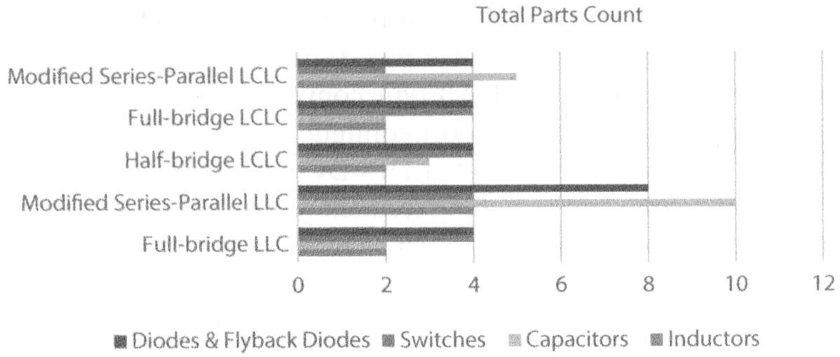

Figure 4.20 Total parts count of unidirectional resonant DC–DC converter.

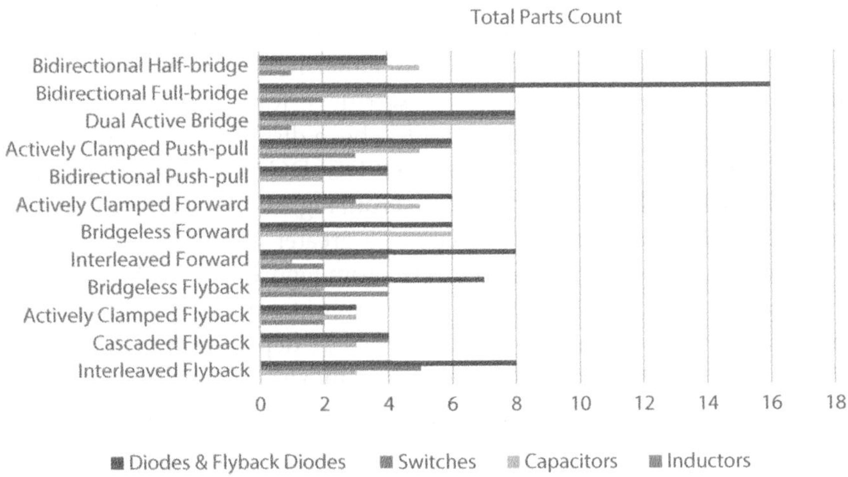

Figure 4.21 Total parts count of bidirectional isolated DC–DC converter.

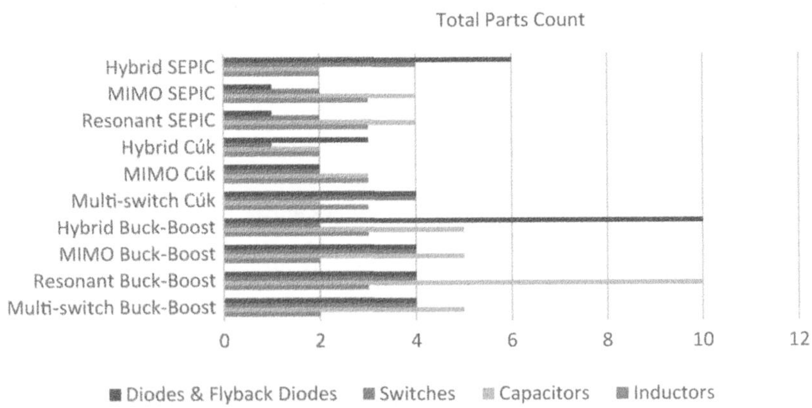

Figure 4.22 Total parts count of bidirectional non-isolated DC–DC converter.

scheme employed and field of application. Tables 4.7, 4.8 & 4.9 depicts a comparative analysis of unidirectional DC-DC converters, bidirectional isolated DC-DC converters and bidirectional non-isolated DC-DC converters respectively; according to their distinct feature, control scheme employed and field of application.

4.6 Conclusion

This chapter presents a comprehensive review of the emerging trends in the classification of DC–DC converters typically used in microgrid applications. This chapter classifies the DC–DC converters into two major categories—unidirectional converters and bidirectional converters—according to their specific usage. There is also a brief-up about the combination of different switching techniques and control schemes implemented for the operation of various DC–DC converters as mentioned in Section 4.4. A comparative analysis is presented to provide a clear-cut idea of the converters with parameters. The main objective is to present the most efficient converter with the lowest voltage stress, maximum switching frequency for specific power ratings, reduced switching and conduction losses, and less circuit complexity. Broadly speaking, every converter in its domain has remarkable operation owing to its application. We cannot compare all the converters on a single parameter and estimate their performance. Therefore, a unique color coding is adopted to group up similar converters. It is essential to focus on those specific groupings and a comparison that has been done on them. However, according to the comparative analysis, full-bridge LLC type is more efficient among unidirectional resonant DC–DC converters, bridge-type converters (especially dual-active bridge DC–DC converters) are most versatile among bidirectional isolated DC–DC converters, and multi-input multi output buck-boost converters are more reliable among bidirectional non-isolated DC–DC converters. The above-mentioned DC–DC converters are mostly preferred as power electronic interfaces in a DC microgrid or a hybrid microgrid. Hence, research advancements are focused on the direction of enhancing the operation of these DC–DC converters to provide reliable, dynamic, and efficient power conversion technology in a microgrid.

Table 4.1 Comparative configuration analysis of unidirectional resonant DC–DC converters.

		Active components		Passive components		
Converters	**Voltage gain**	**Inductors**	**Capacitors**	**Switches**	**Diodes and flyback diodes**	**References**
Full-bridge LLC	$\frac{1}{\sqrt{\left[1-\left(\frac{f_s}{f_r}\right)\right]^2+\left[\frac{f_s}{f_r}\left(\frac{1}{Q}\right)\right]^2}}$	2	2	4	0+4	[11]
Modified series–parallel LLC	$\frac{1}{\sqrt{\left[1+\frac{L_r}{L_m}\left[1-\left(\frac{f_r}{f_s}\right)\right]\right]^2+\left[\sqrt{\frac{L_r}{C_r}}\frac{1}{R}\left(\frac{f_s}{f_r}-\frac{f_r}{f_s}\right)\right]^2}}$	4	10	4	4+4	[16]

(*Continued*)

Table 4.1 Comparative configuration analysis of unidirectional resonant DC–DC converters. (*Continued*)

Converters	Voltage gain	Active components		Passive components		References
		Inductors	Capacitors	Switches	Diodes and flyback diodes	
Half-bridge LCLC	$\frac{1}{\sqrt{\left[1+\frac{L_r}{L_{meq}}-\frac{L_r}{L_{meq}}\left(\frac{f_r}{f_s}\right)^2\right]^2+\left[\frac{\pi^2}{8N^2}Q\left(\frac{f_s}{f_r}-\frac{f_r}{f_s}\right)\right]^2}}$	2	3	4	0+4	[22]
Full-bridge LCLC	$\frac{1}{\sqrt{\left[3-\left(\frac{f_s}{f_r}\right)^2-\left(\frac{f_r}{f_s}\right)^2\right]^2+\left[\frac{1}{Q}\left(\frac{f_r}{f_s}-\frac{f_s}{f_r}\right)\right]^2}}$	2	2	4	0+4	[11]
Modified series–parallel LCLC	------------	4	5	2	2+2	[20]

where f_s = switching frequency; f_r = resonant frequency; $L_{meq}=L_p-\frac{1}{(2\pi f_s)^2C_s}$; $Q=\frac{1}{R}\sqrt{\frac{L_r}{C_r}}$; N = transformation ratio

Table 4.2 Comparative operational analysis of unidirectional resonant DC–DC converters.

Converters	Power rating	Output voltage range	Input voltage range	Switching frequency	Efficiency	References
Full-bridge LLC	2.5 kW	120 V	80–170 V	30–100 kHz	98%	[13]
Modified series–parallel LLC	480 W	24 V	750–800 V	120 kHz	94.8%	[16]
Half-bridge LCLC	500 W	250–400 V	12 V	150 kHz	96.4% at 400V input, 300W	[22]
Full-bridge LCLC	295 W	4.8 V	40 V	520 kHz	>95% for 60%–100% loading	[19]
Modified series–parallel LCLC	5 W	+3.3 to + 5 V	15V–20 V	150 kHz	-------	[20]

Table 4.3 Comparative configuration analysis of bidirectional isolated DC–DC converters.

Converters	Voltage gain	Active components		Passive components		Voltage stress	References
		Inductors	Capacitors	Switches	Diodes and flyback diodes		
Interleaved flyback	Boost mode: $\frac{2D}{(1-D)N}$ Buck mode: $\frac{(4D-1).N}{2D}$	0	3	5	8	-NA-	[23]
Cascaded flyback	$N\frac{D}{(1-D)}$	0	n+1	2n	0+2n	-NA-	[25]
Actively clamped flyback	$N\frac{D}{(1-D)}$	2	3	2	3	$V_{max} = V_{1_{max}} + NV_2 + i_{L_r}(t_3)\sqrt{\frac{L_r}{C_{clamp}}}$	[29]

(*Continued*)

Table 4.3 Comparative configuration analysis of bidirectional isolated DC–DC converters. (*Continued*)

Converters	Voltage gain	Active components		Passive components		Voltage stress	References
		Inductors	Capacitors	Switches	Diodes and flyback diodes		
Bridgeless flyback	$N\frac{\sqrt{2}D}{(1-D)}$	4	2	4	7	-NA-	[31]
Interleaved forward	N D	2	1	4	8	$\frac{V_1}{2(1-D)}$	[33] [35]
Bridgeless forward	N D	0	6	2	4+2	$V_1\frac{2C_{eq1}+C_1}{C_{eq1}+C_1}$	[36]
Actively clamped forward	N D	2	5	3	3+3	$V_1\frac{(1-2D)}{(1-D)}$	[38] [39]
Bidirectional push–pull	N D	0	2	4	0+4	-NA-	[41]

(*Continued*)

Table 4.3 Comparative configuration analysis of bidirectional isolated DC–DC converters. (*Continued*)

Converters	Voltage gain	Active components		Passive components		Voltage stress	References
		Inductors	Capacitors	Switches	Diodes and flyback diodes		
Actively clamped push–pull	N D	3	5	6	0+6	$\frac{V_{1_{max}}}{N}$	[45]
Dual active bridge	N M	1	8	8	0+8	-NA-	[49]
Modular DAB	N D	n	2	2x + 2y	0 + 2x + 2y	-NA-	[52]
Bidirectional full-bridge	$\frac{2N}{(1-D)}$	2	4	8	8+8	$v_{C_r}(t)+\frac{V_2}{2N}$	[57] [59]
Bidirectional half-bridge	ND	1	5	4	0+4	-NA-	[57]

where N = turns ratio, D = duty cycle, n = number of modules connected in cascade, t_3 = switching instant, V_1 = input voltage, V_2 = output voltage, C_{eq1} = equivalent capacitance of input circuit, M = voltage transfer ratio, x = number of switches in left arm, y = number of switches in right arm, NA = not available

Table 4.4 Comparative operational analysis of bidirectional isolated DC–DC converters.

Converters	Power rating	Output voltage range	Input voltage range	Switching frequency	Efficiency	References
Interleaved flyback	500 W	48 V	400 V	50 kHz	93%	[23]
Cascaded flyback	1.4 kW	400 V	50 V	---	---	[26]
Actively clamped flyback	120 W	---	120 V	150 kHz	80%	[29]
Bridgeless flyback	300 W	50 V	90–265 V	50 kHz	92.5%	[31]
Interleaved forward	200 W	12 V	36–75 V	500 kHz	87%–90%	[32]
Bridgeless forward	80 W	90 V	90–265 V	---	90%	[36]
Actively clamped forward	300 W	48 V	200–400 V	100 kHz	92%–93%	[38]
Bidirectional push–pull	1 kW	400 V	48 V	10 kHz	87%	[41]
Actively clamped push–pull	500 W	440 V	30–40 V	50 kHz	97.7%	[45]
Dual active bridge	10 kW	28 V	270 V	100 kHz	92%	[49]
Modular DAB	60 MW	5 kV	30 kV	1 kHz	99%	[53]
Bidirectional full-bridge	10 kW	330 V	260–320 V	10.5 kHz	96.5%	[60]
Bidirectional half-bridge	100 W	5 V	48 V	150 kHz	85%	[61]

Table 4.5 Comparative configuration analysis of bidirectional non-isolated DC–DC converters.

Converters	Voltage gain	Active components		Passive components		References
		Inductors	Capacitors	Switches	Diodes and flyback diodes	
Multi-switch buck-boost	$V_o = \frac{\sum_{i=1}^{N}\left(D_{i_{eff}} V_i\right)}{1-\sum_{i=1}^{N} D_{i_{eff}}}$	2	5	4	0+4	[65]
Resonant buck-boost	$\frac{2}{(1-D)}\frac{\left(X_{L_{R2}}+R_{ac}\right)R_{ac}}{\left(X_{L_{R1}}+X_{C_P}\right)\left(X_{L_{R2}}+R_{ac}\right)+R_{ac}}$	3	10	4	0+4	[68]
MIMO buck-boost	$V_5 = \frac{1}{(1-D_3)}V_3 + \frac{(1-D_1)}{(1-D_3)}(V_1+V_2)$	2	5	4	0+4	[71]
Hybrid buck-boost	$-\frac{(1+5D+2D^2)}{(1-D)}$	3	5	2	8+2	[75]
Multi-switch Cúk	$\frac{D}{1-D}$	3	2	4	0+4	[76]

(*Continued*)

Table 4.5 Comparative configuration analysis of bidirectional non-isolated DC–DC converters. (*Continued*)

Converters	Voltage gain	Active components		Passive components		References
		Inductors	**Capacitors**	**Switches**	**Diodes and flyback diodes**	
MIMO Cúk	$-\frac{D}{(1-D)^2}+\frac{D}{(1-D)}$	3	3	2	2+0	[77]
Hybrid Cúk	$\frac{D}{1-D}\left(1+\frac{D}{N}\right)$	2	2	1	3+0	[80]
Resonant SEPIC	$\frac{D}{1-D}$	3	4	2	1+0	[82]
MIMO SEPIC	$V_o=\frac{D_2V_2+D_{eff}V_1}{1-D_1}$	3	4	2	1+0	[88]
Hybrid SEPIC	$\frac{2D}{(1-D)}$	2	2	4	2+4	[89]

Where $D_{i_{eff}}$ = duty ratio of each switch, R_{ac} = effective AC load resistance, $X_{L_{R1}}$ *and* $X_{L_{R2}}$ = inductive reactance of resonant inductors 1 and 2, X_{C_p} = capacitive reactance of parallel resonant capacitance, D = duty ratio, N = turns ratio of transformer, $D_{eff} = D_2 - D_1$ = Effective duty ratio of MIMO SEPIC converter

Table 4.6 Comparative operational analysis of bidirectional non-isolated DC–DC converters.

Converters	Power rating	Output voltage range	Input voltage range	Switching frequency	Efficiency	Reference
Multi-switch buck-boost	300 W	24 V	48 V	200 kHz	97.5%	[64]
Resonant buck-boost	350 W	380 V	48 V	105 kHz	95.5%	[68]
MIMO buck-boost	1 kW	36–60 V	22–40 V	20 kHz	97.5%	[70]
Hybrid buck-boost	500 W	50 V	-400 V	50 kHz	---	[75]
Multi-switch Cúk	1 kW	360 V	250 V	20 kHz	95.7%	[76]
MIMO Cúk	1 kW	72 V	24 V	---	93.9%	[79]
Hybrid Cúk	50 W	92 V	30 V	50 kHz	87%	[80]
Resonant SEPIC	0.3–3 W	3–9 V	3.6–7.2 V	20 MHz	83.7%	[83]
MIMO SEPIC	10.1 W	52.8 V	12.1 V	32 kHz	91.64%	[88]
Hybrid SEPIC	1 kW	400 V	240 V	50 kHz	---	[89]

Table 4.7 Comparative analysis of the overall configuration of unidirectional resonant DC–DC converter.

Converters	Features	Control scheme	Applications	References
Half-bridge LLC	1. Simple in construction 2. Isolated DC–DC power stage provides wide range of voltage gain 3. Soft switching reduces power loss and EMI	Output voltage control	• Integration of renewables in microgrids • Low-cost switched mode power supplies	[13]
Full-bridge LLC	1. Greater efficiency with switching frequency near resonant frequency 2. Less switching losses with high switching frequency 3. The three criteria to guarantee ZVS are: a) Inductive input impedance b) Sufficient energy stored in resonant tank c) Sufficient dead time	Frequency control	• Server, telecom, consumer electronics • Auxiliary drives in traction	[13] [14]
Modified series–parallel LLC	1. Two LLC resonant circuits are connected in a unique full bridge series-parallel combination 2. Fundamental frequency analysis of converter for finding voltage gain 3. ZVS is easily achieved for switches	Frequency control modulation	• Light rail transit system • Small-scale industries • Supplying residential loads	[16]
Half-bridge LCLC	1. In server and communication systems hold-up problem is solved with an LCLC converter 2. Resonant circuit elements are designed with first harmonic approximation	Asymmetric pulse width modulation	• Computer and telecommunication systems	[22]
Full-bridge LCLC	1. Efficient with high power density, low switching loss, high frequency, and h.v. applications 2. Operate between two resonant frequencies with constant output voltage and input resistance variation 3. ZVS is easily achieved for switches and ZCS for diodes	Pulse frequency modulation	• Lasers, traveling-wave tube amplifiers • X-ray imaging, radio frequency generation, EV integration, etc.	[17] [18] [19]

(*Continued*)

Table 4.7 Comparative analysis of the overall configuration of unidirectional resonant DC–DC converter. (*Continued*)

Converters	Features	Control scheme	Applications	References
Modified series–parallel LCLC	1. Multi-output converter with regulated output voltages and controlled PWM and frequency operation 2. The converter was analyzed for steady-state response and transient start-up conditions 3. A regulation error of <5%, initial overshoot of ~10% with $\zeta = 0.6$ was obtained [20]. This determines a good steady-state stability characteristic	Combination of frequency control and PWM control	• Signal processing applications • Microprocessors and computer systems	[20]

Table 4.8 Comparative analysis on the overall configuration of bidirectional isolated DC–DC converter.

Converters	Features	Control scheme	Applications	References
Interleaved flyback	1. Bidirectional power flow capability 2. Dual transformer shares equally the HV side of microgrid so that voltage stress over switches reduces also reduces the current stress with parallel connection of LV side 3. Converter operates in boundary conduction mode with different loads [23] 4. Type of a multi-input flyback converter 5. Interleaving reduces ripples and EMI with better transient response	Input and output voltage regulation and duty ratio control by central controller DSPIC30F2010	• Emergency telecommunication • Remote area power supply • PV-based relay stations, navigation aids, etc. • Battery backup facility	[23] [24]
Cascaded flyback	1. Dielectric electro-active polymers technology requires high-voltage power electronics (HVPE) to do the conversion that implements a cascaded flyback converter 2. Bidirectional power DC–DC converter exhibited dynamic and accurate voltage control 3. Cascading can provide a wide range of voltage levels	Current control, voltage control (PI controller) [25], fuzzy logic control [26]	• Wave and tidal power generators • DC microgrid with energy source and DC loads	[25] [26]

(*Continued*)

Table 4.8 Comparative analysis on the overall configuration of bidirectional isolated DC–DC converter. (*Continued*)

Converters	Features	Control scheme	Applications	References
Actively clamped flyback	Active clamping is most beneficial in: 1. Reutilizing the stored energy of the high-frequency transformer 2. Clamps the voltage transients during switching 3. Delay time presence due to active clamping has reduced switching losses The designed three-port actively clamped flyback converter could: 1. Accurately track the MPP when there is no droop control 2. Sigma–delta modulator could provide droop control by choosing which output port should receive energy from input switching	Hill climbing MPPT with direct duty cycle perturbation and droop control on sigma–delta modulation	• Switched-mode power supplies • Solar PV systems • LED and LCD lighting systems	[28] [29]
Bridgeless flyback	1. Size of isolating transformer is reduced due to reduced peak magnetizing current of the primary side 2. Efficient power factor correction by completely making the input voltage in phase with input current, i.e., having lower harmonic currents 3. ZVS operation ensures no power loss during turn-on	Peak current mode PFC controller [31]	• Low power lighting • Microgrids with DC loads	[30] [31]

(*Continued*)

Table 4.8 Comparative analysis on the overall configuration of bidirectional isolated DC–DC converter. (*Continued*)

Converters	Features	Control scheme	Applications	References
Interleaved forward	1. High conversion efficiency 2. The increase in the number of switches reduces the voltage and current stress across individual switches. 3. Ripple current cancellation effect and ripple frequency doubling reduce the size of output filter	Phase shift control and duty ratio control [33]	• DC lighting • DC microgrids • Voltage regulation module	[32] [33] [34] [35]
Bridgeless forward	1. Power factor correction in both discontinuous conduction mode (DCM) or boundary conduction mode (BCM) 2. Low output voltage ripple, low voltage stress on switches and low power applications. 3. Harmonics in the line current is reduced by an improved constant on-time control	Constant on-time control	• Power factor correctors	[36] [37]
Actively clamped forward	Multi-switch actively clamped topology (particularly 2SW/4SW) had: 1. Delayed turn-off gate signal and a clamping diode reduced voltage stress 2. Wide input range applications 3. Partial zero voltage switching 4. Boot-strapping technique implemented	Expanded duty cycle control	• High power applications • IC technology	[38] [39] [40]

(*Continued*)

Table 4.8 Comparative analysis on the overall configuration of bidirectional isolated DC–DC converter. (*Continued*)

Converters	Features	Control scheme	Applications	References
Bidirectional push–pull	1. High step-up ratio, isolation prevents combined damage, compact in size and efficient 2. Dead time can enhance bucking by adjusting duty cycle properly 3. Effective in bidirectional power flow	PWM and MPPT technique	• Telecom system • DC microgrids	[41] [42]
Actively clamped push–pull	1. Highly efficient in boost mode and buck mode 2. ZVS is achieved for switches for all possible operations 3. ZCS is achieved for LV devices 4. Radio frequency noise and EMI can be reduced with proper line impedance stabilization 5. Eliminates the snubber circuit requirement	Hardware control with Xilinx FPGA board	• Portable battery system for commercial use • UPS in fuel cell systems	[43] [44] [45]
Dual active bridge	1. Highly efficient with fast dynamic response to transient conditions 2. Wide voltage range, low ripples in output current, and low current stress on switches 3. Reduces switching losses with soft switching technique in clamped configurations 4. Most preferred converter for bidirectional power transfer functions. 5. Highly reliable converter for islanded operations.	Phase shift control [47], virtual DC machine control [48], extended phase shift—direct power control [51]	• Smart grid network • Electric vehicles • UPS • Aerospace microgrids	[46] [47] [48] [49] [50] [51]

Table 4.9 Comparative analysis of overall configuration of bidirectional non-isolated DC–DC converters.

Converters	Features	Control scheme	Applications	References
Multi-switch buck-boost	1. Highly efficient, robust, complex, and compact in structure 2. Ease the bidirectional power flow capability. 3. Perfect voltage balance capability in between a unipolar and bipolar microgrid 4. Incorporates a few passive elements to realize configuration 5. Increase in power loss due to several switches	Tri-state logic [66], digital PI control [64], FPGA controller [67]	Microgrids integrated with • Hybridized energy storage systems • Safety extra low voltage level devices Hybrid microgrids interlinking AC and DC microgrid or unipolar and bipolar microgrid	[64] [65] [66] [67]
Resonant buck-boost	1. ZVS turn-on for switches and ZCS turn-on and turn-off for diodes 2. Reduced voltage stress across switches 3. High conversion ratio 4. Reduced volume/weight ratio of magnetics	Soft switching [68], PWM switching [69]	• UPS • Power supplies in EVs • Power supplies in industries using renewables	[68] [69]
MIMO buck-boost	1. Highly efficient, reduction in unnecessary power conversion stages, reduction in overall size and component count, cheaper and high power or energy density 2. Higher voltage boosting and bidirectional power flow capability 3. Integration with multiple resources, hybrid storage system, and variety of loads	PI and PID control [70], MPP control with cascaded PI loops [71]	• Modern loads in residential and industrial fields • Distributed generation systems • EVs and other energy conversion systems	[70] [71] [72] [73]

(*Continued*)

Table 4.9 Comparative analysis of overall configuration of bidirectional non-isolated DC–DC converters. (*Continued*)

Converters	Features	Control scheme	Applications	References
Hybrid buck-boost	1. High gain or conversion ratio with a low duty cycle 2. Reduced voltage stress across switches allows use of low voltage rating switches 3. Disadvantages of cascaded configuration such as high conduction loss, component rating, and low efficiency can be overcome with this hybrid configuration	Asynchronous sinusoidal PWM + PI control	• Rural electrification • Autonomous industrial electrification	[74] [75]
Multi-switch Cúk	1. It is a highly efficient voltage doubler circuit and uses components with a low voltage rating 2. Reduces size of the filter capacitor due to low ripple current 3. Voltage stress is reduced to half across switches	PWM Control	• Integration of battery energy system in microgrid • Portable electronic devices • Power factor correctors	[76]
MIMO Cúk	1. A voltage balancing circuit suitable for bidirectional power flow 2. It reduces the voltage and current ripples in the output and input 3. It combines the advantages of both Cúk and SEPIC	Duty cycle control	• Power supply for telecommunication system • Electric machine drives • Solar energy system • Multi-level inverters	[77] [78] [79]
Hybrid Cúk	1. High conversion ratio with wide input voltage range 2. Designed to operate for both AC and DC loads 3. Voltage stress reduction across the power switches	Modified unipolar sinusoidal PWM [81]	• Microgrid and nanogrid applications • Switched-mode power supply	[80] [81]

(*Continued*)

Table 4.9 Comparative analysis of overall configuration of bidirectional non-isolated DC–DC converters. (*Continued*)

Converters	Features	Control scheme	Applications	References
Resonant SEPIC	1. Highest frequency of operation with a wide voltage range, compact in size, and highly efficient 2. Good transient response characteristics 3. Counters load disturbances, requires magnetic energy storage, and operates with low power	Fixed frequency PWM with hysteresis override [83]	• Aircraft communication • Computer microprocessors • Modern space communications • Local voltage regulation in microgrids	[82] [83]
MIMO SEPIC	1. Highly efficient with low switching losses 2. Explicit usage in incorporating energy storage system due to buck-boost characteristics of SEPIC converter 3. In isolated versions, it provides low EMI due to low-input ripple current 4. Acts as a voltage equalizer in a multi-stacked converter reducing circuit complexity	Arduino microcontroller based on Atmega 328 [88], duty cycle control [85]	• Hybrid systems • Microgrids interfacing supercapacitors • Rural irrigation purposes • Microgrids with energy storing systems	[84] [85] [86] [87] [88]
Hybrid SEPIC	1. Efficient stepping-up of voltages even at higher duty ratios, reduced filters, and reliable power flow 2. Low voltage stress across switches 3. Low ripple currents in output and input enhance charging and discharging	Quantitative theory based robust controller [90]	• Renewable energy source integration with energy storage • Nanogrid applications in homes • EV auxiliary power supply	[89] [90] [91] [92]

Appendix

A systematic tabular representation of configuration, operation, and overall analysis is presented for each DC–DC converter category. It should be noted that a particular color coding procedure is followed to specify similar types of converters with similar configurations.

References

1. Gellings, C.W., *The smart grid: Enabling energy efficiency and demand response*, CRC Press, Taylor and Francis Group, IEEE, River Publishers, 2009.
2. https://ourworldindata.org/emissions-by-sector
3. https://www.iea.org/data-and-statictics/data-product/world-energy-statistics-and-balances
4. Mumtaz, F., Yahaya, N.Z., Meraj, S.T., Singh, B., Kannan, R., Ibrahim, O., Review on non-isolated DC-DC converters and their control techniques for renewable energy applications. *Ain Shams Eng. J.*, 12, 4, 3747–3763, 2021. Elsevier.
5. Jeremy, L.J., Ooi, C.A., Teh, J., Non-isolated conventional DC-DC converter comparison for a photovoltaic system: A review. *J. Renew. Sustain. Energy*, 12, 1, 1–22, 2020.
6. Gandhi, K. and Gupta, S.K., Operational strategies and electricity market structure of microgrid: A critical review. *Renewable Energy Focus J.*, 39, 163–171, December 2021, Elsevier.
7. Kumar, J., Agarwal, A., Singh, N., Design, operation and control of a vast DC microgrid for integration of renewable energy sources. *Renew. Energy Focus*, 34, 00, 17–36, September 2020, Elsevier.
8. Butt, O.M., Zulqarnain, M., Butt, T.M., Recent advancement in smart grid technology: Future prospects in the electrical power network. *Ain Shams Eng. J.*, 12, 1, 687–695, Elsevier, 2021.
9. Erickson, R.W. and Maksimović, D., *Fundamentals of power electronics*, Second Edition, Kluwer Academic Publishers, Secaucus, NJ, USA, 2000.
10. Yu, S.-Y., Chen, R., Viswanathan, A., Survey of resonant converter topologies. *Texas Instruments Power Supply Design Seminar*, 2018, www.ti.com/psds.
11. Salem, M. and Yahya, K. Resonant power converters, in: *Electric Power Conversion*, London, UK: IntechOpen, 2019.
12. Urno, J., Shoji, H., Kanouda, A., Kaminaga, Y., Hitachi, *Unidirectional DC-DC converters*, United States Patent, US, Jul,7, 2009, 7,557,546 B2.
13. Zubeita Luis, E. and Lehn Peter, W., A high efficiency unidirectional DC/DC converter for integrating distributed resources into DC microgrids, in: *2015 IEEE First International Conference on DC Microgrids (ICDCM)*, IEEE, 2015.
14. Bhajana, V.V.S.K., Drabek, P., Jara, M., Design & analysis of a full bridge LLC DC-DC converter for auxiliary power supplies in traction. *Sadhana*, 43, 1–8, Indian Academy of Sciences, 2018.
15. Paul, A., K J, L., Paul, R.T., Unidirectional DC/DC converters for microgrids. *GRDJE*, 74–76, May 2019.
16. Lin, B.R., Analysis of series-parallel resonant converter for DC microgrid applications. *MDPI Processes*, 9, 542, 2021.
17. Conesa, A., Velasco, G., Martinez, H., Roman, LCLC resonant converter as maximum power point tracker in PV systems, in: *2009 13th European*

Conference on Power Electronics and Applications, Taylor & Francis, European Power Electronics & Drives (EPE-Barcelona), 2009.

18. Xuan, Y., Yang, X., Chen, W., Liu, T., Hao, X., A Novel Three-level CLLC resonant DC-DC converter for bidirectional EV charger in DC microgrids. *IEEE Trans. Ind. Electron.*, 68, 3, 2334–2344, Mar, 2021.
19. Zhao, B., Wang, G., Hurley, W.G., Analysis and performance of LCLC resonant converters for high-voltage high frequency applications. *IEEE J. Emerg. Sel. Topics Power Electron.*, 5, 3, 1272–1286, Sept, 2017.
20. Ang, Y., Bingham, C.M., Foster M, P., Stone, D.A., Analysis and control of dual-output LCLC resonant converters, and the impact of leakage inductance. *IEEE, 7th International Conference on Power Electronics and Drive Systems*, 2007.
21. Dobrucky, B. and Koscelnik, J., Multi-element resonant topology based on LCLC circuit: Theory & application, in: *Proceedings of the 9th International Conference on Circuits, Systems, Signal and Telecommunications (CSST'15)*, WSEAS, 2017.
22. Chen, Y., Wang, H., Hu, Z., Liu, Y.F., Afsharian, J., Yang, Z., LCLC resonant converter for hold up mode operation, in: *2015 IEEE Energy Conversion Congress and Exposition (ECCE)*, IEEE, 2015.
23. Chen, S.M., Liang, T.J., Huang, Y.H., A isolated bidirectional interleaved flyback converter battery backup system application, in: *2013 IEEE International Symposium on Circuits and Systems (ISCAS)*, 2013.
24. Venugopal, K. and Kanaksabapathy, P., Three port multi winding flyback converter for DC microgrid applications. *IEEE, International Conference on Technological Advancements in Power and Energy*, 2017.
25. Eitzen, L., Graf, C., Maas, J., Cascaded bidirectional flyback converter driving DEAP transducers, in: *IECON 2011 - 37th Annual Conference of the IEEE Industrial Electronics Society*, IEEE, 2011.
26. Manikandan, C.T. and Sundarrajan, G.T., Fuzzy logic controller for closed loop cascaded flyback converter fed PMDC-motor system. *Int. J. Power Electron. Drive Syst.*, 11, 4, 1857–1865, Dec 2020.
27. Ruedi, H., Thalheim, J., Garcia, O., *Advantages of advanced active clamping*, Power Electronics, Europe, 8, 2009, www.IGBT-Driver.com.
28. Chub, A., Zinchenko, D., Vinnikov, D., Blinov, A., Three-port flyback converter for photovoltaic module integration in bipolar DC microgrids, in: *2020 IEEE International Conference on Industrial Technology (ICIT)*, 2020.
29. Jailyn, P.A., Dheeraj, A., Rajini, V., Analysis of active clamp flyback converter. *Modern Appl. Sci.*, 9, 1, 12–26, 2015.
30. Chen, X., Jiang, T., Huang, X., Zhang, J., A high efficiency bridgeless flyback PFC converter for adapter application, in: *2013 Twenty-Eighth Annual IEEE Applied Power Electronics Conference and Exposition (APEC)*, 2013.
31. Tseng, S.Y., Huang, P.J., Wu, D.H., Power factor corrector with bridgeless flyback converter for DC loads applications. *Energies*, 11, 1–20, 2018, MDPI.

32. O'Loughlin, M., *200W interleaved forward converter design review using TI's UCC28221 PWM controller*, Texas Instruments, May 2004, Application Report SLUA312. https://www.ti.com/lit/an/slua312/slua312
33. Dasohari, M., Neti, V., Porpandiselvi, S., Manikyavani, A.V.R., An interleaved dual double-ended forward converter based led driver for DC lighting grids. *21st National Power Systems Conference (NPSC)*, IEEE, 2020.
34. Habumugisha, D., Chowdhury, S., Chowdhury, S.P., A DC-DC interleaved forward converter to step-up DC voltage for DC microgrid applications, in: *2013 IEEE Power & Energy Society General Meeting*, 2013.
35. Wu, H. and Xing, Y., Families of forward converters suitable for wide input voltage range applications. *IEEE Trans. Power Electron.*, 29, 11, 6006 – 6017, Nov 2014.
36. Lan, Z., Xie, X., Dong, H., Liu, S., A single-stage high power factor bridgeless forward converter with an improved constant on-time control, in: *2013 1st International Future Energy Electronics Conference (IFEEC)*, IEEE, 2013.
37. Soldano, M. and Salato, M., *Bridgeless bidirectional forward type converter*, US Patent Application Publication, US 2007/0008747 A1, Jan. 11, 2007. https://patents.google.com/patent/US20070008747A1/en.
38. Park, K.B., Moon, G.W., Youn, M.J., Two-switch active-clamp forward converter with one clamp diode and delayed turnoff gate signal. *IEEE Trans. Ind. Electron.*, 58, 10, 4768–4772, Oct 2011.
39. Doutreloigne, J. and Pauw, H.D., An active clamping H-bridge for isolated forward DC-DC converters, integrated in a high-voltage smart-power IC technology. *Proceedings of the World Congress on Engineering and Computer Science*, San Fransisco, USA, vol. II, 24-26 Oct, 2012.
40. Design Guide, *200W active clamp forward DC-DC converter*, Toshiba Electronic Devices & Storage Corporation, Nov, 2021. https://toshiba.semicon-storage.com/ap-en/semiconductor/design-development/referencedesign/detail.RD175.html
41. Solanki, D., Malwadkar, O., Mahajan, S., Nikam, S., Bajaga, R., Design of bidirectional push pull converter for microgrid applications, in: *2017 Second International Conference on Electrical, Computer and Communication Technologies (ICECCT)*, IEEE, 2017.
42. Saravana Prakash, P., Kalpana, R., Singh, B., Solar photovoltaic array fed pushpull buck DC-DC converter for telecom load, in: *2017 14th IEEE India Council International Conference (INDICON)*, 2017.
43. Prathibha, M.R. and Sridhar, H.S., High step-up high frequency push-pull DC-DC converter using mppt with DC motor load. *IEEE, International Conference on Computation of Power, Energy, Information and Communication (ICCPEIC)*, 2017.
44. Bal, S., Rathore, A.K., Srinivasan, D., Naturally clamped snubberless soft-switching bidirectional current-fed three-phase push-pull DC/DC converter for DC microgrid application, in: *2015 IEEE Applied Power Electronics Conference and Exposition (APEC)*, 2015.

45. Schulz, M., Wild, M., Chacon, R., Wunder, B., Marz, M., A bidirectional and isolated DC/DC converter connecting mobile battery systems to a DC grid in commercial buildings, in: *2019 IEEE 4th International Future Energy Electronics Conference (IFEEC)*, 2019.
46. Dinesh, V. and Prasad, E.S., Simulation of dual active converter for energy storage system. *International Journal of Engineering Trends and Technology (IJETT)*, 27, 2, 79–83, Sept 2015.
47. Barone, G., Brusco, G., Burgio, A., Motta, M., Menniti, D., Pinnarelli, A., Sorrentino, N., A dual active bridge DC-DC converter for application in a smart user network. *IEEE, Australasian Universities Power Engineering Conference, AUPEC 2014*, Curtin University, Perth, Australia, 28 Sept–1 Oct 2014.
48. Guo, Y., Meng, J., Wang, Y., Wang, C., A virtual DC machine control strategy for dual active bridge DC-DC converter. *IEEE, PES Innovative Smart Grid Technologies Asia*, 2019.
49. Tarisciotti, L., Costabeber, A., Linglin, C., Walker, A., Galea, M., Evaluation of isolated DC/DC converter topologies for future HVDC aerospace microgrids, in: *2017, IEEE Energy Conversion Congress and Exposition (ECCE), IEEE*, 2017.
50. Le, T.T., Nguyen, M.K., Wang, C., Choi, S., Fault-tolerant control of three-phase bidirectional current-fed dual active bridge DC-DC converter. *IEEE Transportation Electrification Conference & Expo (ITECE)*, 2021.
51. Hou, N., Song, W., Zhu, Y., Sun, X., Li, W., Dynamic and static performance optimization of dual active bridge DC-DC converters. *J. Mod. Power Syst. Clean Energy*, 6, 3, 607–618, Springer, 2018.
52. Hosseini, S.H., Sabahi, M., Sedaghati, F., Gharehpetian, G.B., A new extended topology for dual active bridge DC-DC converter. *IEEE, EPEC*, London, ON, Canada, 2015.
53. Stieneker, M. and Doncker, R.W.D., Dual-active bridge DC-DC converter systems for medium-voltage DC distribution grids, in: *2015 IEEE 13th Brazilian Power Electronics Conference and 1st Southern Power Electronics Conference (COBEP/SPEC)*, IEEE, 2015.
54. Kado, Y., Okutani, S., Katagiri, K., Huang, P.Y., Autonomous DC microgrid consisting of triple active bridge converters, in: *2019 IEEE Third International Conference on DC Microgrids (ICDCM)*, IEEE, 2019.
55. Saha, J., Gorla, N.B.Y., Panda, S.K., Modulation of direct matrix-basd dual-active -bridge (MB-DAB) AC-DC converters. *12th Energy Conversion Congress & Exposition - Asia (ECCE-Asia)*, IEEE, May 2021.
56. Sang, Y., Junyent-Ferré, A., Green, T.C., Operational principles of three-phase single active bridge DC/DC converters under duty cycle control. *IEEE Trans. Power Electron.*, 35, 8, 8737–8750, IEEE, Aug 2020.
57. Deng, F. and Chen, Z., Control of improved full-bridge three-level DC/DC converter for wind turbines in a DC grid. *IEEE Trans. Power Electron.*, 28, 1, 314–324, Jan 2013.

58. Zhang, W., Li, H., Liang, Q., Shi, Z., Zhang, T., Huang, L., Series phase-shifted full-bridge DC-DC converters in a new type of full scale photovoltaic DC power generation system. *IEEE, 2nd International Conference on Power and Renewable Energy*, 2017.
59. Suryadevara, R. and Parsa, L., Full-bridge zcs-converter-based high-gain modular DC-DC converter for pv integration with medium-voltage DC grids. *IEEE Trans. Energy Convers.*, 34, 1, 302–312, Mar 2019.
60. Ikeda, S., Kajiwara, K., Tsuji, K., Kurokawa, F., Efficiency improvement of isolated bidirectional boost full bridge DC-DC converter. *7th International Conference on Renewable Energy Research and Application*, Paris, France, Oct, 2018.
61. RD174-RGUIDE-01, *Half-bridge DC-DC power supply basic simulation circuit-reference guide*, TOSHIBA Electronics Devices & Storage Corporation, Jul, 2019. https://toshiba.semicon-storage.com/ap-en/semiconductor/design-development/referencedesign/detail.RD174.html.
62. Jing, P., Wang, C., Jiang, W., Zhang, G., Performance analysis of isolated three-level half-bridge bidirectional DC/DC converter. *IEEE 7th International Power Electronics and Motion Control Conference - ECCE*, Asia, Harbin, China, Jun, 2012.
63. Phetphimoon, W. and Bhumkittipich, K., Modelling and simulation of bidirectional half bridge DC-DC converter, in: *2016 13th International Conference on Electrical Engineering/Electronics, Computer, Telecommunications and Information Technology (ECTI-CON)*, IEEE, 2016.
64. Schulz, M., Schleippmann, N., Gosses, K., Wunder, B., März, M., Four switch buck/boost converter for DC microgrid applications. *EPE'20 ECCE Europe, European Power Electronics & Drives Association*, IEEE, 2020.
65. Pires, V.F., Foito, D., Cordeiro, A., Pires, A.J., A bidirectional DC-DC converter to interlink unipolar and bipolar DC microgrids. *9th IEEE International Conference on Smart Grid*, Jul, 2021.
66. Choubey, A.K. and Lopes, L.A.C., *A tri-state 4-switch bi-directional converter for interfacing supercapacitors to DC microgrids*, IEEE, 2017.
67. Jiya, I.N., Khang, H.V., Kishor, N., Ciric, R., Four quadrant switch based multiple-input DC-DC converter. *IEEE 12th Energy Conversion Congress & Exposition – Asia (ECCE-Asia)*, 2021.
68. Patil, D., Rathore, A.K., Srinivasan, D., A non-isolate bidirectional soft switching current fed LCL resonant dc/dc converter interface energy storage in DC microgrid, in: *2015 IEEE Applied Power Electronics Conference and Exposition (APEC)*, IEEE, 2015.
69. Litvani, L. and Hamar, J., New five level resonant DC/DC buck-boost converter, in: *2018 International Conference on Smart Energy Systems and Technologies (SEST)*, IEEE, 2015.
70. Akar, F., Tavlasoglu, Y., Ugur, E., Vural, B., Aksoy, I., A bidirectional non-isolated multi-input DC-DC converter for hybrid energy storage systems in electric vehicles. *IEEE Trans. Veh. Technol.*, 65, 10, 7944–7955, Oct 2016.

71. Saafan, A.A., Khadkikar, V., El Moursi, M.S., Zeineldin, H.H., A new multiport DC-DC converter for DC microgrid applications. *IEEE Industry Application Society Annual Meeting (IAS)*, 2021.
72. Kumaravel, S., Karthikeyan, V., Kumar, G.G., Jithin, T.J., Seshagiri Rao, V., Dual-input dual-output DC-DC converter for DC microgrid applications, in: *2018 IEEE International Conference on Power Electronics, Drives and Energy Systems (PEDES)*, IEEE, 2018.
73. Liang, Y., Zhang, H., Du, M., Sun, K., Parallel coordination control of multiport DC-DC converter for stand-alone photovoltaic-energy storage systems. *CPSS Trans. Power Electron. Appl.*, 5, 3, 235–241, Sept 2020.
74. Bagewadi, M.D. and Dambhare, S.S., A novel hybrid converter derived from buck-boost circuit for standalone applications. *EPE'17 ECCE Europe, European Power Electronics and Drives Association & the Institute of Electrical and Electronics Engineers (IEEE)*, 2017.
75. Gupta, N., Almakhles, D., Sagar Bhaskar, M., Sanjeevikumar, P., Holm-Nielsen, J.B., Mitolo, M., Novel hybrid high gain converter: Combination of cuk and buck-boost structures with switched inductor for DC microgrid. *2020 2nd Global Power, Energy and Communication Conference (IEEE GPECOM2020)*, Oct, 20-23, 2020.
76. de Almeida, A.B. and Font, C.H.I., DC-DC bidirectional 4-switch cuk converter with voltage-doubler concept for interfacing batteries in microgrids. *13th IEEE International Conference on Industry Applications*, 2018.
77. Marjani, J., Imani, A., Hekmati, A., Afjei, E., A new dual output DC-DC converter based on sepic and cuk converter. *International Symposium on Power Electronics, Electrical Drives, Automation and Motion*, IEEE, 2016.
78. Biswas, S., Dhople, S., Mohan, N., A three-port bidirectional DC-DC converter with zero-ripple terminal currents for PV/microgrid applications, in: *IECON 2013 - 39th Annual Conference of the IEEE Industrial Electronics Society*, IEEE, 2013.
79. Asthana, P. and Verma, V., Dual channel bidirectional interleaved cuk converter for battery source interfaced with LVDC μGrid, in: *2019 IEEE 4th International Future Energy Electronics Conference (IFEEC)*, IEEE, 2019.
80. Lica, S., Pop-Calimanu, I.M., Lascu, D., Popescu, S., Tomoroga, M., Ciresan, A., A new hybrid cuk converter, in: *2018 International Symposium on Electronics and Telecommunications (ISETC)*, IEEE, 2018.
81. R, S. and Kanakasabapathy, P., Hybrid converter based on cuk topology to supply both AC and DC loads, in: *IEEE International Conference on Technological Advancements in Power & Energy*, 2015.
82. Ikeda, S., Tsuji, K., Kajiwara, K., Kurokawa, F., Active Clamp Resonant SEPIC Converter for Local Voltage Regulation in 48V DC Based Energy Management Systems. *7th International Conference on Renewable Energy Research & Applications*, Paris, France, IEEE, Oct, 2018.

83. Hu, J., Sagneri, A.D., Rivas, J.M., Han, Y., Davis, S.M., Perreault, D.J., High-frequency resonant SEPIC converter with wide input & output voltage ranges. *IEEE Trans. Power Electron.*, 27, 1, 189–200, Jan 2012.
84. Tomar, A., Nguyen, P.H., Mishra, S., SEPIC-MISO Converter based PV water pumping system – an improved performance under mismatching conditions, in: *2020 IEEE 9th Power India International Conference (PIICON)*, IEEE, 2020.
85. Mohanty, G., SEPIC based multi input DC-DC converter. *5th International Conference for Convergence in Technology (I2CT)*, Pune, India, Mar 29-31, 2019.
86. K, N., Sailaja, V., Deepa., K., Manjunath, H.V., A DC-DC multi-output SEPIC converter for suburban power application. *International Conference on Electronics, Communication and Computational Engineering (ICECCE)*, IEEE, 2014.
87. Uno, M. and Tanaka, K., Single-switch cell voltage equalizer based on multistacked SEPIC for series-connected energy storage cells, in: *2011 IEEE 33rd International Telecommunications Energy Conference (INTELEC)*, IEEE, 2011.
88. Mahmud, T., Ahmed Rakeen, S.M., Araf, A.U., Ahmed, N., Azad, A.K.M A. M., Design and accuracy assessment of a multi-input single ended primary inductor converter (SEPIC) for highly efficient output from hybrid sources of renewable energy. *IEEE International Systems Conference (SysCon)*, 2022.
89. Sebaje, A.S., Martins, M. L.D. S., Illa Font, C.H., A hybrid bidirectional DC-DC converter based on a SEPIC/zeta converter with a modified switched capacitor cell. *Brazilian Power Electronics Conference (COBEP)*, IEEE, 2021.
90. Eate, V.K. and Veerachary, M., Analysis of Two-Input Switched Inductor-Capacitor Hybrid Buck-SEPIC DC-DC Converter. *IEEE Transportation Electrification Conference (ITEC-India)*, 2017.
91. Arun Shastry, P.S., Suresh, K.V., Vinayaka, K.U., Hybrid wind-solar systems using cuk-SEPIC fused converter with quasi-Z-source inverter. *IEEE Power, Communication and Information Technology Conference (PCITC)*, Siksha 'O' Anusandhan University, Bhubaneswar, India, 2015.
92. Maurya, R. and Patle, P., Quadratic SEPIC derived hybrid output converter for photo-voltaic applications, in: *2017 6th International Conference on Computer Applications In Electrical Engineering-Recent Advances (CERA)*, IEEE, 2017.
93. Chincholkar, S.H., Jiang, W., Chan, C.Y., A modified hysteresis-modulation-based sliding mode control for improved performance in hybrid DC-DC boost converter. *IEEE Trans. Circuits Syst. II: Express Br.*, 65, 11, 1683–1687, Nov 2018.
94. Abdel-Rahim, O. and Wang, H., A new high gain DC-DC converter with model predictive control based MPPT technique for photovoltaic system. *CPSS Trans. Power Electron. Appl.*, 5, 2, 191–200, Jun 2020.

95. Chen, Y., Zhang, Y., Liu, Y., Distributed Model predictive control for multi-port DC-DC converter wind-solar complementary system. *6th IEEE International Conference on Predictive Control of Electrical Drives and Power Electronics*, 2021.
96. Kapat, S. and Krein, P.T., A tutorial and review discussion of modulation, control and tuning of high-performance DC-DC converters based on high-performance DC-DC converters based on small-signal and large-signal approaches. *IEEE Open J. Power Electron.*, 1, 339–371, Sept 2020.
97. Giraldo, O.D.M., Ruiz, A.G., Ortega-Velazquez, I., Espinosa-Perez, G., Passivity-based control for battery charging/discharging applications by using a buck-boost DC-DC converter. *IEEE Green Technologies Conference*, 2018.
98. Xu, Q., Vafamand, N., Chen, L., Dragicevic, T., Xie, L., Frede, B., Review on advanced control technologies for bidirectional DC/DC converetrs in DC microgrids. *IEEE J. Emerging Sel. Top. Power Electron.*, 9, 2, 1205–1221, Apr 2021.
99. Jafari, M., Malekjamshidi, Z., Lu, D.D.C., Zhu, J., Development of a fuzzy-logic-based energy management system for a multiport multioperation mode residential smart microgrid. *IEEE Trans. Power Electron.*, 3283–3301, 34, 4, Apr 2019.
100. Pardhi, P.K. and Sharma, S.K., High gain non isolated DC converter employed in single-phase grid-tied solar photovoltaic supply system. *IEEE Trans. Ind.y Appl.*, 57, 5, . 5170–5182, Sep/Oct 2021.
101. Dong, W., Li, S., Fu, X., Li, Z., Fairbank, M., Gao, Y., Control of a buck DC-DC converter using approximate dynamic programming and artificial neural networks. *IEEE Trans. Circuits Syst. I Regul. Pap.*, 68, 4, 1760–1768, Apr 2021.
102. Hajihosseini, M., Andalibi, M., Gheisarnejad, M., Farsizadeh, H., Khooban, M.H., DC/DC power converter control-based deep machine learning techniques: Real-time implementation. *IEEE Trans. Power Electron.*, 35, 10, 9971–9977, Oct 2020.
103. Shi, X., Chen, N., Wei, T., Wu, J., Xiao, P., A reinforcement learning-based online-training AI controller for DC-DC switching converters. *IEEE, The 6th International Conference on Integrated Circuits and Microsystems*, 2021.
104. Buccella, C., Cecati, C., Siano, P., A fuzzy-logic-controlled resonant converter for renewable energy sources applications. *IEEE, 37th Annual Conference of IEEE Industrial Electronic Society*, Nov 2011.
105. Yu, K., Zhuo, F., Wang, F., Jiang, X., Deep-learning-based steady-state modelling and model predictive control for CLLC DC-DC resonant converter in DC distribution system. *IEEE Applied Power Electronics Conference and Exposition (APEC)*, 2022.
106. Ghosh, S. and Batarseh, I., Grid-tied two PV LLC converter with dual mppt algorithm based on the adaptive neuro fuzzy interface system (ANFIS). *IEEE Energy Conversion Congress and Exposition (ECCE)*, 2021.

107. Anitha, T. and Nachammai, N., Simulation of single-input fuzzy logic control of multi-output zero-voltage switching push-pull quasi resonant converter. *IEEE, 11th International Conference on Advanced Computing (ICoAC)*, 2019.
108. Selvaperumal, S., Ranjan, C., Christober, A., Murlidharan, S., Stability and performance investigation of a fuzzy-controlled LCL resonant converter in an RTOS environment. *IEEE Trans. Power Electron.*, 28, 4, 1817–1832, Apr 2013.
109. Gorji Saman, A., Sahebi Hosein, G., Ektesabi, M., Rad, A.B., Topologies and control schemes of bidirectional DC-DC power converters: An overview. *IEEE Access*, 7, 117997–118019, Sept, 2019, Special Section on Advanced Energy Storage Technologies and Their Applications.
110. Jeung, Y.C. and Lee, D.C., Voltage and current regulations of bidirectional isolated dual-active-bridge DC-DC converters based on a double-integral sliding mode control. *IEEE Trans. Power Electron.*, 34, 7, 6937–6946, Jul 2019.
111. Xie, Y., Ghaemi, R., Sun, J., Freudenberg, J.S., Model predictive control for a full bridge DC/DC converter. *IEEE Trans. Control Syst. Technol.*, 20, 1, 164–172, Jan 2012.
112. Zhao, D., Shen, K., Chen, L., Wang, Z., Liu, W., Yang, T., Wheeler, P., Improved active damping stabilization of DAB converter interfaced aircraft DC microgrids using neural network-based model predictive control. *IEEE Trans. Transp. Electrif.*, 8, 2, 1541–1552, Jun 2022.
113. Tang, Y., Hu, W., Xiao, J., Chen, Z., Huang, Q., Chen, Z., Blaabjerg, F., Reinforcement learning based efficiency optimization scheme for the DAB DC–DC converter with triple-phase-shift modulation. *IEEE Trans. Ind. Electron.*, 68, 8, 7350–7361, Aug 2021.
114. Kayaalp, R.I., Demirdelen, T., Tumay, M., A novel fuzzy logic control for bidirectional DC DC converter and comparison with dual phaseshift control method in medium voltage applications, in: *2016 IEEE International Conference on Computational Intelligence and Virtual Environments for Measurement Systems and Applications (CIVEMSA)*, IEEE, 2016.
115. Hou, N., Song, W., Wu, M., Minimum-current-stress scheme of dual active bridge DC–DC converter with unified phase-shift control. *IEEE Trans. Power Electron.*, 31, 12, 8552–8561, Dec 2016.
116. Askarian I. Bagawade, S., Pahlevani, M., Knight, A.M., Bakhshai, A., Robust digital nonlinear control system for dual active bridge (DAB) DC/DC converters with asymmetric half-cycle modulation. *IEEE J. Emerging Sel. Topics Ind. Electron.*, 1, 2, 123–132, Oct 2020.

5

Analysis of PWM Techniques on Multiphase Multilevel Inverter for PV Applications in Microgrids

B. Hemanth Kumar[1]*, Kavali Janardhan[1], R. Seshu Kumar[2], J. Ramesh Rahul[3] and P. Sanjeevi Kumar[4]

[1]*Department of EEE, Sree Vidyanikethan Engineering College, Tirupathi, India*
[2]*Department of EEE, KPR Institute of Engineering and Technology, Coimbatore, India*
[3]*Department of EEE, Gayatri Vidya Parishad College of Engineering, Visakhapatnam, India*
[4]*Department of Business Development and Technology, Aarhus University, Herning, Denmark*

Abstract

Due to the growth in consumers over the past few years and the advent of high-power industries, the energy grid has had to supply much more energy. This has caused the production of conventional energy to significantly raise its worldwide emissions. As a result, the incursion of renewable energy sources (RES) in the power grid has increased significantly. The most often used resources these days are photovoltaic systems. Power electronics are a crucial component of photovoltaic generation; the need for effective power electronic converters is growing as a result of falling energy costs, rising energy conversion efficiency, increasing power density, and adhering to strict grid code requirements. Since they control the power flows from both the generators and energy storage devices, inverters are also the primary actuator in the regulation of AC microgrids.

Many academics now prefer multilayer inverters (MLIs) over two-level inverters due to their capability to provide reduced higher DC link voltages, electromagnetic interference, and superior efficiency. The multiphase multilevel inverters are fervently chosen in high-power areas because of their redundant structure, reliability with high fault tolerant capabilities, flexibility to operate as independent

**Corresponding author*: hemub09@gmail.com

P. Sivaraman, C. Sharmeela and P. Sanjeevikumar (eds.) Microgrids for Commercial Systems, (133–156) © 2024 Scrivener Publishing LLC

single-phase systems, and reduced rating of power devices. Additionally, devices with voltage limitations can reach large power ratings thanks to multilevel converter technology. In order to utilize multilevel multiphase converters, new modulation techniques must be required.

The findings of a study of various pulse width modulation (PWM) techniques on cascaded H-Bridge (CHB) inverters with five, six, and seven phases are presented in this work. A number of modulation control approaches may be used to control MLIs. The contribution of the common mode waveform towards the modulating signal determines the type of the modulation approach in carrier-based PWM. The impact of various PWM schemes on a variety of metrics is explored regarding the harmonic distortion and DC bus utilization of the inverter output. The simulation tool MATLAB/SIMULINK was used to conduct simulations on a five-phase, six-phase, and seven-phase five-level CHB inverter running in a symmetric mode.

Keywords: Multilevel inverter, pulse width modulation, microgrid, harmonic distortion, multiphase inverter

5.1 Introduction

Several distinct multilayer inverter (MLI) structures have been described in the literature throughout the years. MLI is categorized into two categories depending on the number of separate DC sources incorporated within its structure. The usefulness of these converters varies according to their modularity, control complexity, and number of components. MLIs were first introduced in 1981 and have gained quickly in popularity during the subsequent two decades. When compared to standard two-level voltage source inverters (VSIs), MLI delivers reduced harmonic distortion, lower device ratings, lower dv/dt stress on the devices, lower switching frequency, and so on [1, 2]. Furthermore, the utilization of two-level VSIs for medium- and high-power applications necessitates the use of semiconductor switches in series and/or parallel to survive the high voltage and/or high current requirements. This leads to current/voltage sharing issues due to inherent variances in system characteristics, turn-off and turn-on periods, and so on. MLIs are a superior option for higher-power applications and are hence often favored [3–5].

The fundamental concept of the MLI is to generate a sinusoidal voltage commencing a variety of voltage levels, which are generally taken from the capacitor's voltage sources. Inverters are referred to be multilevel inverters when their output-phase voltage levels are more than or equal to three [6].

The usage of multilevel inverters has risen and emerged as a demanding alternative for power conversion and high-power applications since its structure alone decreases harmonics. Drive systems and AC power supply are two applications that employ MLIs. The output voltage is produced from multiple levels of input DC voltage to produce a higher output voltage and thus reduce the voltage stress. Multi-level inverters reduce switching losses, and as the DC side voltage levels are increased, more steps are added to the output, resulting in a waveform that is closer to a sinusoidal wave with less harmonic distortion [7, 8].

Multilevel inverters have lately been prominent in medium-voltage and high-power industrial applications owing to attributes such as decreased low-power dissipation on IGBTs, harmonic distortion, lower CMV, lower voltage stress, fewer switching losses, and greater DC bus usage. MLIs are divided into three types, as shown in Figure 5.1. The renowned and developed topologies are flying capacitor, diode neutral point clamped, and cascaded H-bridge (CHB) inverters [9].

DCMLI and FCMLI both need a single DC source to function. The architectures of DCMLI and FCMLI are nearly identical; the main variation is that DCMLI uses clamping diodes instead of flying capacitors in FCMLI. While DCMLI and FCMLI need distinct DC sources depending on the number of cells utilized, CHB MLI requires no additional components such as diodes and capacitors.

The neutral point MLI [10, 11] is regarded as one of the earliest three-level inverters, sometimes known as a neutral point clamped inverter, since it has two capacitor voltage sources connected in series at its midway. Each leg of the three-level inverter is outfitted with two pairs of serial switching devices. To clamp the neutral point to the center of every unit pair, clamping diodes are utilized. The output of the three-level inverter yields the quasi-square waveform.

The diode-clamp that emerged is found for greater inverter levels. While an inverter's level increases, further voltage steps are added en route for the resulting output waveform, resulting in a stairs wave comparable to the sinusoidal wave with reduced harmonic content. The output voltage waveform may ultimately reach an unlimited amount of voltage levels while exhibiting no harmonic distortion. Regrettably, the amount of voltage

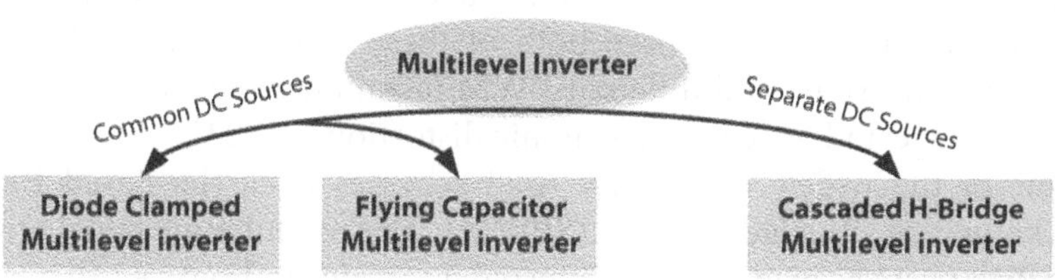

Figure 5.1 Classification of multilevel inverters.

levels that are achievable is rather limited, not only owing to worries about unbalanced voltage but also due to circuit architecture, voltage clamping conditions, and packing constraints [12].

CHB multilevel inverters are extensively used for the following reasons: they are more dependable, have fault tolerant capabilities, and can be readily regulated since they need independent DC sources and have a modular construction. The required topology of the inverter is determined by the specific application. In general, several modulation methods are worn to modulate the output voltage of MLI and decrease the harmonic distortion [13, 14].

Multiphase converters are three-phase converter extensions with extra legs [15]. They go through more than three stages. The only significant variation between a three-phase and a multiphase converter is the leg number. Despite this, more degrees of redundancy are possible, and control becomes more complicated and adaptive. The usage of multiphase power converters has expanded in the past 10 years due to advancements in high-power switching technology and computing platforms, for instance, field-programmable gate arrays and digital signal processors [16, 17]. Concerns about building unsoiled transportation systems, renewable energy generating systems (particularly wind energy systems), and variable-speed high-reliability drive systems are claimed to be the primary motivators for the multiphase converter sector. Multiphase converters are utilized in a variety of applications depending on whether they convert DC to AC, AC to DC, DC to DC, or AC to AC. DC to AC converters are mostly used in variable-speed drives, while multiphase renewable-energy-generating systems utilize AC to AC and AC to DC converters. In the automobile sector, multiphase DC–DC converters are used [18–20].

Multiphase inverters have become more popular during the last decade due to their reliability with fault-tolerant capacity and redundant and are widely used in high-power electric drive areas like five-, six-, and seven-phase induction motors and some renowned applications like propulsion in ships and traction applications (both in electric and hybrid) vehicles, whereas conventional inverters have limitations like harmonic current losses, less reliability, and DC-link current harmonics. Consequently, multi-phase inverters that have both of these benefits are combined in multi-level inverters, and different PWM techniques are used to regulate output voltage and lessen the harmonic distortion [21–23].

One of the widely extensively used schemes is carrier-based PWM, also known as sinusoidal PWM (SPWM). The modulating signal within the SPWM technique is often a sinusoidal waveform; however, this is not always the case. A triangular wave is often used as the carrier signal. PWM approaches are classified into three types: phase disposition, phase opposition disposition, and alternative phase opposition disposition. These are based on the carrier pulse alignment for MLIs [24–26]. Discontinuous PWM and continuous PWM are two more categorization schemes depending on the switching times that occur in a basic cycle. Discontinuous PWM, sometimes known as continuous PWM, is switching that occurs constantly throughout the course of one cycle. Space vector PWM (SVPWM) is another PWM approach that is often used to control MLIs; however, employing this technique at higher voltage levels will be more difficult [27–35].

The electrical grid system is constantly undergoing remarkable changes, mostly as a result of significant technological advances. It is becoming more decentralized as a result of the increasing penetration of RES and technological breakthroughs in energy storage, for example. These results are occurring at all the levels of the grid system, from utility grids and transmission to distributed grids, and many businesses have already predicted that a return to a solely centralized grid system would not occur. Alternatively, at the dispersed network level, DC microgrids, which were previously only utilized in telecommunication/data center applications, are now commonly suggested in the areas where power is generated from RES. Aside from a few exceptions, these microgrids were intended to operate in mutually grid-tied and islanded modes and are sometimes referred to as hybrid DC/AC microgrids. It is distinguished by a mixture of dispersed generating units, storage systems, and loads that are linked to either an alternating current or direct current grid. Both grids are linked to the primary distributed AC grid through a multiport power electronics converter that can handle various energy conversions. Such hybrid microgrids have the potential to significantly improve essential applications such as commercial buildings [36].

Due to the numerous benefits of distributed generation (DG), its penetration on the primary grid has expanded significantly in recent years. It is primarily based on RES, which lowers pollution. Environmental effect is reduced, and local energy sources can be exploited and harvested. The increasing presence of DG necessitates an examination of its effect on the power grid consecutively to reduce line loadings, losses, and reactive power requirements [37]. However, due to the intermittent nature, uncertainty, and randomness caused by meteorological conditions, renewable energy sources are difficult to integrate directly into the power system. In this regard, because microgrids incorporate distributed and RES, energy storage devices, and a wide range of loads, they serve as an effective interface between power grid and the DG [38].

Microgrids are local distribution systems that incorporate generating, storage, and load capacities, and they may operate independently or in conjunction with the main power grid [39]. This capability to operate both connected to and detached from the mains grid enhances the dependability and power quality of the devices' connected users. In addition, almost zero energy buildings may be achieved if these microgrids are designed according to an ecodesign [40]. Microgrids are capable of supporting either alternating current (AC) or direct current (DC). Each solution has unique characteristics that indicate various benefits and drawbacks that must be considered. The control, protection, and power loss comparisons between the two kinds of microgrids are described in [41–43]. In the same vein, this article gives a comprehensive analysis of AC and DC microgrids, highlighting the primary features of each microgrid's components. In addition, the paper covers and analyzes technical, economic, and regulatory concerns, including available power converter designs, relevant directive, monitoring systems, and economic analysis. The primary goals of this analysis are as follows: describe the benefits and downsides of each technology. A microgrid is required for identifying and categorizing possible combinations and designs. The aforementioned goals become a tool for selecting the best appropriate microgrid that meets the requirements of a particular circumstance. In addition, this study provides a snapshot of the present status of microgrids and suggests research directions for addressing their existing demands and further their development. The primary goal of this chapter is to analyze several PWM algorithms for multi-phase MLIs which are useful in commercial microgrids for DC–AC power conversion.

5.2 Cascaded H-Bridge Multiphase Multilevel Inverter

The first MLI topology for semiconductors was created and specified by Baker and Bannister in 1975 [8]. It had a cascaded topology, which is a single-phase serial connection. CHB converters are currently recognized as CHB. CHB MLIs are made up of two or more single-phase HB inverters in a row. Two legs of the voltage source step make up each H-bridge, and line–line voltage serves as the inverter's output voltage.

As a consequence, each HB inverter may produce three separate voltage levels. Every leg has just two alternative switching states to prevent DC-link capacitor short circuits. For 1-φ H-bridge inverters, there are four unique switching states. Two of the four switching states are duplicated switching states.

The CHB converter is based on the basic H-Bridge block structure. Four switches make up a 1-φ HB. Each switch is a combination of transistors and antiparallel diodes. A single HB is also called a cell or module. The CHB converter consists of a number of cells that are linked in a sequence combination. In this bridge, each cell gives three voltage levels 0, V, and -V. If the MLI has n number of cells, then it produces $2n + 1$ voltage levels in the phase voltage.

There is a unique DC source for every H-Bridge. The DC sources in CHB MLI are inversely associated to the number of cells used. In a CHB converter, two modes of operation are likely:

- Symmetric mode—where the DC sources in all cells are equal.
- Asymmetric mode—where the DC sources in all cells are not equal.

There will be more output voltage levels achieved in an asymmetric mode of operation than in a symmetric mode of operation. This study takes into account CHB operation in symmetric mode. Each phase for a five-level system needs two H-bridge modules connected in cascade. Table 5.1 shows the switching sequence for five-level CHB MLI per phase. Figures 5.2, 5.3, and 5.4 successively depict the five-level inverter in five-phase, six-phase,

Table 5.1 Switching sequence for five-level CHB MLI per phase.

S_{1x}	S_{2x}	S_{3x}	S_{4x}	S_{5x}	S_{6x}	S_{7x}	S_{8x}	Voltage level
1	1	0	0	1	1	0	0	+2E
1	1	0	0	0	1	0	1	+E
0	1	0	1	1	1	0	0	+E
0	1	0	1	0	1	0	1	0
1	0	1	0	1	0	1	0	0
0	0	1	1	1	0	1	0	-E
1	0	1	0	0	0	1	1	-E
0	0	1	1	0	0	1	1	-2E

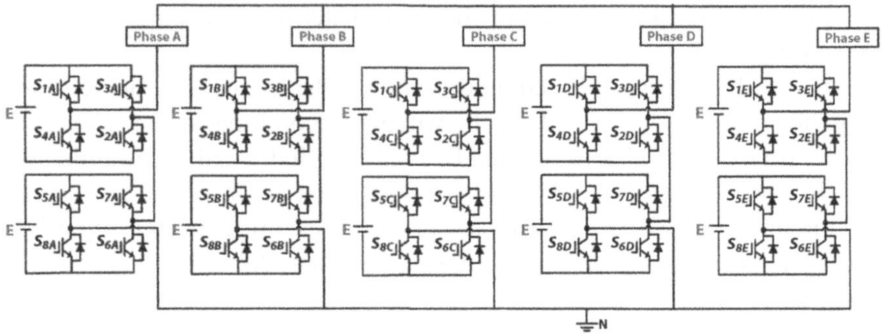

Figure 5.2 Structure of five-phase five-level MLI.

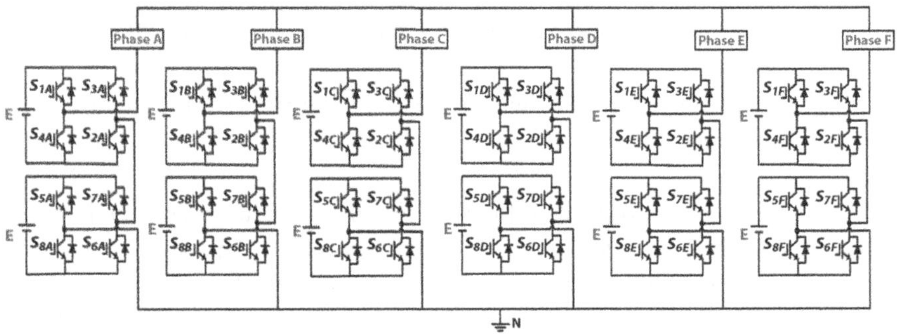

Figure 5.3 Structure of six-phase five-level MLI.

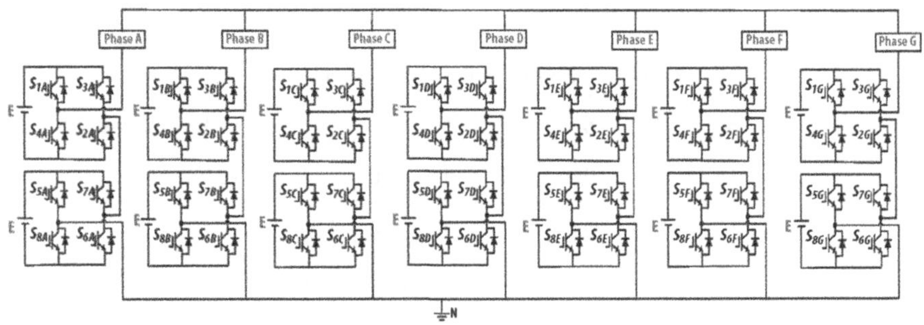

Figure 5.4 Structure of seven-phase five-level MLI.

and seven-phase configurations. In 5-φ inverter, the phase shift amid two AC output voltages is 72°, whereas in 6-φ inverter, the phase shift between two output voltages is 60°, and in 7-φ inverter, the phase shift among two output voltages is 51.4°.

5.3 Modulation Techniques for Multilevel Inverter

Different modulation methods are used by multilevel inverters to get a better output voltage responsiveness with the least amount of harmonic distortion. Fundamental switching frequency modulation and high switching frequency PWM, which is seen in Figure 5.5, are the two main categories of approaches. The analysis in this chapter uses a carrier-based PWM approach.

5.3.1 High Switching Frequency PWM Technique

In the two-level PWM example, consecutively to generate a sinusoidal output wave, a multilevel pulse width modulation technique uses carrier waves with a much higher switching frequency than the reference signal. Harmonic distortions in the output voltage waveform can be mitigated *via* phase-shifting schemes.

The carrier-based PWM techniques can be generally classified as detailed in the following discussions.

5.3.1.1 Phase-Shifted Modulation (PSM)

In the PSM technique, the peak-to-peak amplitude and frequency of all triangle carrier waves should be identical. Two neighboring carrier waves,

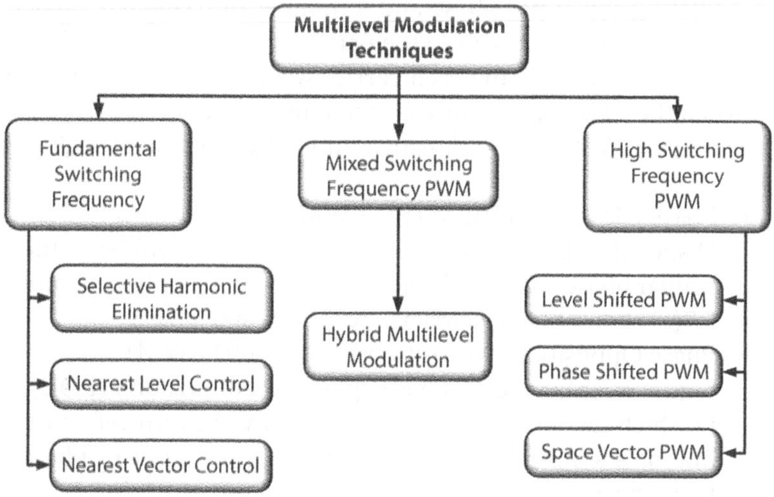

Figure 5.5 Multilevel modulation techniques.

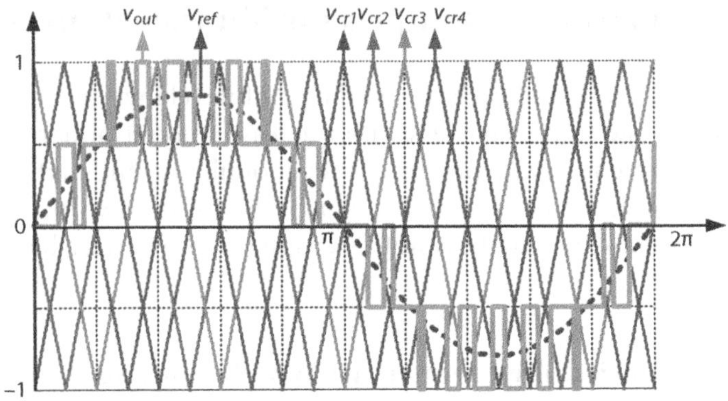

Figure 5.6 Phase-shifted carrier waveforms, modulating waveform, regions, and nature of output voltage of a five-level converter.

however, will have different phases. Figure 5.6 shows the modulating signal superimposed over a phase-shifted carrier waveform for a five-level converter.

5.3.1.2 Level-Shifted Modulation (LS-PWM)

It uses carrier signals that are vertically offset from one another. Multilevel converters may use the same PWM methods that are often used for converters of the classic kind. In this instance, the methods are known as multicarrier PWM. The main common modulation methods for MPI involve one modulating signal per phase and a number of triangular carrier signals. The command pulses sent to the parts of the similar leg of the bridge are determined by comparing each carrier waveform and the modulating signal when the CHBMI structure is taken into account.

Level-shifted PWM can be classified into three types of disposition:

- Phase Disposition PWM (PD PWM):
  Disposition in phases PWM has carriers that will be in the similar phase above and below the zero reference line. Figure 5.7(a) shows that all carriers are in the same phase.
- Phase Opposition Disposition PWM (POD PWM):
  Figure 5.7(b) depicts a phase opposition disposition (POD), in which carrier signals beyond and below the zero-reference line of a sinusoidal reference signal are 180° out of phase with one another.

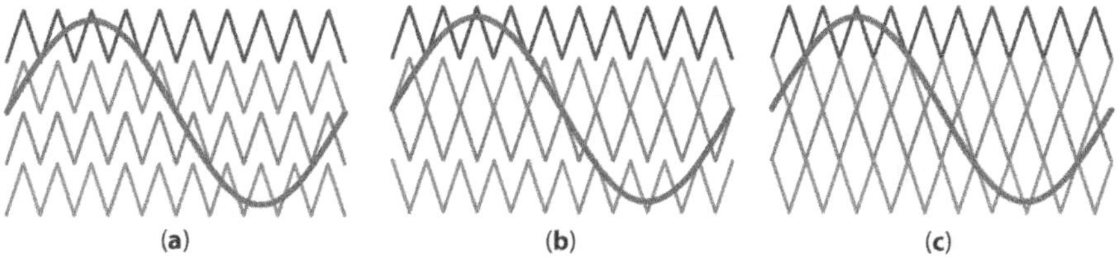

Figure 5.7 Carrier (triangular) and reference (sinusoidal) waveform for five-level converter: (a) PD, (b) POD, and (c) APOD.

- Alternative Phase Opposition Disposition PWM (APOD PWM):
  Figure 5.7(c) illustrates the APOD modulation used for voltage balancing where every carrier signal is 180° out of phase with its neighboring carrier.

For an MLI with several levels identical to N, the amount of carrier signals N_c is specified through the subsequent equation:

$$N_C = N - 1 \tag{5.1}$$

As a function of *N*, the modulation index *m* is as follows:

$$M = \frac{A_M}{A_C}(N-1) \tag{5.2}$$

A_M is the modulating signal's amplitude, while A_C is the carrier signal's amplitude.

Additionally, the ratio among the frequency of the carrier waveform F_c and the output frequency of fundamental waveform F_1 determines the frequency modulation index M_f:

$$M_F = \frac{F_C}{F_1} \tag{5.3}$$

Figure 5.8 shows the reference, carrier, and output voltage waveform of a five-level inverter. For the five-level inverter, four carriers (v_{Cr1}, v_{Cr2}, v_{Cr3}, and v_{Cr4}) are used. By comparing the sinusoidal waveform (v_{ref}) through the triangular waveforms, the switching instants are provided to the inverter to obtain the required output voltage as follows:

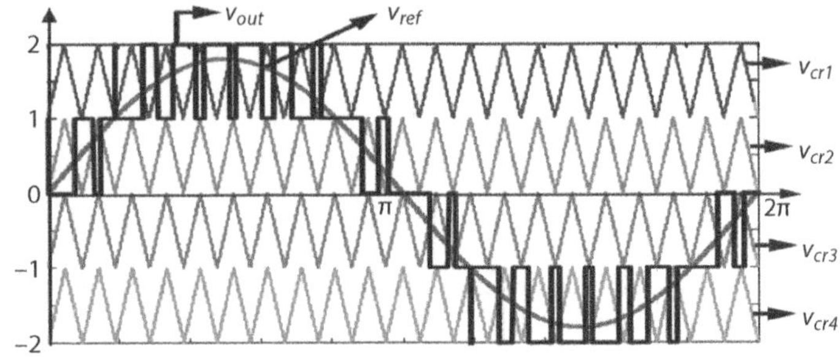

Figure 5.8 Output voltage of a five-level converter with PD-LSPWM.

$$v_{out} = \begin{cases} \dfrac{V_{DC}}{2} & : \quad v_{ref} > v_{cr1} \\ \dfrac{V_{DC}}{4} & : \quad v_{cr1} > v_{ref} > v_{cr2} \\ 0 & : \quad v_{cr2} > v_{ref} > v_{cr3} \\ -\dfrac{V_{DC}}{4} & : \quad v_{cr3} > v_{ref} > v_{cr4} \\ -\dfrac{V_{DC}}{2} & : \quad v_{cr4} > v_{ref} \end{cases} \tag{5.4}$$

The following methods can be obtained depending on the waveform of the modulating signal.

5.3.2 Sinusoidal PWM

To determine the switching states for each pole in the inverter, sinusoidal PWM (SPWM) compares sinusoidal AC V_{ref} by means of a triangular high-frequency carrier wave in real time.

A three-phase sinusoidal wave can be generated using a carrier-based sinusoidal PWM method:

$$V_{AM} = V_M \,\mathrm{Sin}(\mathrm{wt}) \tag{5.5}$$

$$V_{BM} = V_M \,\mathrm{Sin}\left(\mathrm{wt} - \frac{2\pi}{3}\right) \tag{5.6}$$

$$V_{CM} = V_M \, \text{Sin}\left(\text{wt} + \frac{2\pi}{3} \right) \tag{5.7}$$

The three signals are comparable to a high-frequency triangle wave and are utilized for modifying or controlling signals. All three phases are compared with a single common triangle wave.

5.3.3 Harmonic Injection

Using the SPWM technique, the maximum output voltage is 0.5 V DC. It is possible that the available voltage will be insufficient when using the SPWM method in motor driving applications. The machine has to be derated so that it produces less torque in this application. By injecting third harmonics into the modulating signal, the PWM inverter's output voltage may be increased by carrier-based modulation. It is shown that the peak of the reference signal may be reduced by the addition of an acceptable third harmonic component to the basic reference signal. Therefore, it is achievable to increase the inverter's output voltage by increasing the reference value of the generated modulating signal over 0.5 V DC. The third harmonic of the reference signal is canceled out in the legs and is not seen in the phase voltages of the output. This means the absence of low-order harmonics in the output voltage. It is possible to utilize the modulating signals to find the sweet spot for injecting third harmonics.

$$V_{AM} = V_M \, \text{Sin(wt)} + \text{V}_{\text{M3}} \, \text{Sin(3wt)} \tag{5.8}$$

$$V_{BM} = V_M \, \text{Sin}\left(\text{wt} - \frac{2\pi}{3} \right) + \text{V}_{\text{M3}}\text{Sin(3wt)} \tag{5.9}$$

$$V_{CM} = V_M \, \text{Sin}\left(\text{wt} + \frac{2\pi}{3} \right) + \text{V}_{\text{M3}}\text{Sin(3wt)} \tag{5.10}$$

5.3.4 Switching Frequency Optimal

As shown in Figure 5.6, by the addition of an offset to the sinusoidal reference signals, the output voltage of a three-phase voltage source inverter may be raised. Like the third harmonic injection system, the offset

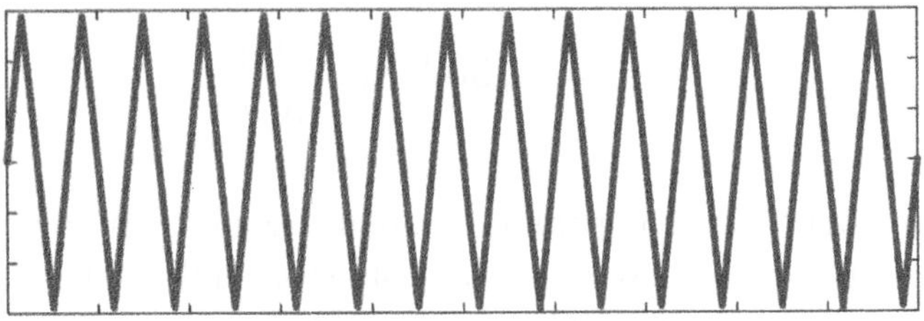

Figure 5.9 Offset signal.

accumulation PWM method works by dipping the peak of the modulating signal to increase the value of *m*.

The SFO PWM modulating signal is achieved by subtracting the voltage-offset V_{offset} shown in Figure 5.9 from a sinusoidal reference signal, which is specified as follows:

$$V_{offset} = (Max(V_a, V_b, V_c) + Min(V_a, V_b, V_c))/2 \tag{5.11}$$

where V_a, V_b, and V_c are the reference voltage's instantaneous values. As a result, the reference signals' instantaneous values are given by:

$$V_{a,\,sfo} = V_a - V_{offset} \tag{5.12}$$

$$V_{b,\,sfo} = V_b - V_{offset} \tag{5.13}$$

$$V_{c,\,sfo} = V_c - V_{offset} \tag{5.14}$$

Although sinusoidal modulating signals are common in industrial commercial inverters, the THI PWM and SFO PWM approaches allow for a more efficient utilization of the DC link voltage at higher modulation indices.

5.4 Simulation Results

The MATLAB/Simulink tool is used to simulate a five-level, five-phase, six-phase, and seven-phase CHB inverter running in symmetric mode with a R-L load, allowing the PWM techniques to be tested on a variety of inverter parameters. Table 5.2 shows the list of variables used in the simulations.

Table 5.2 Simulation parameters.

Parameter	Values
DC supply (E)	50 V
Resistance (R)	110 Ω/ph
Inductance (L)	120 mH (X_L= 37.68 Ω/ph)
Carrier frequency (f_c)	2,500 Hz

Figure 5.10 illustrates the output line voltage and phase voltage of a five-phase, five-level CHB employing sinusoidal PWM at unity modulation index. Figure 5.11 illustrates the output phase voltage and line voltage of a six-phase, five-level CHB employing sinusoidal pulse width modulation at unity modulation index. Figure 5.12 illustrates the output phase voltage and line voltage of a seven-phase, five-level cascaded H-bridge employing sinusoidal pulse width modulation at unity modulation index.

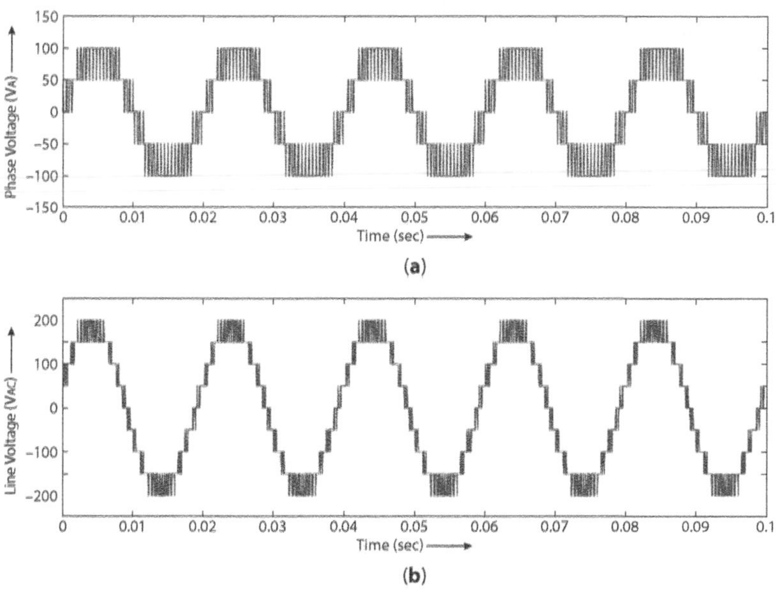

Figure 5.10 Five-phase five-level CHB inverter phase voltage and line voltage for the SPWM technique.

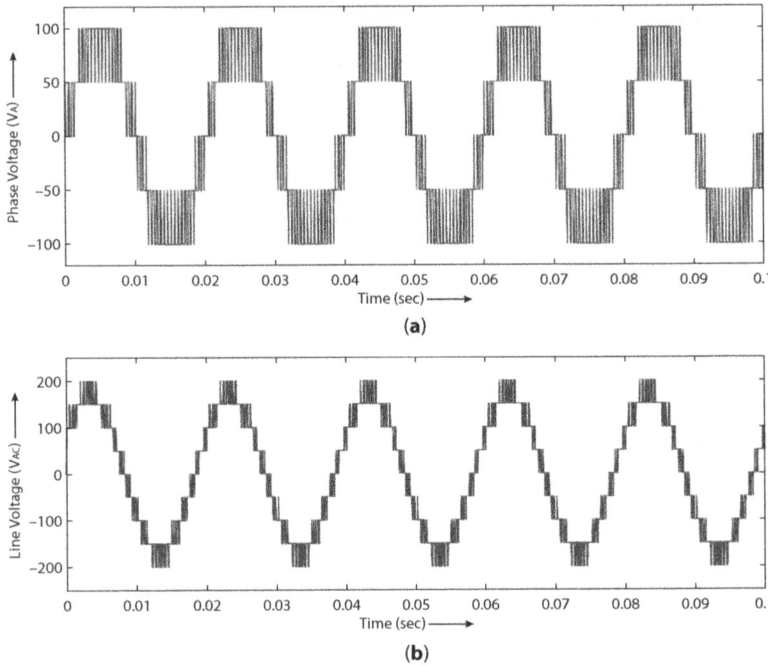

Figure 5.11 Six-phase five-level CHB inverter phase voltage and line voltage for the SPWM technique.

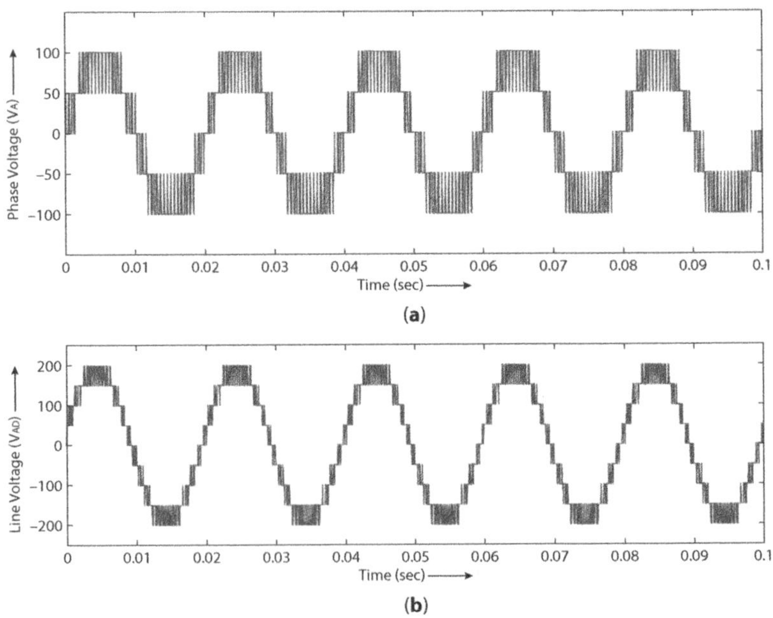

Figure 5.12 Seven-phase five-level CHB inverter phase voltage and line voltage for the SPWM technique.

Figure 5.13 illustrates the output phase and line voltage of a five-phase, five-level CHB employing harmonic injection PWM at unity modulation index. Figure 5.14 illustrates a harmonic injection pulse width modulated six-phase, five-level cascaded H-bridge with unity M output phase voltage and line voltage. Figure 5.15 illustrates a seven-phase, five-level cascaded harmonic injection pulse width modulated H-bridge output line voltage and phase voltage with a unity modulation index.

Figure 5.16 depicts the output line and phase voltage of a five-phase, five-level CHB employing switching frequency optimized pulse width modulation approach at 1.0 M. Figure 5.17 illustrates a six-phase, five-level cascaded H-bridge with the output phase voltage and line voltage modulated at the switching frequency's ideal pulse width. Figure 5.18 illustrates a seven-phase five-level CHB with optimum PWM at unity modulation index and output phase voltage and line voltages.

Different PWM approaches have been used in the inverter output voltages, and the results have been analyzed. As a result, when comparing all PWM approaches, the SFO-PWM scheme in a multiphase multilevel inverter provides greater DC bus usage and lower harmonic distortion than the SPWM and THI PWM schemes.

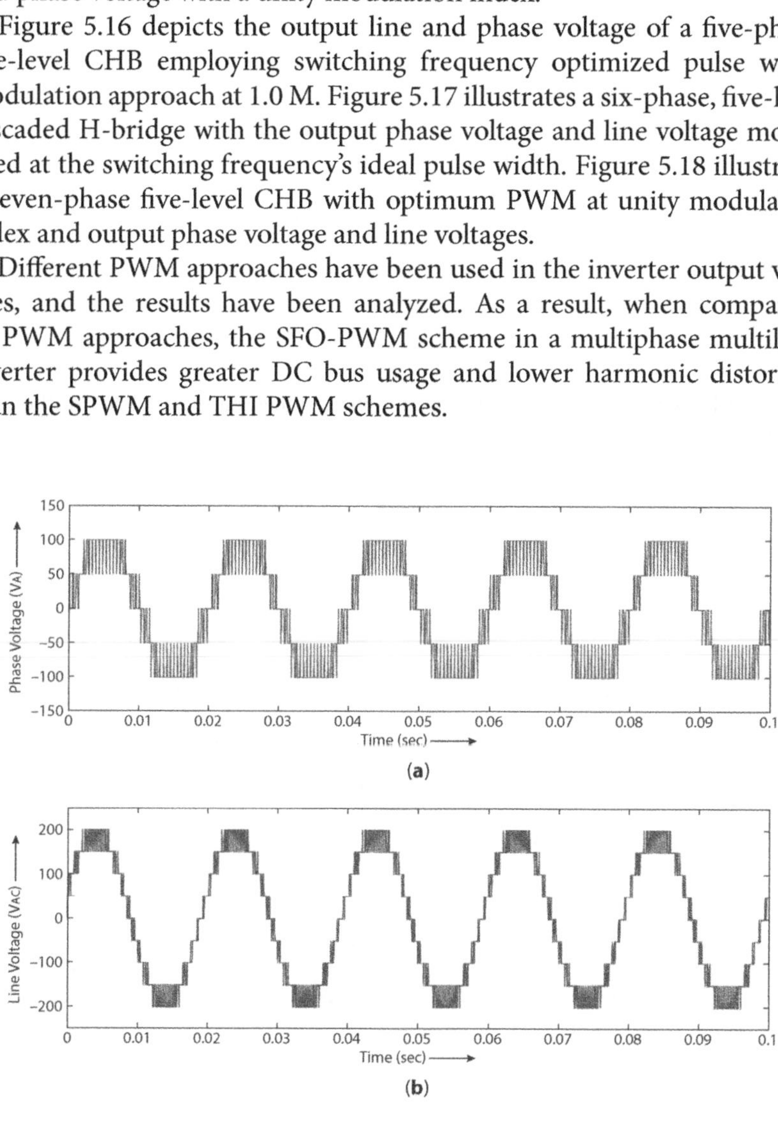

Figure 5.13 Five-phase five-level CHB inverter phase voltage and line voltage for the HIPWM technique.

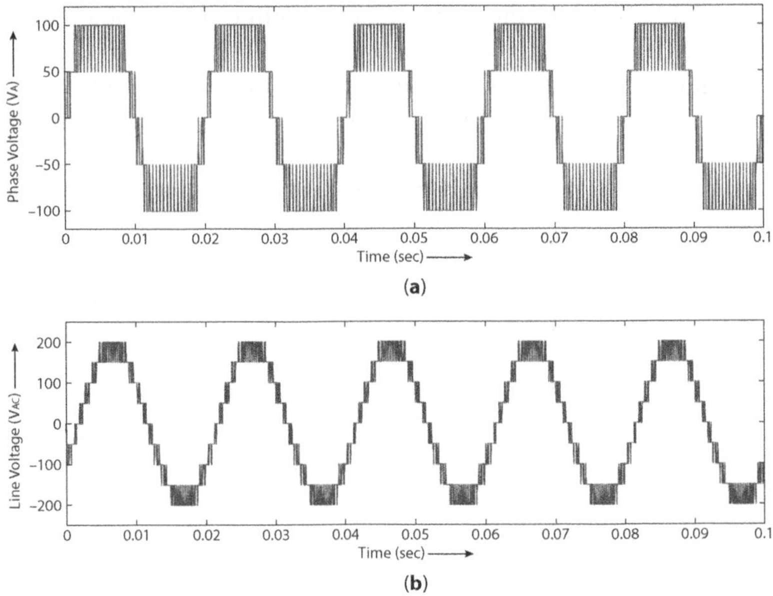

Figure 5.14 Six-phase five-level CHB inverter phase voltage and line voltage for the HIPWM technique.

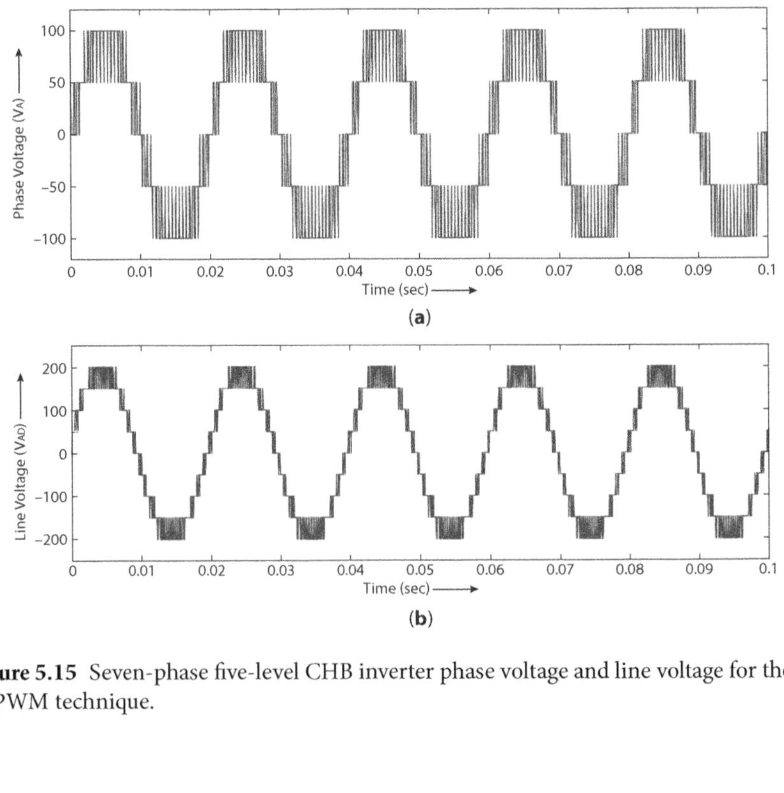

Figure 5.15 Seven-phase five-level CHB inverter phase voltage and line voltage for the HIPWM technique.

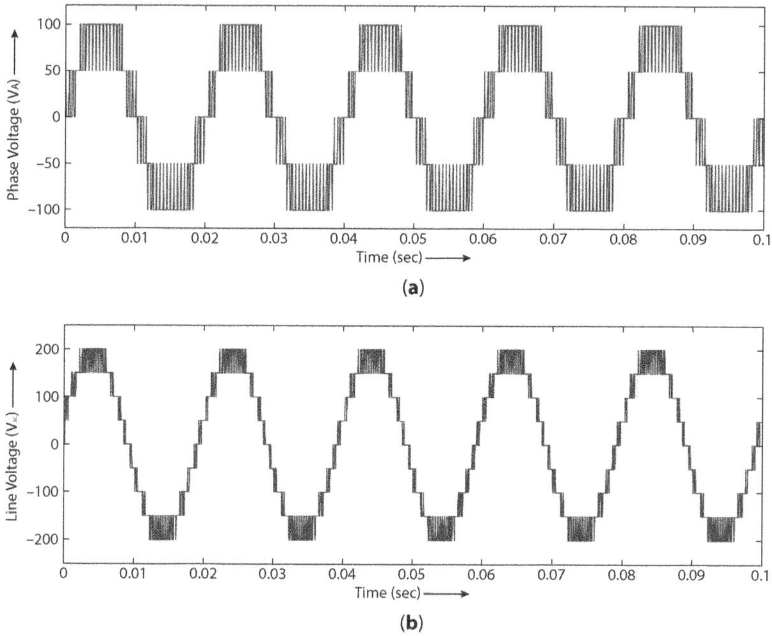

Figure 5.16 Five-phase five-level CHB inverter phase voltage and line voltage for the SFOPWM technique.

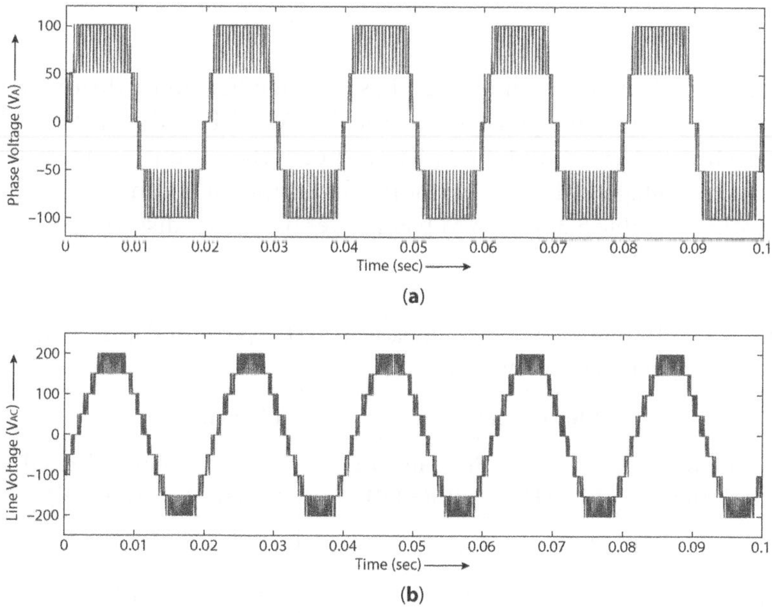

Figure 5.17 Six-phase five-level CHB inverter phase voltage and line voltage for the SFOPWM technique.

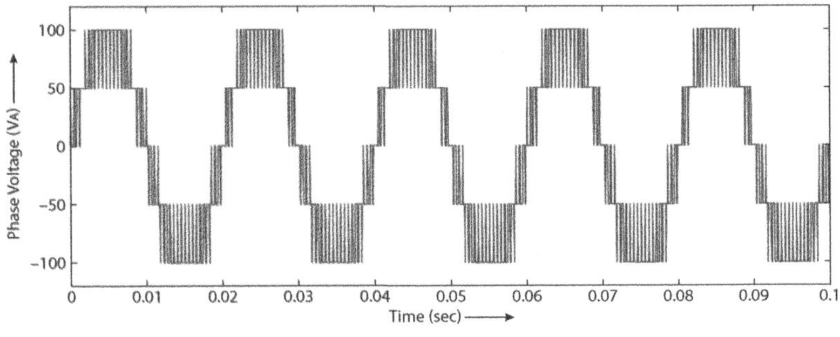

(a)

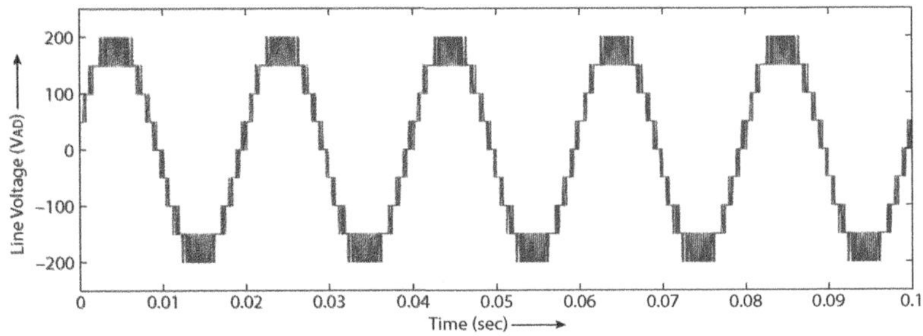

(b)

Figure 5.18 Seven-phase five-level CHB inverter phase voltage and line voltage for the SFO PWM technique.

Table 5.3 displays the line and phase voltage harmonic distortion and RMS value for the five-phase inverter's SPWM, HIPWM, and SFO PWM approaches. Table 5.4 and Table 5.5, respectively, exhibit comparisons of phase, line voltage RMS, and harmonic distortion for six and seven phases. According to Table 5.3, HIPWM has a lower harmonic distortion for phase

Table 5.3 THD and RMS value comparison for the phase and line voltages of a five-phase five-level CHB inverter.

Modulation techniques	Phase voltage		Line voltage	
	%THD	Fundamental RMS (V)	%THD	Fundamental RMS (V)
SPWM	29.35	69.58	15.79	132.2
HIPWM	27.34	72.75	14.57	138.3
SFO PWM	27.65	72.53	14.43	137.9

Table 5.4 THD and RMS value comparison for the phase and line voltages of a six-phase five-level CHB inverter.

Modulation techniques	Phase voltage		Line voltage	
	%THD	Fundamental RMS (V)	%THD	Fundamental RMS (V)
SPWM	29.35	69.58	18.06	120.7
HIPWM	29.06	78.97	15.82	136.4
SFO PWM	30.25	80.3	15.62	138.3

Table 5.5 THD and RMS value comparison for the phase and line voltages of a seven-phase five-level CHB inverter.

Modulation techniques	Phase voltage		Line voltage	
	%THD	Fundamental RMS (V)	%THD	Fundamental RMS (V)
SPWM	29.35	69.58	15.86	136.1
HIPWM	27.99	70.92	14.87	138
SFO PWM	27.9	71.01	14.86	138.2

voltage than SPWM, and for line voltage, HIPWM has a lower harmonic distortion. In a five-phase, five-level CHB inverter, HIPWM also has a higher DC bus usage. According to Table 5.4, the SPWM for line voltage in a six-phase, five-level CHB inverter has a high DC bus utilization, whereas offset PWM for phase voltage has a reduced harmonic distortion.

According to Table 5.5, offset PWM has a high DC bus utilization for both the line and phase voltages in a seven-phase, five-level CHB inverter, whereas HIPWM has a reduced harmonic distortion for both the phase and line voltages. Seven-phase CHB five-level inverters provide greater DC bus usage and lesser THD to five, six, and seven phases.

5.5 Conclusion

Different modulation techniques such as SPWM, harmonics injected PWM, and switching frequency optimal-based PWM techniques have been investigated in this work on the multiphase (five, six, and seven phases) multilevel (five-level) inverter. A multilevel CHB inverter is under consideration. According to the simulation results, switching frequency optimal technique produces a better inverter output voltage profile in terms of fundamental RMS value, DC bus utilization, and harmonic content than sinusoidal PWM and harmonics-injected PWM all phases of the inverter. It was also established that the dependability of the inverter rises as the number of inverter phases grows. To investigate the inverter output voltage profile for various modulation strategies, simulation results on a five-level CHB multiphase inverter working in symmetric mode have been presented.

References

1. Lai, J.S. and Peng, F.Z., Multilevel converters – A new breed of power converters. *IEEE Trans. Ind. Appl.*, 32, 3, 509–517, May/June 1996.
2. Rodriguez, J., Franquelo, L.G., Kouro, S., Leon, J.I., Portillo, R.C., Prats, M.A.M., Perez, M.A., Multilevel converters: An enabling technology for high-power applications. *Proc. IEEE*, 97, 11, 1786–1817, 2009.
3. Wu, B., *High-power converters and AC drives*, IEEE Press, Piscataway, New Jersey, USA, 2006.
4. Tolbert, L., Peng, F.Z., Habetler, T., Multilevel converters for large electric drives. *IEEE Trans. Ind. Appl.*, 35, 1, 36–44, Jan./Feb. 1999.
5. Rodriguez, J., Lai, J.S., Peng, F.Z., Multilevel inverters: A survey of topologies, controls, and applications. *IEEE Trans. Ind. Electron.*, 49, 4, 724–738, Aug. 2002.
6. Walker, G. and Ledwich, G., Bandwidth considerations for multilevel converters. *IEEE Trans. Power Electron.*, 14, 1, 74–81, 1999.
7. Gupta, K.K., Ranjan, A., Bhatnagar, P., Sahu, L.K., Jain, S., Multilevel inverter topologies with reduced device count: A review. *IEEE Trans. Power Electron.*, *31*, 1, 135–151, 2015.
8. Poorfakhraei, A., Narimani, M., Emadi, A., A review of multilevel inverter topologies in electric vehicles: Current status and future trends. *IEEE Open J. Power Electron.*, *2*, 155–170, 2021.
9. Panagis, P., Stergiopoulos, F., Marabeas, P., Manias, S., Comparison of state of the art multilevel inverters, in: *2008 IEEE Power Electronics Specialists Conference*, IEEE, pp. 4296–4301, 2008, June.
10. Rodriguez, J., Bernet, S., Steimer, P.K., Lizama, I.E., A survey on neutral-point-clamped inverters. *IEEE Trans. Ind. Electron.*, *57*, 7, 2219–2230, 2009.
11. Nabae, A., Takahashi, I., Akagi, H., A new neutral-point-clamped PWM inverter. *IEEE Trans. Ind. Appl.*, (5), IA-17, 518–523, 1981.
12. Zaragoza, J., Pou, J., Ceballos, S., Robles, E., Jaen, C., Corbalan, M., Voltage-balance compensator for a carrier-based modulation in the neutral-point-clamped converter. *IEEE Trans. Ind. Electron.*, *56*, 2, 305–314, 2008.

13. Malinowski, M., Gopakumar, K., Rodriguez, J., Perez, M.A., A survey on cascaded multilevel inverters. *IEEE Trans. Ind. Electron., 57*, 7, 2197–2206, 2009.
14. Wu, B. and Narimani, M., *Cascaded H-bridge multilevel inverters*, Wiley-IEEE Press, Piscataway, New Jersey, USA, 2017.
15. Kelly, J.W., Strangas, E.G., Miller, J.M., Multi-phase inverter analysis, in: *IEMDC 2001. IEEE International Electric Machines and Drives Conference (Cat. No. 01EX485)*, IEEE, pp. 147–155, 2001, June.
16. Guo, H., Wang, R., Kong, W., Qu, R., Design of multiphase inverter system for variable-speed applications based on DSP and FPGA, in: *2019 22nd International Conference on Electrical Machines and Systems (ICEMS)*, IEEE, pp. 1–4, 2019, August.
17. Hou, L., Su, Y., Chen, L., DSP-based indirect rotor field oriented control for multiphase induction machines, in: *IEEE International Electric Machines and Drives Conference, 2003. IEMDC'03*, vol. 2, IEEE, pp. 976–980, 2003, June.
18. Jones, M., Vukosavic, S.N., Levi, E., Parallel-connected multiphase multi-drive systems with single inverter supply. *IEEE Trans. Ind. Electron., 56*, 6, 2047–2057, 2009.
19. Tahaa, W., Azerb, P., Callegaro, A.D., Emadi, A., Multiphase traction inverters: State-of-the-art review and future trends. *IEEE Access*, 10, 4580–4599, 2022.
20. Levi, E., Multiphase electric machines for variable-speed applications. *IEEE Trans. Ind. Electron., 55*, 5, 1893–1909, 2008.
21. Brazhinkov, A.V. and Dovzhenko, N.N., Control potentials and advantages of multiphase AC drives, in: *PESC 98 Record. 29th Annual IEEE Power Electronics Specialists Conference (Cat. No. 98CH36196)*, vol. 2, IEEE, pp. 2108–2114, 1998, May.
22. Brazhnikov, A.V. and Belozerov, I.R., Non-traditional control and advantages of multiphase AC inverter drives, in: *2011 International Conference on Energy, Automation and Signal*, IEEE, pp. 1–6, 2011, December.
23. López, Ó., Álvarez, J., Doval-Gandoy, J., Freijedo, F.D., Multilevel multiphase space vector PWM algorithm. *IEEE Trans. Ind. Electron., 55*, 5, 1933–1942, 2008.
24. Holtz, J., Pulsewidth modulation for electronic power conversion. *Proceedings of the IEEE Conference*, vol. 82, 8, pp. 1194–1214, Aug. 1994.
25. Holmes, D.G. and Lipo, T.A., *Pulse width modulation for power converters*, Wiley, New York, 2003.
26. B.P., McGrath, Holmes, D.G., Multicarrier PWM strategies for multilevel inverters. *IEEE Trans. Ind. Electron.*, 49, 4, 858–867, Aug 2002.
27. Hemanth Kumar, B. and Lokhande, M.M., Investigation of switching sequences on a generalized SVPWM algorithm for multilevel inverters. *J. Circuits, Syst. Comput., 28*, 02, 1950036, 2019.
28. Kumar, B.H. and Subburaj, V., Integration of RES with MPPT by SVPWM Scheme, in: *Intelligent Renewable Energy Systems*, pp. 157–178, 2022.

29. Nandhini, E. and Sivaprakasam, A., A review of various control strategies based on space vector pulse width modulation for the voltage source inverter. *IETE J. Res.*, 68, 5, 1–15, 2020.
30. Kumar, B.H. and Lokhande, M.M., An enhanced space vector PWM for nine-level inverter employing single voltage source. *2017 IEEE Transportation Electrification Conference (ITEC-India)*, pp. 1–6, 2017, https://doi.org/10.1109/ITEC-India.2017.8333711.
31. Busireddy, H.K., Lokhande, M.M., Karasani, R.R., Borghate, V.B., A modified space vector PWM approach for nine-level cascaded H-bridge inverter. *Arab. J. Sci. Eng.*, 44, 3, 2131–2149, 2019.
32. Kumar, B.H., Lokhande, M.M., Reddy, K.R., Borghate, V.B., An improved space vector pulse width modulation for nine-level asymmetric cascaded H-bridge three-phase inverter. *Arab. J. Sci. Eng.*, *44*, 3, 2453–2465, 2019.
33. Hemanth Kumar, B. and Lokhande, M.M., Investigation of switching sequences on a generalized SVPWM algorithm for multilevel inverters. *J. Circuits Syst. Comput.*, 28, 02, 1950036, 2019.
34. Hemanth Kumar, B., Bhavani, A., Jeevithesh, C.V., Padmanaban, S., Subburaj, V., A new series-parallel switched capacitor configuration of a DC–DC converter for variable voltage applications, in: *Electric Vehicles*, pp. 247–270, Springer, Singapore, 2021.
35. Hemanth Kumar, B., Prabhu, S., Janardhan, K., Arun, V., Vivekanandan, S., A switched capacitor-based multilevel boost inverter for photovoltaic applications. *J. Circuits Syst. Comput.*, 2350057, 32, 04, 2022.
36. Parhizi, S., Lotfi, H., Khodaei, A., Bahramirad, S., State of the art in research on microgrids: A review. *IEEE Access*, 10, 3, 890–925, 2015 Jun.
37. Lopes, J.P., Hatziargyriou, N., Mutale, J., Djapic, P., Jenkins, N., Integrating distributed generation into electric power systems: A review of drivers, challenges and opportunities. *Electr. Power Syst. Res.*, 77, 9, 1189–1203, 2007 Jul 1.
38. Justo, J.J., Mwasilu, F., Lee, J., Jung, J.W., AC-microgrids versus DC-microgrids with distributed energy resources: A review. *Renew. Sust. Energy Rev.*, 24, 387–405, 2013 Aug 1.
39. Jiayi, H., Chuanwen, J., Rong, X., A review on distributed energy resources and microgrid. *Renew. Sustain. Energy Rev.*, 12, 9, 2472–2483, 2008 Dec 1.
40. Martirano, L., Habib, E., Parise, G., Greco, G., Manganelli, M., Massarella, F., Parise, L., Demand side management in microgrids for load control in nearly zero energy buildings. *IEEE Trans. Ind. Appl.*, 53, 3, 1769–1779, 2017 Feb 22.
41. Planas, E., Gil-de-Muro, A., Andreu, J., Kortabarria, I., de Alegría, I.M., General aspects, hierarchical controls and droop methods in microgrids: A review. *Renew. Sustain. Energy Rev.*, 17, 147–159, 2013 Jan 1.
42. Bayati, N., Hajizadeh, A., Soltani, M., Protection in DC microgrids: A comparative review. *IET Smart Grid*, 1, 3, 66–75, 2018 Oct.
43. Faisal, M., Hannan, M.A., Ker, P.J., Hussain, A., Mansor, M.B., Blaabjerg, F., Review of energy storage system technologies in microgrid applications: Issues and challenges. *IEEE Access*, 6, 35143–35164, 2018 May 28.

6

Mathematical Modeling and Analysis of Solar PV–Electrolyzer–Fuel Cell-Based Power Generation System

R. Aruna* **and K. Punitha**

Department of EEE, P. S. R. Engineering College, Sivakasi, Tamil Nadu, India

Abstract

The energy need is increased due to changes in modern lifestyle. In a survey, approximately 60% of the population in India are using vehicles and commercial electronic products. The use of electrical energy increases, leading to a demand in fossil fuel and raising the greenhouse gas effect. The increases in energy demand go ahead to a search in different power and energy sources to meet out the energy requirement. The renewable energy resources are paving a way to meet out the demand, but these have a main drawback being the seasonal change in energy production. To beat out the losses of weather-dependent resources, various energy storage techniques are used. On using batteries, periodic maintenance is required and has a short lifespan. For long-term storage, research works are moving towards the hydrogen-based energy storage system. On proper storing of hydrogen gas, the gas is further used for electrical energy production using fuel cell technology. The fuel cell can be used in both vehicle application and also for electrical energy production. This chapter presents the environment-friendly solar PV fuel cell-based power generation system as a micro-grid for commercial application. The fuel cell-based power generation system was integrated with solar PV system for the production of hydrogen gas. The whole power generation system comprises of solar PV system, electrolyzer, hydrogen gas and oxygen gas storage tank, and fuel cell with suitable converters. Initially, the characteristics of each component in the system are analyzed by developing the simulink model of the system using MATLAB. Each component is interconnected with a suitable converter device. To enhance the performance of the system, a suitable energy management system is likewise developed. The intention behind the development

**Corresponding author*: arunamuthukrishnan@gmail.com

P. Sivaraman, C. Sharmeela and P. Sanjeevikumar (eds.) Microgrids for Commercial Systems, (157–186) © 2024 Scrivener Publishing LLC

of the energy management system is to not waste the renewable energy sources and to meet out the energy demands.

Keywords: Solar PV panel, PEM electrolyzer, hydrogen tank, PEM fuel cell, DC-DC converter, inverter

6.1 Introduction

Increasing emphasis on delivering the energy demand through green energy sources has led to a quantum development of renewable energy sources. These resources are hindered by their intermittent and unpredictable nature. Nowadays, the generation of electricity from solar energy is an on-hand system. In Tunisia Research and Technology Centre of Energy, a solar PV power plant is installed and measured the meteorological data for 1 year. As a result, it was concluded that, in the solar PV energy generated system, it was able to save 4,500 kWh of energy, and the CO_2 emission is reduced by approximately 2,600 kg per year [1]. The drawback on using a solar power plant is that the power output from the solar PV plant is intermittent, and in order to get the regular power output from a solar PV plant, the energy storage system is widely used. Generally, battery energy storage systems are commonly used for this purpose, and they will also store the excess amount of solar energy. The batteries have some demerits in terms of periodic maintenance, short lifespan, increase in cost, and environmental impacts [2]. To resolve those problems, a hydrogen energy storage system is used as an emission-free storage system with a long-term storage capacity. The review articles [3] give a detailed report on the renewable power system with various hydrogen storage techniques to store the energy for a long period. The utilization of hydrogen gas using a fuel cell and its thermal management are presented. The modeling of renewable power generation systems comprises of PV panel/wind turbine/electrolyzer/hydrogen tank/fuel cell showing the integrative action performed by various resources for the proper production of effective energy is presented [4–9]. In order to increase the efficiency of the power-generating system, a detailed mathematical model is required. The mathematical model of a solar PV cell is presented [10–12] to enhance the performance of the system. The maximum power point tracking technique is combined with a power electronic circuit which provides an interface with load [13, 14].

In order to support an eco-friendly environment, the use of hydrogen is one of the better ways. The hydrogen can be produced from both fossil and non-fossil resources. Thus, hydrogen becomes a substitutable energy of fossil fuel which is a light element, portable, with no emission of harmful gases, and supportive for sustained power generation. Water electrolysis is one of the methods for hydrogen production. Various types of electrolyzers are available in market. Among them, the polymer electrode membrane (PEM) electrolyzer is more efficient, compact, and easy to maintain. The mathematical model of the PEM electrolyzer is presented with their characteristics in [15–19]. The generated hydrogen gas from the electrolyzer needs to be stored carefully. The hydrogen gases may be stored in the form of compressed gas, liquid, and metal [1, 22].

The stored hydrogen gas is utilized for power generation using a fuel cell. The fuel cell has high efficiency, ultra-low emissions, noise-free operation, and superior reliability and durability in space, automotive, and standalone applications [20]. The fuel cell generates electricity by a chemical reaction which produces power and heat [21, 22]. The PEM fuel cell uses polymer as solid electrolyte and platinum as catalyst. It will not produce any crack or leak. The above-mentioned systems are interconnected to utilize the energy from various resources. The suitable energy management system will reduce the losses during various load changing conditions [23]. For the real-time implementation, the sizing calculation of solar PV, PEM electrolyzer, hydrogen gas storage tank, and PEM fuel cell are presented in the literature [24]. A management strategy for the power generation system comprises of solar cell, fuel cell, and super capacitor as storage element [25]. The developed management system overcomes the sudden change in demand and provides permanent supply to the load.

A battery management system is designed for a standalone system consisting of photovoltaic, wind generation system, fuel cell, and storage batteries using artificial intelligence techniques [26]. For improving the power quality of solar PV, wind power generation system, and fuel cell, a suitable management system is presented to mitigate all the power quality-related problems [27]. Power and energy management strategies are developed for the hybrid PV and wind power generating system with lithium ion battery storage from the real generation profiles in Denmark. As a result, it was concluded that the proper use of hybrid power generating system will resolve the losses, and the cost will be reduced [28]. In the literature [29], the prospects of integrated solar PV and fuel cell are discussed. The key features and the wide range of applications of integrated solar PV and fuel cell are presented. The weakness/challenges are noted as the efficiency of the system, the slow response of the fuel cell, and the performance degradation with time of the fuel cell and electrolyzer need to be improved. Furthermore, a power management strategy is introduced in a fuel cell/battery hybrid vehicle [30] for fuel consumption as well as for the identification of degradation of fuel cell and battery. From the aforementioned literature, it is noted that, for the proper use of resources, energy management is needed for the reliable flow of electricity. The proposed management system will turn on the fuel cell at the time of low irradiance level to meet the energy demand.

6.2 Hybrid Renewable Energy Storage System

The hybrid renewable energy storage (HRES) system associates with two electricity generation systems that are managed through power electronics circuits. Solar photovoltaic (PV) system and fuel cell power generation system are combined together to provide increased system efficiency and also to balance the energy supply [4]. It is comprised of photovoltaic panels, PEM electrolyzer, compressed hydrogen tank, and PEM fuel cell with suitable converter circuits to meet out the energy demand. The block diagram of the HRES system is shown in Figure 6.1.

6.3 Modeling of the Hybrid Renewable Energy Storage System

6.3.1 PV Panels

The solar PV panel is a semiconductor device which can produce direct current without any effect on the environment. The solar PV system contains solar cells to convert light energy to electrical energy [9]. The photons in sunlight hit the solar cells; the cells absorb the photons and conduct the

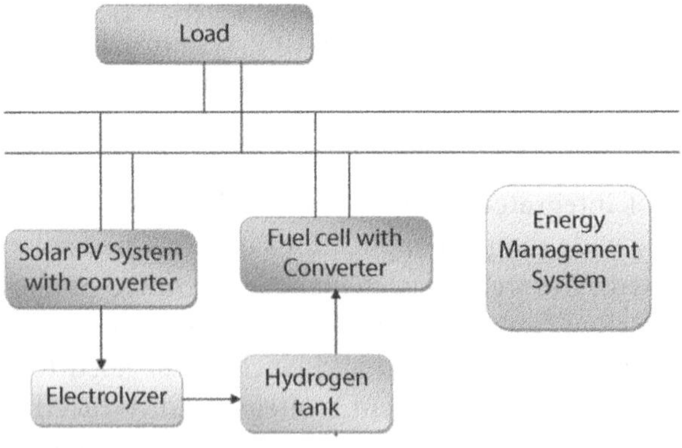

Figure 6.1 Block diagram of hybrid renewable energy storage system.

electrons. The equivalent circuit of a PV cell is shown in Figure 6.2. It is a single diode model.

The mathematical model of a 60-W PV panel [10–12] is given by the following Equations 6.1–6.4.

The module photo-current I_{PH} is as follows:

$$I_{PH} = [I_{SCr} + K_i(T - 298)] \times \lambda/1000 \tag{6.1}$$

The subsystem 1 in Figure 6.3 shows the simulink model of a photo current. The irradiance λ W/m^2 is variable 1, and the operating temperature is 35°C, with a reference temperature of 25°C. The short circuit current (I_{SCr}) at reference temperature is 3.8 A, while the temperature coefficient K_i is 0.0022A/°C.

The module reverse saturation current I_{rs} subsystem 2 simulink model is shown in Figure 6.4.

$$I_{rs} = I_{SCr} / \left[\exp^{(qV_{Oc}/N_s kAT)} - 1\right] \tag{6.2}$$

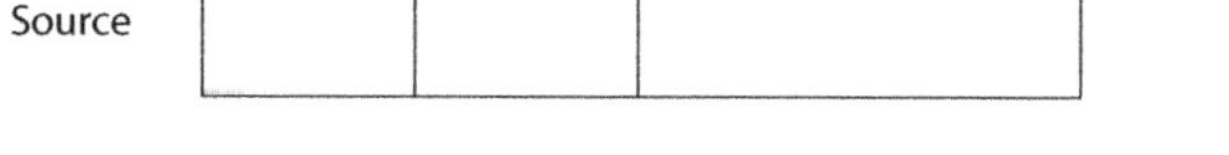

Figure 6.2 Equivalent circuit of a PV cell.

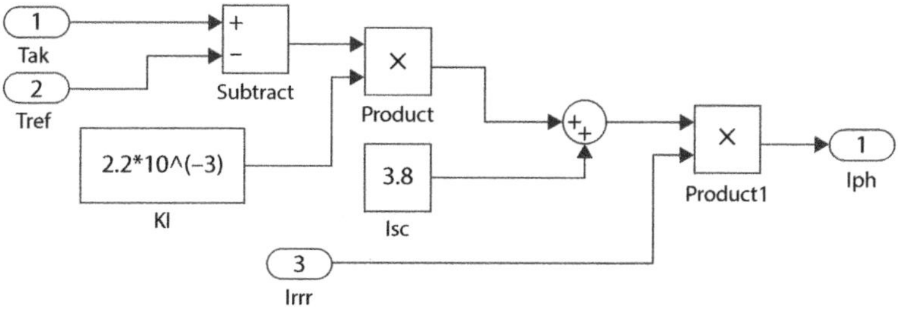

Figure 6.3 Subsystem 1—simulink model of the photo current.

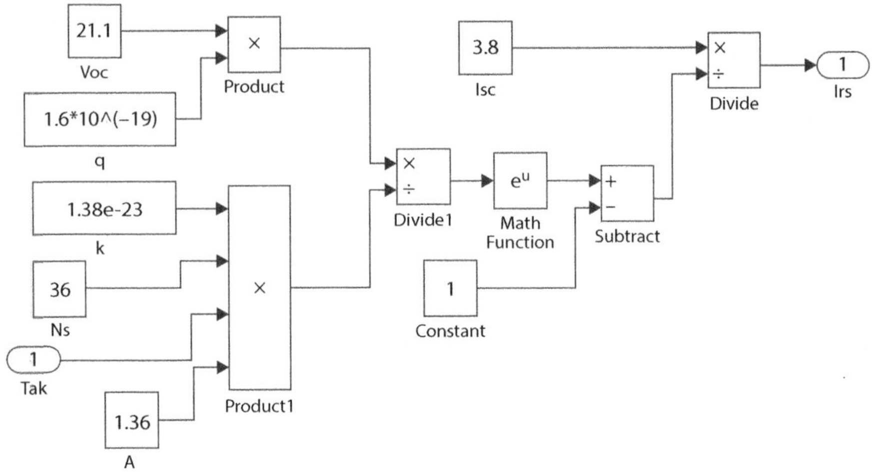

Figure 6.4 Subsystem 2—simulink model of the reverse saturation current.

where the open circuit voltage V_{oc}(V) is 21.1V, q is electron charge at 1.6×10^{-9} C, N_s is the number of cells connected in series, which is 36, k is Boltzmann constant of 1.3805×10^{-23} J/K, and A is the ideality factor at 1.36.

The module saturation current I_d subsystem 3 is shown in Figure 6.5; it varies with the cell temperature, which is given as follows:

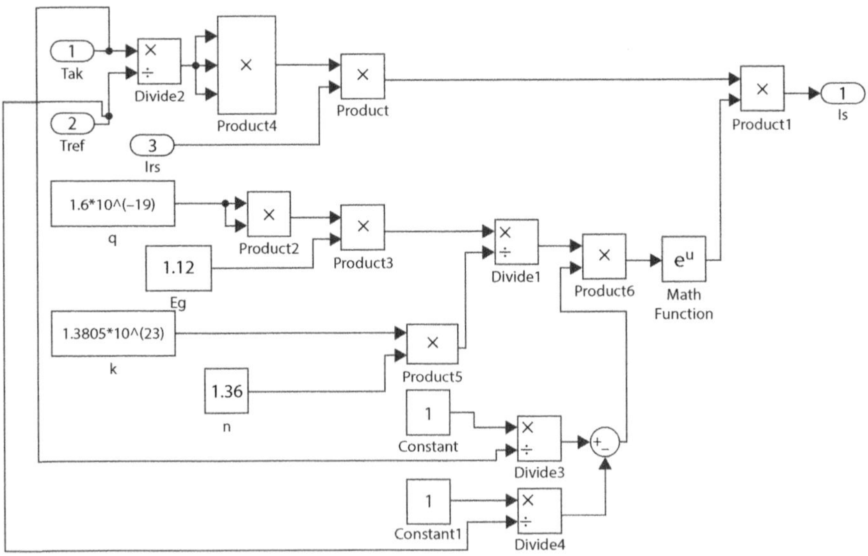

Figure 6.5 Subsystem 3—simulink model of the saturation current.

$$I_d = I_{rs}\left[\frac{T}{T_r}\right]^3 \exp\left[\frac{q \times E_{go}}{Bk}\left\{\frac{1}{T_r} - \frac{1}{T}\right\}\right] \tag{6.3}$$

where E_{go} band gap for silicon is 1.12 eV, and B is the ideality factor.

The current output I of PV module as subsystem 4 based on Equation 6.4 is as follows:

$$I = N_p \times I_{PH} - N_p I_d\left[\exp\left[\frac{q \times (V_{pv} + I_{pv} R_s)}{N_s A k T}\right] - 1\right] - I_{sh} \tag{6.4}$$

and

$$I_{sh} = \frac{V \times \dfrac{N_p}{N_s} + I \times R_s}{R_{sh}} \tag{6.5}$$

where the output voltage of a PV module (V) $V_{pv} = V_{oc}$, I_{sh} is shunt current, N_p is the number of cells connected in parallel, which is considered as 1, series resistance R_s is 0.18 Ω, and parallel resistance R_p is 360Ω based on the Equations 6.1–6.5. The interconnected subsystem of PV panel is shown in Figure 6.6.

The DC–DC boost converter [13] is used to produce a controlled DC output at a desired voltage level. Applying DC voltage across the inductor causes the current to flow through it and the energy to be magnetically stored; then, upon switching off this voltage, the stored energy is transferred as output voltage. By adjusting the on/off time of switches, the output voltage is regulated. The simulink model of the boost converter is shown in Figure 6.7. The capacitor allows ripple to output voltage.

The maximum power point technique (MPPT) method is used to extract the maximum output power automatically by finding the maximum current or maximum voltage under the given temperature and irradiance. The MPPT algorithm works based on the derivative of output power with respect to the panel voltage which is equal to zero at maximum power point [14]. The most fundamental algorithm is the perturb-and-observe (P&O) method, in which the module voltage is periodically given a perturbation, and the corresponding output power is compared with a previous perturbing cycle. Gate pulse to converter switches is provided by P&O, which leads to extract maximum power from the solar module. The

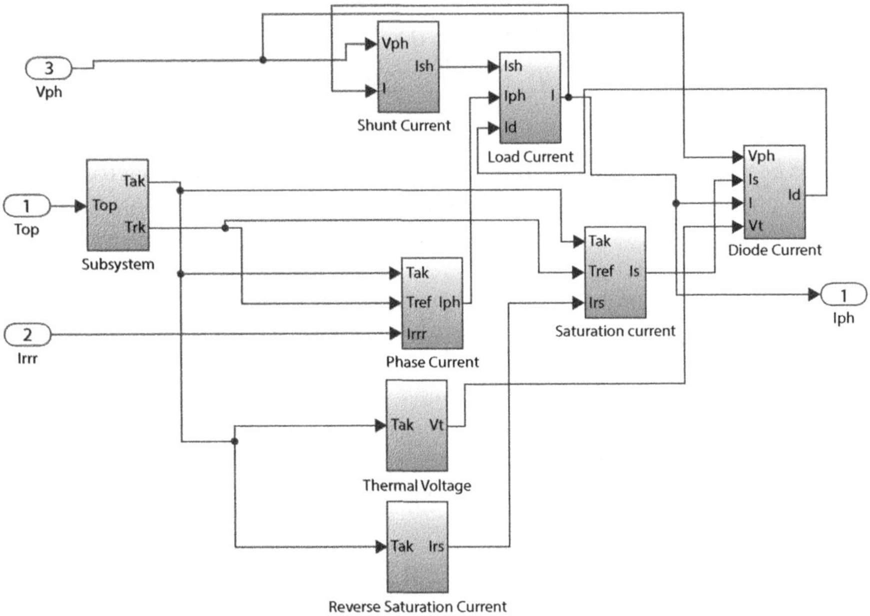

Figure 6.6 Interconnected subsystems of the PV panel.

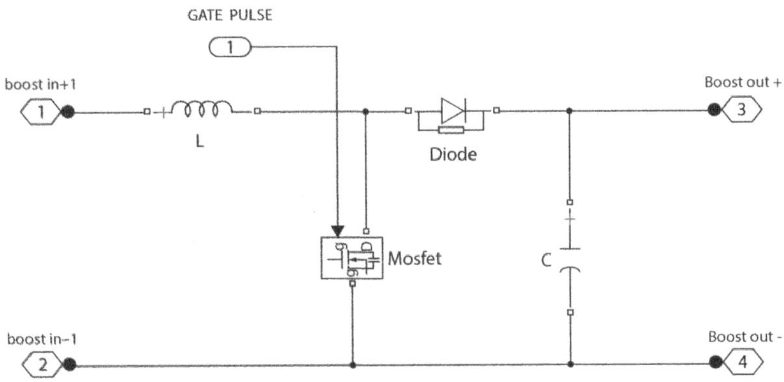

Figure 6.7 Simulink model of the boost converter.

three-phase inverter is employed to modulate the output voltage from the boost converter.

6.3.2 PEM Electrolyzer

Generally, water electrolysis is used to produce hydrogen by dissociation of water molecules into hydrogen and oxygen in combination with a renewable system. The PEM electrolysis is ecologically clean and compact compared with the traditional electrolysis methods [15–19]. The dynamic model is developed based on the electrochemical reaction which is given as follows:

$$H_2O \rightarrow H_2 + 0.5O_2$$

The anode and the cathode sides of an electrolyzer correspond to oxygen and hydrogen, respectively. The reaction rate of hydrogen and oxygen is calculated as follows:

$$H_{2,g} = \frac{nI}{2F}\eta_F, O_{2,g} = \frac{nI}{4F}\eta_F \tag{6.6}$$

The internal current for the travel of hydrogen from the cathode to the anode is generally 99% [16]. The voltage required for the PEM electrolyzer can be expressed as follows:

$$V_{el} = E + V_{act} + V_{ohm} \tag{6.7}$$

According to the Nernst equation, the equilibrium voltage simulink model in Figure 6.8 is expressed as follows:

$$E = 1.484 - 0.85 \times 10^{-3}(T - 298) + 2.3\frac{RT}{2F}\left[\log\left(\frac{P_{H_2}\sqrt{P_{O_2}}}{P_{H_2O}}\right)\right] \tag{6.8}$$

where the universal gas constant R is 8.3145 J mol^{-1} K^{-1}, the Faraday constant F is 9.6485 × 10^4 Cmol^{-1}, the absolute temperature T is 70°C, and the partial pressure of hydrogen, oxygen, and water is P_{H_2}, P_{O_2}, and P_{H_2O}, respectively.

The simulink model of the ohmic over-potential in Figure 6.9 is obtained by the transportation of hydrogen ions obtained as follows:

$$V_{ohm} = JR_{ohm} \tag{6.9}$$

where J is the current density (A/m^2), and the overall ohmic resistance R_{ohm} is expressed as follows:

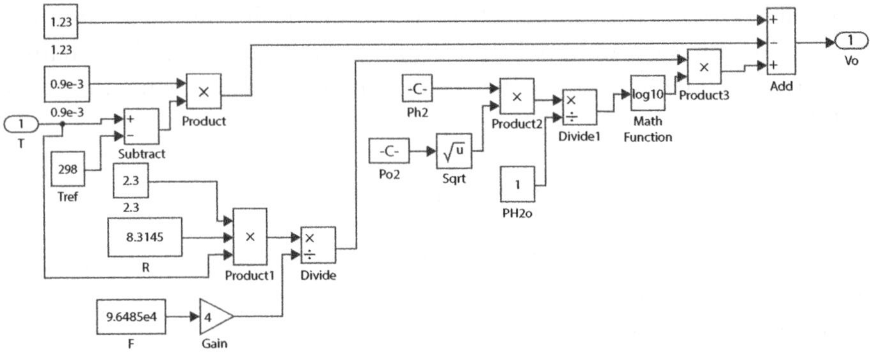

Figure 6.8 Simulink model of the equilibrium voltage.

$$R_{ohm} = \int_0^L \frac{dx}{\sigma[\lambda(x)]} \tag{6.10}$$

where the local ionic conductivity σ(x) of the membrane is expressed as follows:

$$\sigma[\lambda(x)] = [0.5139\lambda(x) - 0.326]\exp\left[1268\left(\frac{1}{303} - \frac{1}{T}\right)\right] \tag{6.11}$$

where $\lambda(x)$ as the water content at location in the membrane is expressed as follows:

$$\lambda(x) = \frac{\lambda_a - \lambda_c}{L} x + \lambda_c \tag{6.12}$$

where membrane thickness L is 50×10^{-6}, and water content at the anode membrane and the cathode membrane is λ_a and λ_c, respectively.

By electrochemical reaction, the activity of the electrodes is varied and measured to be expressed as activation over-potential as shown in Figure 6.10 and expressed as follows:

$$\eta_{act,i} = \frac{RT}{F}\sinh^{-1}\left(\frac{J}{2J_{o,i}}\right), i = a, c \tag{6.13}$$

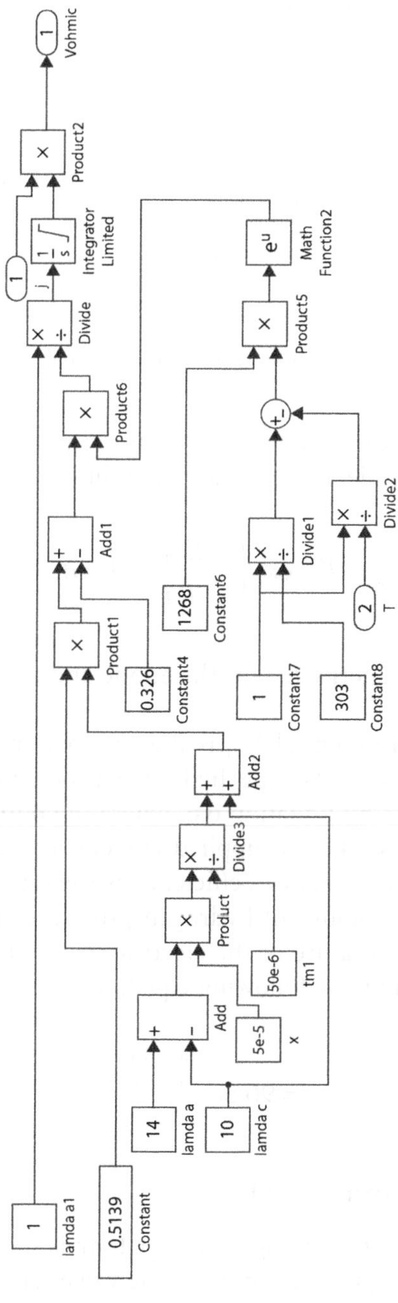

Figure 6.9 Simulink model of the ohmic over-potential.

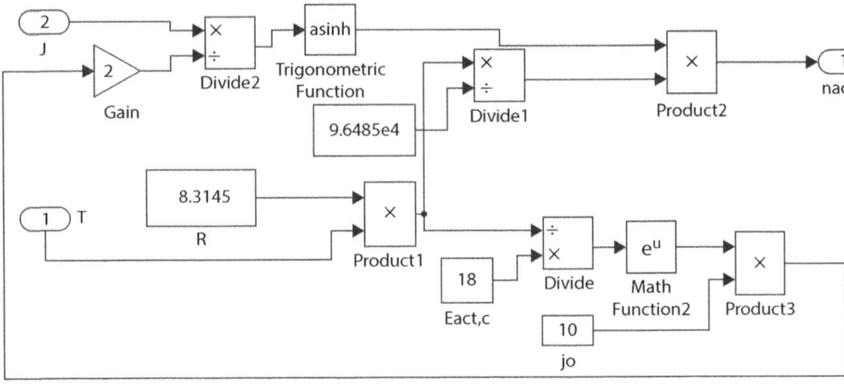

Figure 6.10 Simulink model of the cathode activation potential.

where $J_{0,i}$ is exchange current density with subscripts a and c representing the anode and the cathode. It is expressed as follows:

$$J_{o,i} = J_i^{ref} \exp\left(-\frac{E_{act,i}}{RT}\right), i = a, c \tag{6.14}$$

Here the activation energy E_{act} for the anode and the cathode is 76 and 18, respectively.

The total activation potential V_{act} is the summation of the anode and cathode activation potentials. The hydrogen production rate from the electrolyzer is obtained by Equation 6.6, where the number of PEM electrolyzer cell stacks n is 24, I is supplied current to the electrolyser, and Faraday's efficiency is η_F. Faraday's efficiency is the ratio of the actual and theoretical maximum amount of hydrogen produced in the electrolyzer. Here the working temperature of the electrolyser is 60°C; thus, Faraday's efficiency is calculated by the following equation

$$\eta_F = 96.5e^{\left(\frac{0.09}{I} - \frac{75.5}{I^2}\right)} \tag{6.15}$$

6.3.3 Hydrogen Storage Tank

The simulink model of the hydrogen storage tank is shown in Figure 6.11; in the model energy required for pumps, the valves are ignored. The storage pressure is obtained as follows:

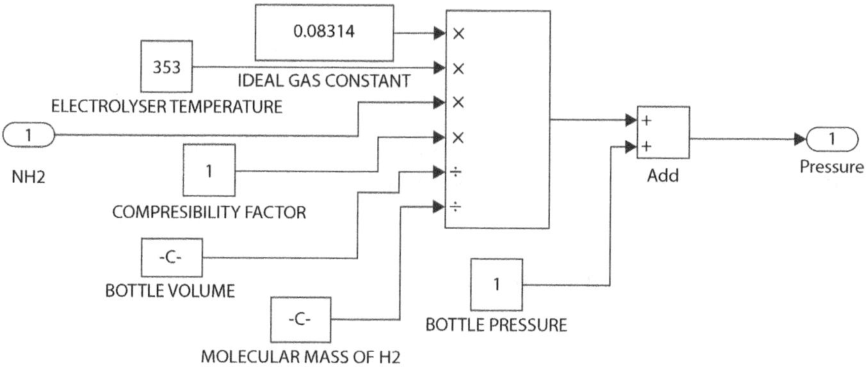

Figure 6.11 Simulink model of the hydrogen pressure.

$$P_b = \frac{zN_{H_2}RT}{M_{H_2}V} + P_i \tag{6.16}$$

where the compressibility factor of hydrogen is 1, the hydrogen flow rate from the electrolyzer is N_{H2}, the molecular mass of H_2 as M_{H2} is 1.0072, the volume of tank is V, and the initial pressure of the tank P_i is 1 bar.

6.3.4 PEM Fuel Cell

The fuel cell produces electricity by an electrochemical reaction. It has two electrodes: one positive anode and one negative cathode. The electricity produced by the reactions takes place at the electrodes. The basic fuel is hydrogen, but oxygen is also required [22]. Thus, fuel cells generate electricity with very less pollution and form a harmless by-product, namely, water.

The hydrogen is oxidized at the anode and yields electron and hydrogen ions [23].

$$H_2 \rightarrow 2H^+ + 2e^-$$

The yielded electrons flow from the anode to the cathode the external circuit. The hydrogen ions travel through the electrolyte and reach cathode. At the cathode, the oxygen gas reacts with the incoming electrons, and the hydrogen ions produce water as a by-product.

$$\frac{1}{2}O_2 + 2H^+ + 2e^- \leftrightarrow H_2O$$

The flows of electrons provide power to the load. The number of cells are connected in series to make a stack and produce DC voltage.

The fuel cell depends on a number of factors such as current density, cell temperature, membrane humidity, and partial pressure of reactants. The potential of a single cell V_{cell} is found using Equation 6.17 [21].

$$V_{cell} = E - V_{act} - V_{ohm} - V_{conc} \tag{6.17}$$

E—Nernst equation is an open-circuit cell potential as a function of cell temperature and partial pressures as shown in Figure 6.12, and the equation is expressed as follows:

$$E = E_o + \frac{RT}{2F}\ln\left(\frac{P_{H_2O}}{P_{H_2} \times P_{O_2}{}^{1/2}}\right) \tag{6.18}$$

where E_0 represents the reference potential at unity, which is 0.6 V, *R* is the universal gas constant, and *F* is the Faraday constant. P_{H_2}, P_{O_2}, and P_{H_2O} are the vapor pressure of hydrogen, oxygen, and water.

V_{act}—Activation loss takes place due to the slowness of reaction on the surface of electrodes. It is analyzed by Tafel's equation, and the model is outlined.

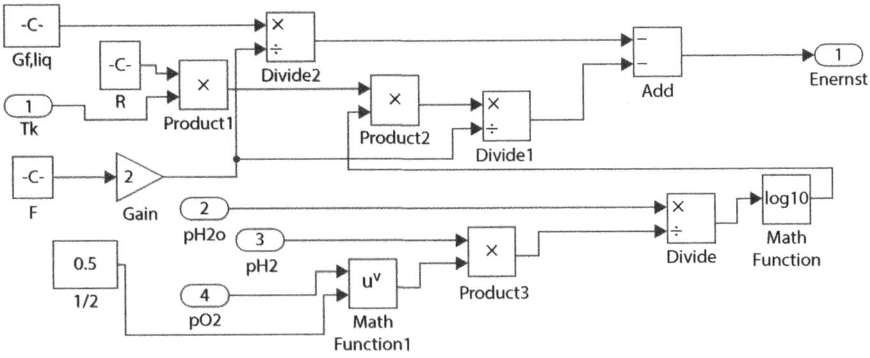

Figure 6.12 Simulink model of the Nernst equation.

$$E_{act} = \xi_1 + \xi_2 + \xi_3\, T\,(\ln\,(CO_2)) + \xi_4\, T\,(\ln\,(i)) \tag{6.19}$$

where ξ_1 to ξ_4 represent constant parametric coefficients, which are 0.9514, 0.00312, 0.000187, and 7.4×10^{-5}, respectively; i is current density, and CO_2 is the oxygen concentration given as a function of stack temperature.

$$CO_2 = \frac{PO_2}{5.08 \times 10^6 e^{(-498/T)}} \tag{6.20}$$

Since the activation overvoltage appears as a voltage drop in the above-mentioned equations and is negative throughout the whole range,

$$V_{act} = -\,E_{act} \tag{6.21}$$

V_{ohm}—At the intermediate current density, the voltage drop is almost linear and ohmic in nature. The membrane resistance R_{mem} is found by dividing the thickness t_m by the conductivity σ.

$$V_{ohm} = IR_{mem} \tag{6.22}$$

$$R_{mem} = \frac{t_m}{\sigma} \tag{6.23}$$

V_{conc}—At a higher current density, the cell potential decreases rapidly due to mass transport limitation. This nonlinearity is termed as concentration over potential, which is given in Equation 6.24.

$$V_{conc} = ae^{(bi)} \tag{6.24}$$

Here the coefficients a (V) and b (cm^2 mA^{-1}) vary with temperature and are given as

$$a = 1.1 \times 10^{-4} - 1.2 \times 10^{-6}\,(T - 273) \tag{6.25}$$

$$b = 8 \times 10^{-3} \tag{6.26}$$

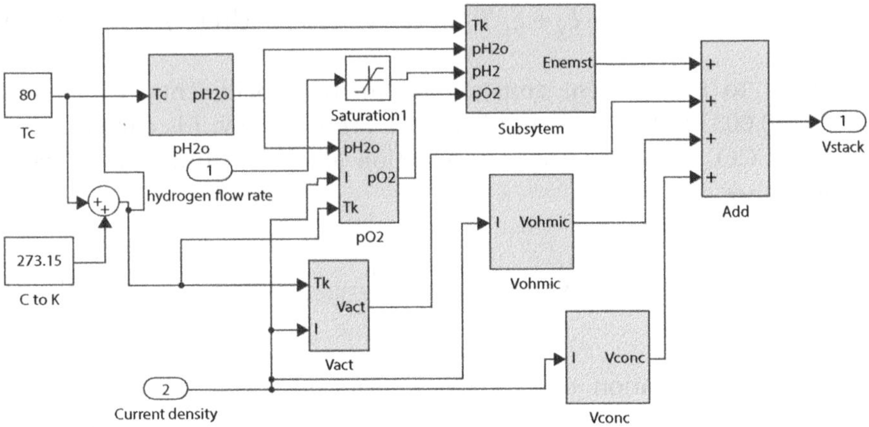

Figure 6.13 Simulink model of the fuel cell.

Thus, from the above-mentioned equations, the cell potential V_{cell} can be solved by the simulink model shown in Figure 6.13. To obtain the stack output, the product of cell potential, and the number of cells N:

$$V_{stack} = V_{cell} \times N \tag{6.27}$$

The output power of PEMFC stack is as follows:

$$P_{stack} = V_{stack} I \tag{6.28}$$

6.4 Characteristic Study of Each Component of the Hybrid Renewable Energy Storage System

6.4.1 Solar PV Panel

The simulation result of the 450-W solar PV panel is simulated using MATLAB. The PV and IV characteristics of the solar PV panel for various irradiance and temperature values are shown in Figures 6.14 and 6.15.

From Figure 6.14 (a and b), it can be observed that, upon increasing the irradiance level, the current is increased, leading to an increase in power value at a constant temperature.

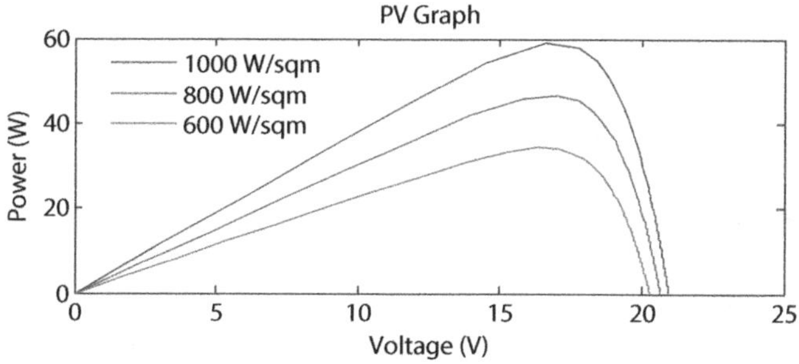

(a)

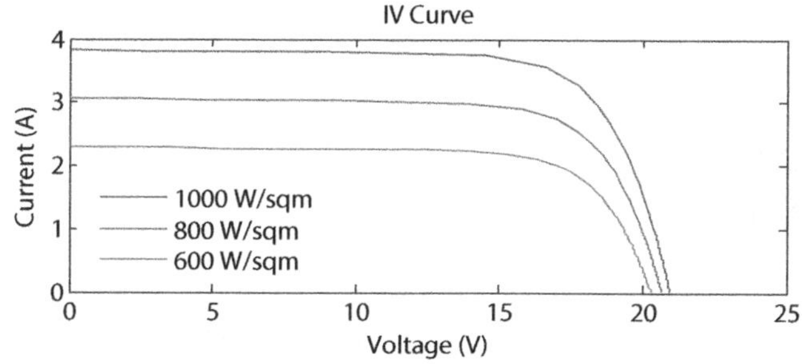

(b)

Figure 6.14 (a) PV characteristics of the solar panel at a constant temperature of 35°C with varying irradiance. (b) IV characteristics of the solar panel at a constant temperature of 35°C with varying irradiance.

From Figure 6.15 (a and b), it can be observed that, upon increasing the temperature, the current is increased at initial voltage value and drops quickly upon increasing the voltage. With the suitable converter circuit, the output of the solar PV is increased to meet the demand.

6.4.2 PEM Electrolyzer

The characteristic curves of the PEM electrolyzer for different temperatures show that as the current density is increased the corresponding cell voltage also increased. At a higher current density, the water molecules separated into a greater number of hydrogen ions and oxygen, and hence a high voltage was it produced. As the temperature increases, the cell voltage is reduced because of Gibbs free energy reduction. Hence, it increases the cell performance and energy conversion. The J–V characteristics of the PEM electrolyser for different temperatures is shown in Figure 6.16.

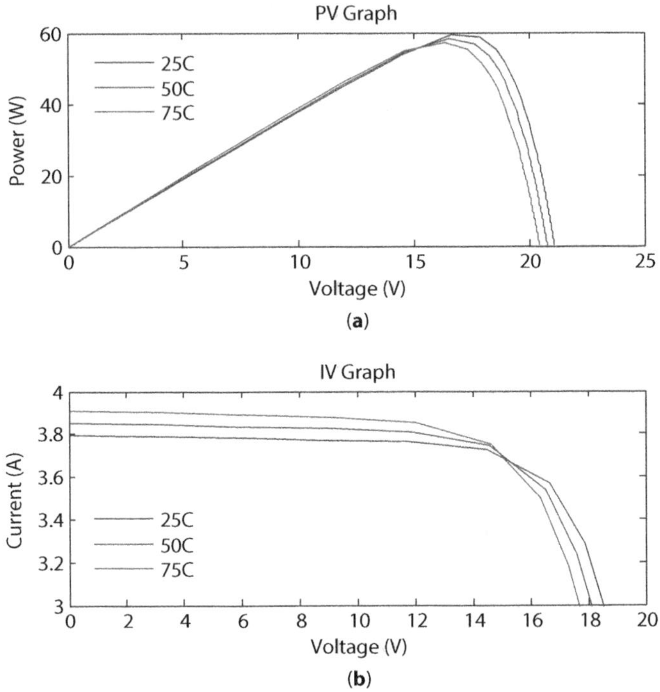

Figure 6.15 (a) PV characteristics of the solar panel at a constant irradiance of 1,000 W/m^2 for varying temperatures. (b) IV characteristics of solar panel at a constant irradiance of 1,000 W/m^2 with varying temperatures.

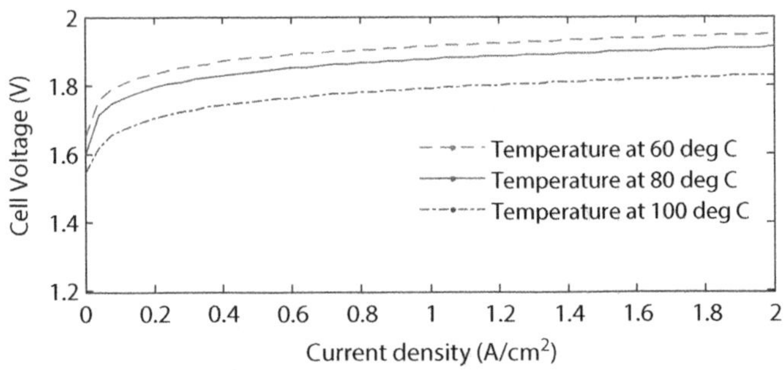

Figure 6.16 J–V characteristics of the PEM electrolyzer.

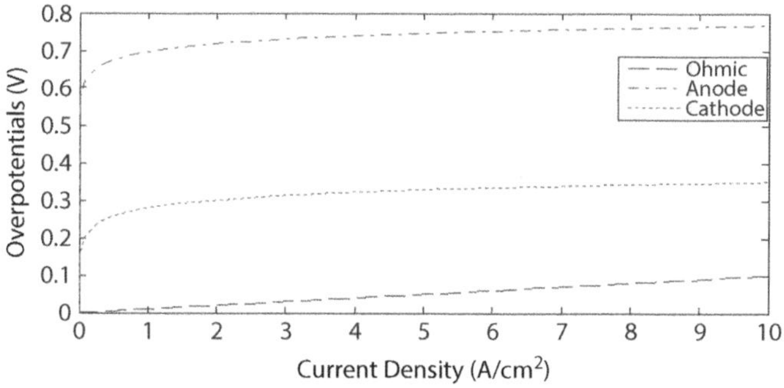

Figure 6.17 Ohmic, anode, and concentration potential of the PEM electrolyzer cell.

Figure 6.17 shows the very small ohmic over-potential that increased steadily with current density. It was due to the high membrane ionic conductivity at typical humidity ($\lambda = 14$) and temperature at 353 K. The cathode activation potential is smaller than the anode activation over-potential due to a fast reduction in kinetics at the cathode surface. However, the anode activation over-potential increased abruptly at a current density which later increased steadily compared with the results obtained in [17].

6.4.3 PEM Fuel Cell

The simulation was carried out by assuming a single-cell PEM fuel cell with a maximum current density of 1.5 A/cm^2. On increasing the operational temperature from 333 to 373 K, it improves the performance due to the increase in the reaction rate kinetics of fuel at the electrode. This leads to an increase in the size of the heat exchanger, coolant, radiator, etc. The corresponding simulation result is shown in Figure 6.18.

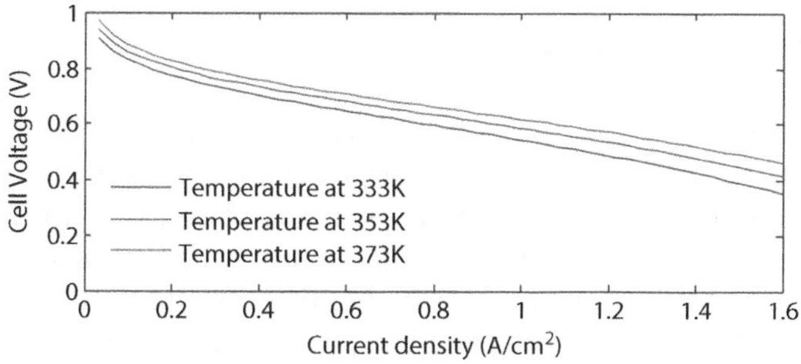

Figure 6.18 J–V characteristic curves obtained for PEM fuel cell at different temperatures.

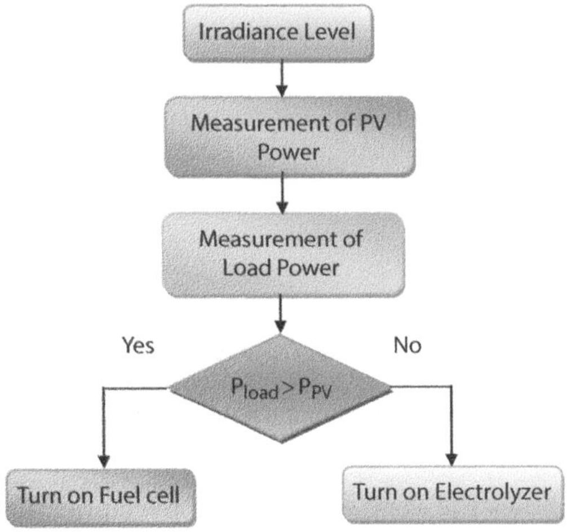

Figure 6.19 Flow chart for power regulation.

6.5 Energy Management System

A regulatory circuit is used to switch on either electrolysis or fuel cell according to the output power of the solar PV panel. The load power is compared with power available from a solar PV panel, and the surplus of energy is diverted to an electrolyzer for producing hydrogen. At the same time, it completely turns off the fuel cell. When the irradiance level starts to decrease, the regulator turned on to the fuel cell as shown in the flow chart of Figure 6.19.

The simulink model of the hybrid renewable storage system in Figure 6.20 shows that all blocks are combined together with the three-phase inverter circuit. It is used to produce a desirable output voltage.

6.6 Result and Discussion

The efficiency of the solar PV is low; it is necessary to track the peak point to operate the PV. Hence, the highest power can be supplied to the load at any condition. A boost converter which is placed next to the PV for extracting maximum power is the maximum power point tracking, the perturb-and-observer method. It matches the impedance by changing the duty cycle of a boost converter. The applied irradiance level is presented in Figure 6.21. The simulated output voltage and current of solar PV with

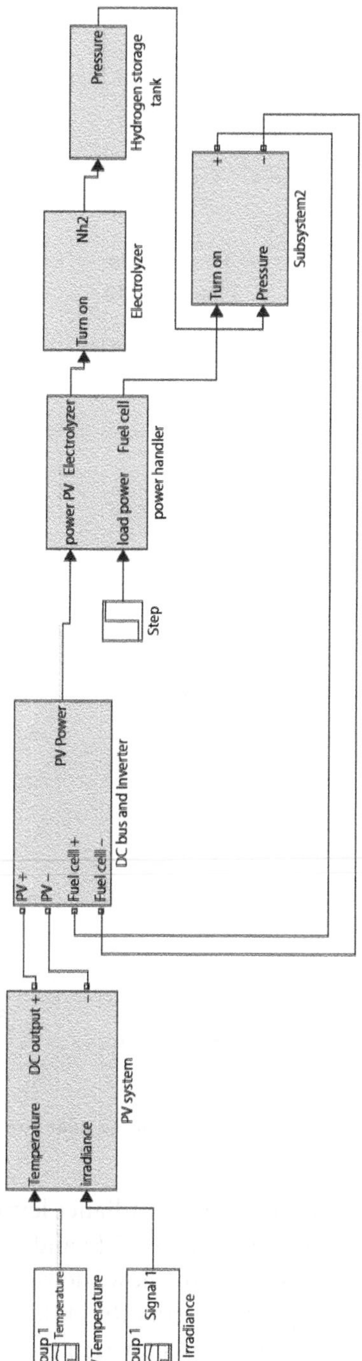

Figure 6.20 Simulink model of a hybrid renewable storage system.

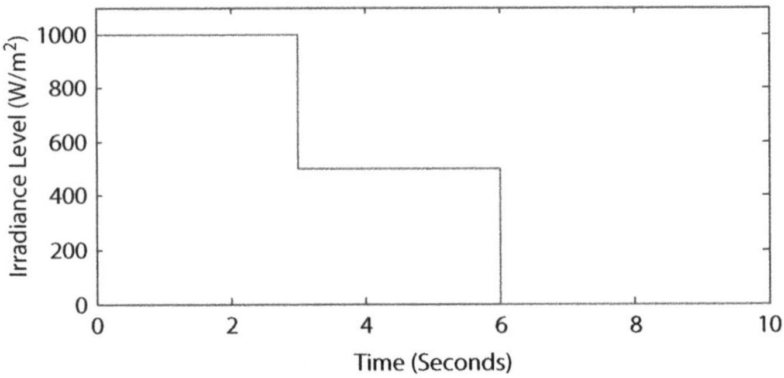

Figure 6.21 Input of solar PV at various irradiance levels.

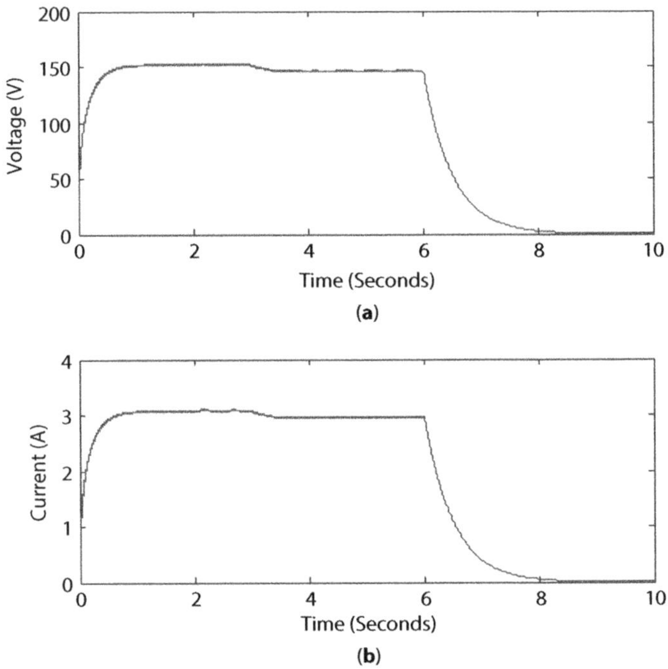

Figure 6.22 (a and b) Output voltage and current of the solar power system.

P&O MPPT boost converter for various irradiance levels at a constant temperature of 35°C [13] is shown in Figure 6.22 (a and b).

A constant load of 200 W is taken to show the variation of energy consumption. The following graph represents the difference in power value, which is delivered to the electrolyzer as shown in Figure 6.23 and required

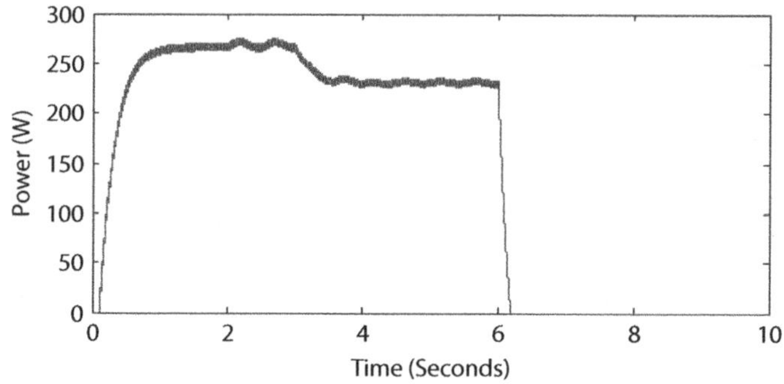

Figure 6.23 The excess power delivered to the electrolyzer.

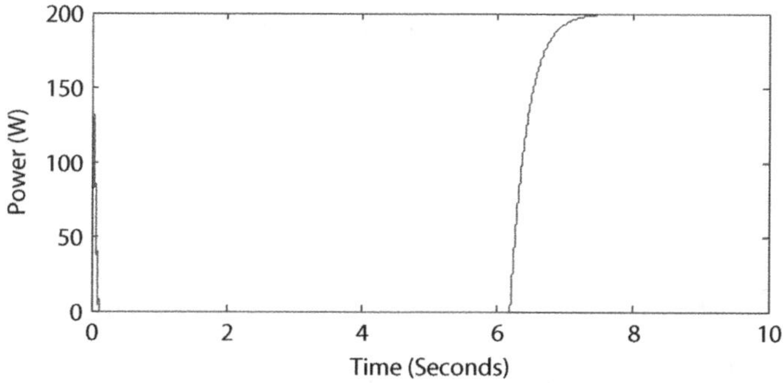

Figure 6.24 The demand of power to turn on a fuel cell.

the power of a fuel cell during the variation in irradiance level as shown in Figure 6.24.

The electrolyzer required a potential for splitting water, and the excess voltage is applied to the electrodes to initiate the electrochemical reactions. When the excess power is delivered to the electrolyzer, it is converted to current density based on the area of the electrolyzer. The current density of electrolyzer is differing due to the change in irradiance level. It was found that, when the current density is increased, then the cell potential is increased steadily with current density. The separated hydrogen gas is stored in the compressed hydrogen storage tank. The generated hydrogen gas partial pressure is synchronized with the current density transients [5]. The following result shows the flow rate of hydrogen as exhibited in Figure 6.25. The increase in current density leads to an increase in the bottle pressure of the hydrogen tank.

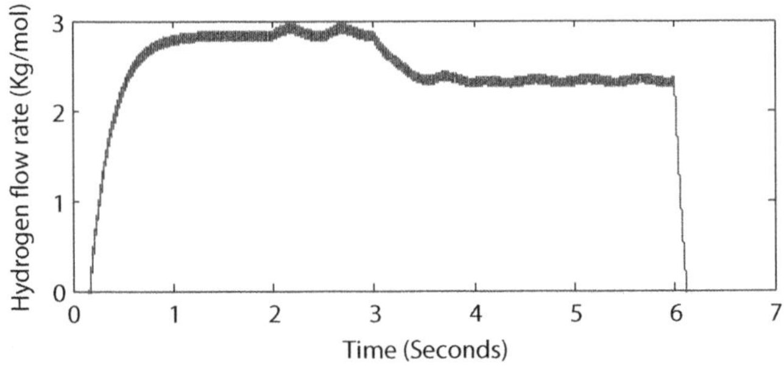

Figure 6.25 Hydrogen flow rate from an electrolyzer.

When the irradiance starts to decrease its level, the hydrogen starts to flow, and the fuel cell produces voltage to meet the load demand. The hydrogen gas was drawn from the compressed gas tank at a constant

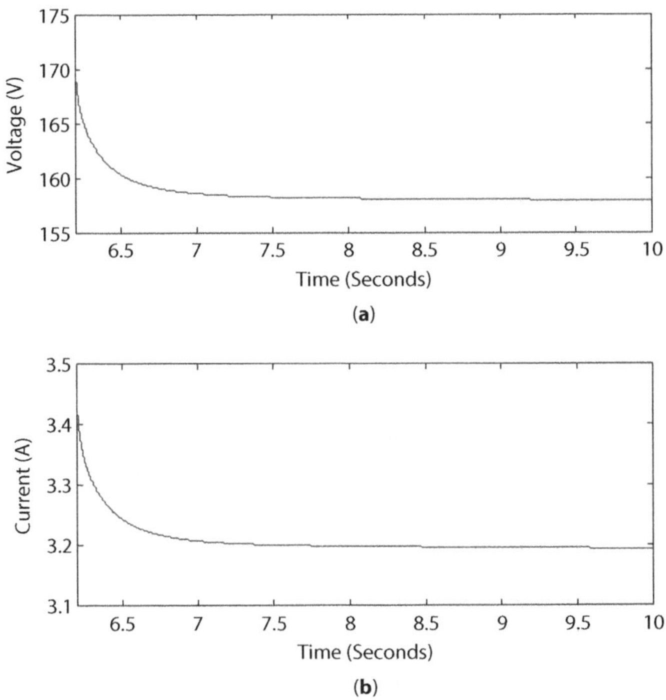

Figure 6.26 (a) DC output voltage of a PEM fuel cell and (b) DC output current of a PEM fuel cell.

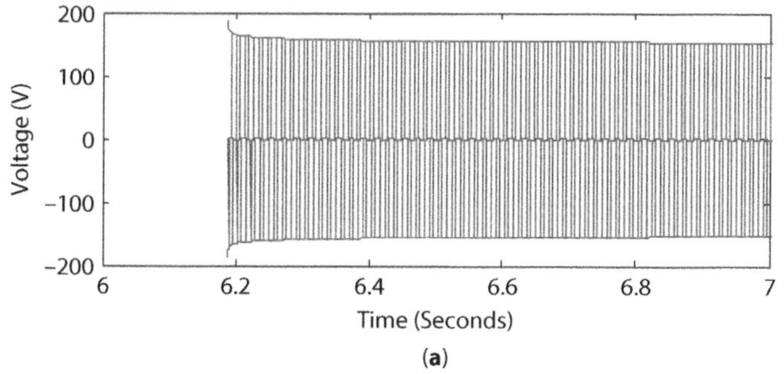

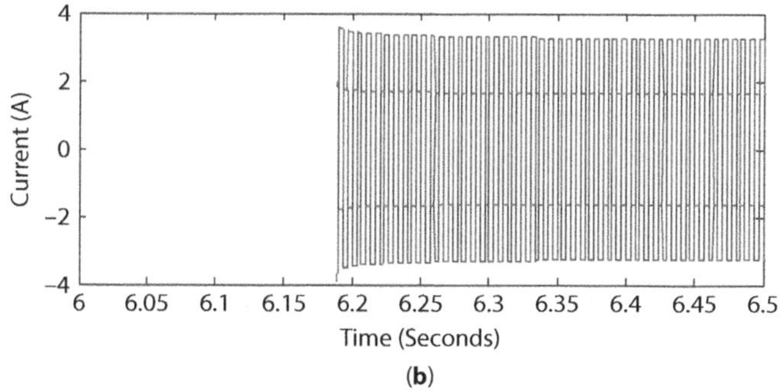

Figure 6.27 (a) Inverter output voltage of a PEM fuel cell and (b) inverter output current of a PEM fuel cell.

pressure, and the DC output voltage and the current of the fuel cell are presented in Figure 6.26. The inverter output voltage and the current are given in Figure 6.27.

The inverter output voltage and the current of the solar PV panel is shown in Figure 6.28. The fuel cell was turned on when the irradiance level decreased and did not meet the demand load. At the same time, it was turned off and the electrolyzer was turned on when excess power was generated in the solar PV system. Similarly, when the irradiance level reached zero, the fuel cell was turned on again to meet the need. The changes in irradiance level values and its corresponding solar PV voltage and current are tabulated in Table 6.1. In addition, the flow rate of hydrogen during the corresponding irradiance level and the corresponding pressure values are mentioned.

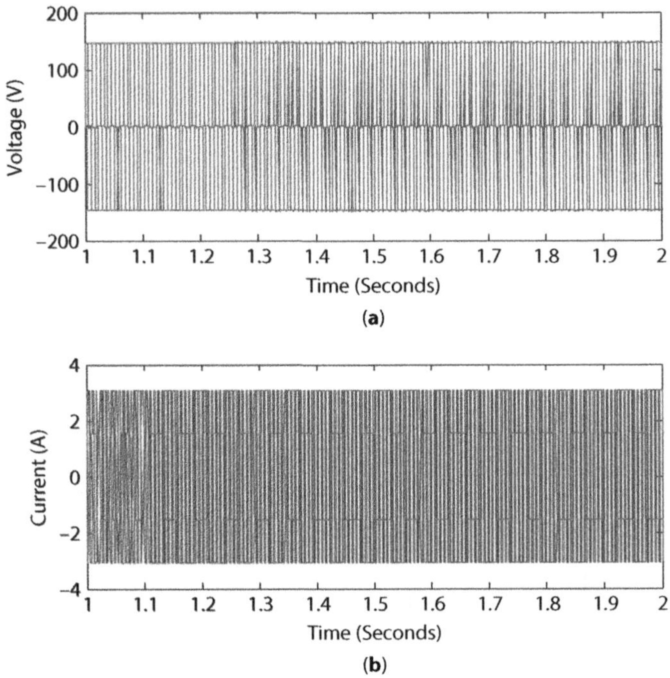

Figure 6.28 (a and b) Inverter output voltage and current of a solar PV panel.

Table 6.1 Various output values for the change in the irradiance level of the hybrid renewable energy storage system.

Irradiance level (W/m^2)	Solar PV voltage (V)	Solar PV current (A)	Hydrogen flow rate (mol/h)	Pressure value in tank (bar)	Fuel cell output power (W)
400	144	2.92	0.06	1.035	0
1,000	153	3.09	0.075	1.045	0
600	148	2.99	0.07	1.04	0
0	0	0	0	0	200

6.7 Summary and Future Scope

In this book chapter, the performance of a hybrid renewable storage system with energy management system was analyzed for various inputs.

Each component characteristic was analyzed with the efficient mathematical model of the systems using MATLAB. The obtained simulation result shows the major limitation on the use of a solar PV power plant that can be overcome by using a hydrogen energy storage system. In the hydrogen energy storage system, the PEM electrolyzer converts the excess solar energy into hydrogen gas and stores it in a compressed hydrogen tank. When the irradiance level is below the required load, the fuel cell turns on. The stored hydrogen gas, at a constant pressure, delivered to the fuel cell. The outputs of the hybrid renewable energy storage system for various irradiance levels are discussed. As a future enhancement, the suitable model of hydrogen energy storage system can be recognized by using a genetic algorithm for system identification. Optimization technique can be used to control the hybrid renewable energy storage system to increase the efficiency of the system's performance. Other renewable energy-based power generation systems can be interconnected to the hydrogen energy storage system.

References

1. Sahlaoui, K. *et al.*, Experimental study of the combined res-based generators and electric storage systems for public buildings, special issue on emerging and renewable energy: Generation and automation. *Energy Procedia*, 162, 275–284, 2019.
2. Kebede, A.A., Kalogiannis, T., Mierlo, J.V., Berecibar, M., A comprehensive review of stationary energy storage devices for large scale renewable energy sources grid integration. *Renew. Sustain. Energy Rev.*, 159, 112213, 2022.
3. Deshmukh, S.S. and Boehm, R.F., Review of modeling details related to renewably powered hydrogen systems. *Renew. Sustain. Energy Rev.*, 12, 2301–2330, 2008.
4. Arabul, F.K., Arabul, A.Y., Kumru, C.F., Boynuegri, A.R., Providing energy management of a fuel cell-battery-wind turbine-solar panel hybrid off grid smart home system. *Int. J. Hydrog. Energy*, 42, 43, 26906–26913, 2017.
5. Dursun, E., Acarkan, B., Kilic, O., Modeling of Hydrogen production with a stand-alone renewable hybrid power system. *Int. J. Hydrog. Energy*, 37, 3098–3107, 2013.
6. Khan, M.J. and Iqbal, M.T., Dynamic modeling and simulation of a small wind-fuel cell hybrid energy system. *Renew. Energy*, 30, 421–439, 2005.
7. Kyriakarakos, G., Dounis, A.I., Arvanitis, K.G., Papadakis, G., A fuzzy logic energy management system for polygeneration microgrids. *Renew. Energy*, 41, 315–327, 2012.

8. Khan, M.J. and Iqbal, M.T., Analysis of a small wind-hydrogen stand-alone hybrid energy system. *Appl. Energy*, 86, 2429–2442, 2009.
9. Lajnef, T., Abid, S., Ammous, A., Modeling, control, and simulation of a solar hydrogen/fuel cell hybrid energy system for grid-connected applications. *Adv. Power Electron.*, 352765, 2013.
10. Pandiarajan, N. and Muthu, R., Mathematical modeling of photovoltaic module with simulink. *Int. Conf. Electr. Energy Syst. (ICEES 2011)*, 3–5, Jan 2011, IEEE, 978-1-61284-379-7/11$26.00 ©.
11. Salmi, T., Bouzguenda, M., Gastli, A., Masmoudi, A., MATLAB/simulink based modelling of solar photovoltaic cell. *Int. J. Renewable Energy Res.*, 2, 2, 213–218, 2012.
12. Bellia, H., Youcef, R., Fatima, M., A detailed modeling of photovoltaic module using matlab. *NRIAG J. Astron. Geophys.*, 3, 53–61, 2014.
13. Bataineh, K.M. and Hamzeh, A., Efficient maximum power point tracking algorithm for PV application under rapid changing weather condition. *ISRN Renew. Energy*, 2014, 13, 673840, 2014.
14. Purti, R., Wibowo, S., Rifa, M., Maximum power point tracking for photovoltaic using incremental conductance method. *2nd International Conference on Sustainable Energy Engineering and Application, ICSEEA*, vol. 68, Energy Procedia, pp. 22–30, 2014.
15. Uzunoglu, M. and Alam, M.S., Dynamic modeling, design, and simulation of a combined PEM fuel cell and ultra capacitor system for stand-alone residential applications. *IEEE Trans. Energy Convers.*, 21, 3, 767–775, 2006.
16. Zini, G. and Tartarini, P., Wind-hydrogen energy stand-alone system with carbon storage: Modeling and simulation. *Renew. Energy*, 35, 2461–2467, 2010.
17. Lee, B., Park, K., Kim, H.M., Dynamic simulation of PEM water electrolysis and comparison with experiments. *Int. J. Electrochem. Sci.*, 8, 235–248, 2013.
18. Ni., M., Leung, M.K.H., Leung, D.Y.C., Electrochemistry modeling of proton exchange membrane (PEM) water electrolysis for hydrogen production. *WHEC 16*, Lyon France, 13–16 June 2006.
19. Zhang, H., Su, S., Lin, G., Chen, J., Efficiency calculation and configuration design of a PEM electrolyzer system for hydrogen production. *Int. J. Electrochem. Sci.*, 7, 4143–4157, 2012.
20. Caliskan, H., Dincer, I., Hepbasli, A., Energy, exergy and sustainability analyses of hybrid renewable energy based hydrogen and electricity production and storage systems: Modeling and case study. *Appl. Therm. Eng.*, 61, 784–798, 2013.
21. Saeed, W. and Warkozek, G., Modeling and analysis of renewable PEM fuel cell system. *Energy Procedia*, 74, 87–101, 2015.
22. Gou, B., Na, W.K., Diong, B., *Fuel cells modeling, control, and applications*, CRC Press, Baco Raton, FL, USA, 2017.

23. Bhuyan, K.C., Sao, S.K., Mahapatra, K., Fuel cell connected to grid through inverter, in: *2012 Students Conference on Engineering and Systems*, pp. 1–4, 2012, 978-1-4673-0455-9/12/$31.00 ©.
24. Aruna, R., Jaya Christa, S.T., Jeyapaul, P.P., Modeling analysis and design of Solar PV based hydrogen energy storage system for residential applications. *IJATEE*, 9, 91, 788–806, 2021.
25. Ferahtia, S., Djerioui, A., Zeghlache, S., Houari, A., A hybrid power system based on fuel cell, photovoltaic source and supercapacitor. *SN Appl. Sci.*, 2, 940, 2020.
26. Bendary, A.F. and Ismail, M.M., Battery charge management for hybrid PV/wind/fuel cell with storage battery, special issue on emerging and renewable energy: Generation and automation. *Energy Procedia*, 162, 107–116, 2019.
27. Samal, S. and Hota, P.K., Design and analysis of solar PV-fuel cell and wind energy based microgrid system for power quality improvement. *Cogent Eng.*, 4, 1, 1402453, 2017.
28. Stroe, D.I., Zaharof, A., Iov, F., Power and energy management with battery storage for a hybrid residential PV-wind system - a case study for Denmark. *12th International Renewable Energy Storage Conference*, Energy Procedia, vol. 155, pp. 464–477, 2018.
29. Ogbonnaya, C., Abeykoon, C., Nasser, A., Turan, A., Ume, C.S., Prospects of integrated photovoltaic-fuel cell systems in a hydrogen economy: A comprehensive review. *Energies*, 14, 6827, 2021.
30. Wang, Y., Moura, S.J., Advani, S.G., Prasad, A.K., Power management system for a fuel cell/battery hybrid vehicle incorporating fuel cell and battery degradation. *Int. J. Hydrog. Energy*, 44, 8479–8492, 2019.

7

Design of DC EV Charging Infrastructure in a Commercial Building Using the Solar PV System

J. Nishanthy*, N. Sivakumar, S. Charles Raja and S. Arockia Edwin Xavier

Department of EEE, Thiagarajar College of Engineering Madurai, Tamil Nadu, India

Abstract

Abundant renewable energy resources such as solar, wind, hydro, and biomass are available in India, particularly in Tamil Nadu. Furthermore, the global adoption of electric vehicles (EVs) is on the rise. In light of this, the current study aims to design a DC microgrid for a commercial building using a solar PV panel system to meet its energy demands. The first step involves designing the microgrid for the selected shopping complex, D-M Art, located in Madurai, Tamil Nadu. Currently, this building lacks EV charging facilities in its parking lot. However, to anticipate that in the future, the availability of EVs may increase significantly and possibly reach up to 90%. This projection underscores the necessity of incorporating a DC microgrid for EV charging within the chosen building. Additionally, the commercial shopping complex faces challenges related to peak energy demand. Lighting loads and heavy loads, such as air conditioners, are currently powered by a combination of generators and utility supply, but this setup does not provide adequate satisfaction during peak periods. Moreover, to address these issues, the design of a DC microgrid and an EV charging infrastructure using solar PV panels should be with the capability for on-grid extension and energy storage to store excess power during power outages and blackouts. Furthermore, we plan to investigate and implement control strategies to optimize the microgrid's performance. This includes simulations conducted using the Hybrid Optimization of Multiple Energy Resources software. Additionally, the combination of a solar on-grid battery system appears to be the most effective solution. Therefore, conducting

**Corresponding author*: jnishanthy123@gmail.com

P. Sivaraman, C. Sharmeela and P. Sanjeevikumar (eds.) Microgrids for Commercial Systems, (187–206) © 2024 Scrivener Publishing LLC

techno-economic and emission studies demonstrates the feasibility and benefits of the proposed DC microgrid design.

Keywords: On-grid solar PV, EV charging infrastructure, techno-economic analysis, emission analysis, HOMER software

7.1 Introduction

In recent times, microgrid design concepts have become more flexible and reliable, thanks to the utilization of renewable energy resources. This shift aims to create environmentally eco-friendly solutions. There is also a growing focus on enhancing research in designing DC microgrids to meet specific load demands. Additionally, there is a global surge in the adoption of electric vehicles (EVs) as a means to produce electrical energy with lower emissions. Battery electric vehicles are at the forefront of sustainable and low-emission transportation. This is especially true when paired with solar-powered charging stations. Furthermore, there are exciting prospects for batteries, with announcements extending to both industry and academic institutions. These opportunities arise from the implementation of charging stations [1].

The focus is on the importance of implementing EV smart charging to mitigate negative power impacts from the grid. This approach not only lowers the charging costs but also reduces grid operational expenses [2]. To tackle the challenges related to power architecture design, storage techniques, energy management optimization, and microgrid control strategies when installing EV charging stations, control techniques and energy management designs are essential for achieving optimal performance [3]. The combination of solar, wind, and battery technologies within a grid-connected system is being examined for feasibility and reliability in economic and environmental aspects. This integrated approach makes EV charging systems highly effective due to increased renewable energy use and reduced carbon emissions [4]; thus, the design is a system using wind, an electrolyzer, a bio-generator, batteries, and a converter. This system can function both as a stand-alone and a grid-connected system, and we have employed the Hybrid Optimization of Multiple Energy Resources (HOMER) tool for its design. It is worth noting that this proposed system is highly suitable for various locations as it adapts to different weather conditions [5]. Through the use of various optimization algorithms, we have developed charging stations. These stations rely on both stand-alone and on-grid renewable energy systems to mitigate peak load intermittency [6]. The integration of a DC microgrid with grid extension is not only feasible but also advantageous for remote areas, rural regions, and small buildings. This DC microgrid, designed for DC loads, aims to reduce protection issues [7].

The development of EVs now includes vehicle-to-grid and grid-to-vehicle capabilities. The focus is on using EVs as a minimum-level energy storage system. We aim to estimate energy absorption from the grid to the EV using a preferred algorithm and then make comparisons [8]. Considering renewable sources such as wind and PV with only grid connections, we find that it is not feasible for charging electric vehicles. Therefore, the most effective solution involves a combination of solar, wind, and battery systems to optimize the EV charging station potential. To enhance system performance and investigate control strategies, we employ MATLAB/Simulink [9].

Evaluating various scenarios for DC fast-charging EVs at highway service centers involves using solar PV, power generation, and battery storage systems to assess the economic feasibility. Our findings indicate a decrease in demand costs thanks to battery storage and reduced consumption costs [10]. Using the HOMER tool, we compare two solar trackers: single-axis and dual-axis trackers. The goal is to identify the most cost-effective system in terms of renewable fraction and emission reduction. The hybrid system successfully meets load demand without shortages, and the dual-axis tracker emerges as the feasible choice [11].

In integrating solar PV with a battery energy system and a diesel generator to investigate the performance of a smart microgrid, our goal is to maximize renewable energy capacity to meet the demand when the grid is unavailable. Additionally, we examine electricity pricing scenarios based on energy costs and selling prices to the grid, with the aim of improving electrical tariff policies [12]. The increasing global adoption of electric vehicles has driven the need for more energy production while reducing energy inefficiency. To address this, we combine solar and wind-based charging stations for electric vehicles, integrated with the grid. This approach helps overcome major challenges in sustainable energy sourcing by leveraging renewable resources. Using HOMER software, we have designed a system to find the optimal solution. Our findings indicate that solar energy output is significantly higher than wind energy when compared to a grid-only system [13].

The growing number of electric vehicles, coupled with limited charging station availability, has prompted an investigation into two methods for managing charging in urban areas. The first method involves mathematical modeling to minimize charging time, while the second employs a charging scheduling algorithm to address charging-related issues. As a solution, we propose using a swarm optimization-based EV scheduling approach to analyze both methods effectively [14]. We designed systems that combine solar PV, wind, and the grid as well as solar PV, wind, grid, and a diesel generator using the HOMER energy tool. Furthermore, we have conducted a technical and economic analysis considering factors such as net present cost, cost of energy, and renewable fraction. The study's results show that the solar-based system is more cost-effective compared to the wind energy system, primarily due to lower initial costs [15].

In light of environmental concerns, we propose a mini-grid system with a capacity of 1.4 MW to serve remote control areas. This system caters to 2,500 households and 120 electric vehicles. We have conducted technical and economic analyses based on voltage, current, and power. The results indicate that solar PV is the most effective in terms of reducing emissions [16]. We designed a system that includes hydrogen load, a hydrogen tank, solar PV, and a wind turbine, all supported by battery storage for electric loads. Using HOMER software for this design, our system modeling and optimization show that the levelized cost of hydrogen is low, and energy production using solar and wind is higher for single-family houses. Consequently, in both utility and renewable energy-based systems, the primary source yields lower emissions and the lowest component costs [17].

We developed a model predictive control-based smart charging strategy for scheduling both conventional EVs and smart charging. This strategy helps manage uncertainties related to EV arrival times and energy demand levels. Moreover, to address increased electrical demand during peak times, we have planned to design EV charging facilities, thereby avoiding grid imbalances [18]. We designed a solar PV–wind–diesel generator-based system with various batteries for EVs, including lead acid, flow-zinc–bromine (ZnBr), and lithium ion. Our aim was to conduct a techno-economic analysis, including factors like net present cost (NPC) and cost of energy (COE). Through various scenarios, we investigated the best battery performance and the most cost-effective battery system. Our findings indicate that ZnBr offers the best battery efficiency among the three different locations integrated with the solar–wind hybrid system [19]. At a university campus, we developed a renewable-based charging integration using HOMER energy software. We considered three scenarios: one without renewable energy, relying solely on a grid-based system; another with renewable and grid-based systems, including solar and wind; and a scenario involving renewable energy with a vehicle-to-grid operation scheme. The simulation results show that the vehicle-to-grid system with solar and wind is the most cost-effective and achieves maximum renewable energy penetration [20].

Based on the gaps identified in previous literature, the primary challenges in charging electric vehicles in commercial buildings using microgrids are energy management strategies and achieving emission-free solutions. This study aims to address these challenges by designing a DC microgrid EV infrastructure to meet the commercial load demand.

Furthermore, the integration of renewable and grid systems for the commercial building is illustrated in Figure 7.1. Additionally, the contributions of the proposed microgrid charging system for the commercial loads are described below:

- Collecting EV specification data and collecting real-time load statistics for within the commercial building
- Designing the DC microgrid electric vehicle charging infrastructure in a commercial building at Madurai, Tamil Nadu, India
- Estimating charging facilities for electrical loads such as lighting air conditioning and electric vehicles in the D-Mart shopping complex
- Employing HOMER software, analyzing the EV microgrid architecture for techno-economic evaluations such as NPC, COE, and emission reduction.

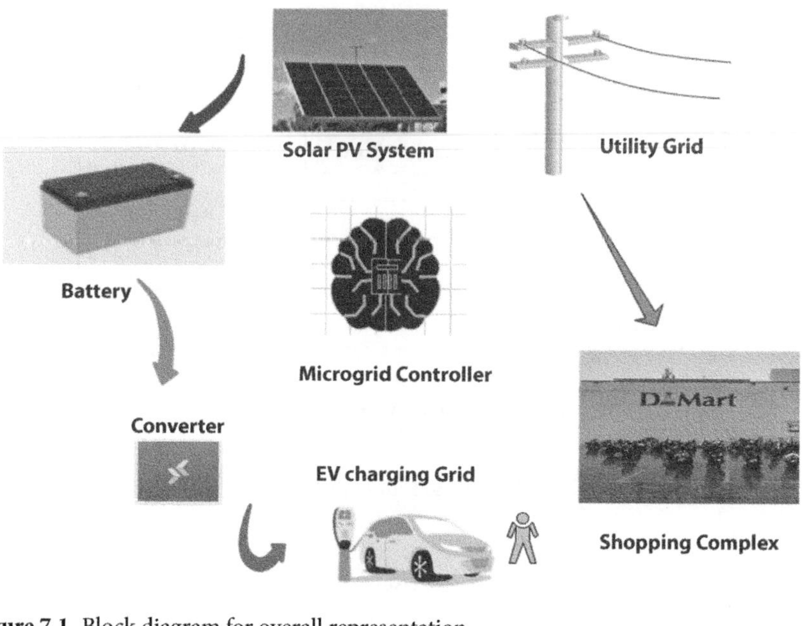

Figure 7.1 Block diagram for overall representation.

7.2 Methodological Analysis

The design of EV DC charging infrastructure for commercial loads, utilizing solar renewable energy, is implemented in a selected location within a shopping complex. This implementation involves analyzing the performance of the commercial building. The proposed work is designed using HOMER software, which requires load details, location maps, cost ratings, and vehicle parameters. The software's process layout is illustrated in

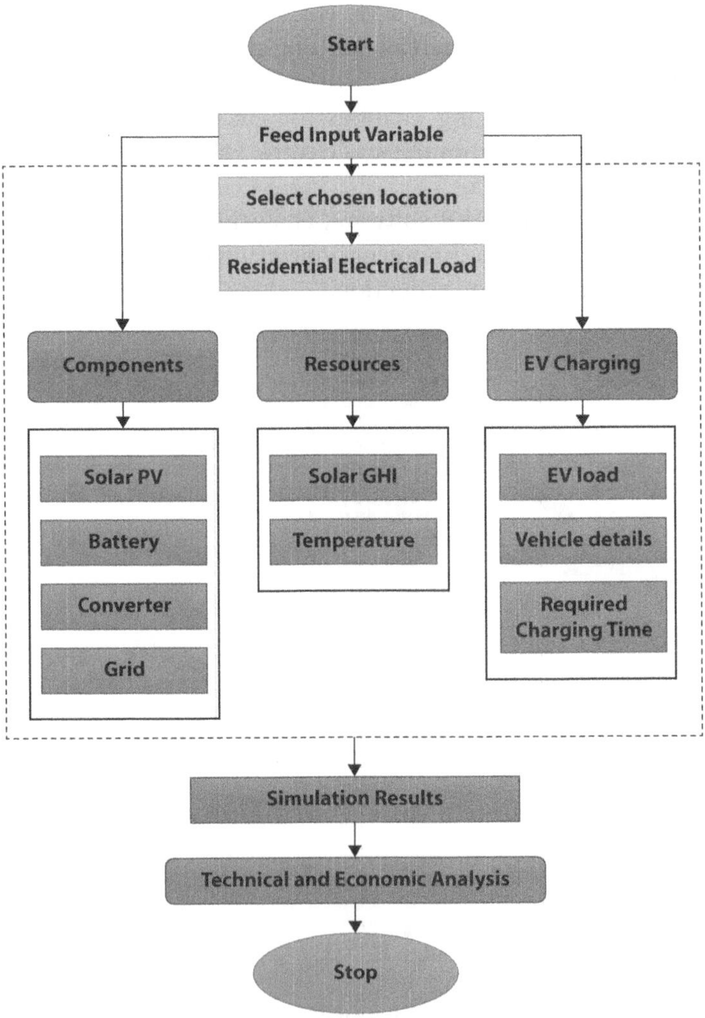

Figure 7.2 Structural outline for the proposed system in HOMER.

Figure 7.2. The subsequent steps in the proposed microgrid design include system description, location details, commercial load, EV parameters, and environmental considerations for solar energy.

7.2.1 System Configuration

Consider the proposed microgrid solar system for the selected commercial shopping complex building. This building boasts 5,500 square feet of space on a single floor, with ample room both in front of the building and in the parking area. This extensive space makes it ideal for installing a solar PV system and optimizing the charging potential for both EVs and commercial loads in an efficient manner [21].

In this building, approximately 300 conventional vehicles arrive each day, while there are about 10 EVs. The total number of vehicles, including both EVs and conventional engine vehicles, is detailed in Table 7.1.

7.2.2 Site Location in Environmental Aspects

The sunlight radiation data from weather forecasting plays a pivotal role in sizing the solar EV charging infrastructure. This sizing depends on solar irradiation and temperature data provided by NREL within the HOMER tool. The selected building offers ample space to accommodate a solar PV array, and its location enjoys high solar energy utilization capabilities. Implementing such a design within the shopping complex yields economic and environmental benefits, including NPC, COE, and emission reduction. Using this software, we can determine the optimization of cost ratios compared to the base case default system, ultimately identifying the most cost-effective solution. Solar power serves as the primary resource during grid unavailability.

Table 7.1 List of the arrival of vehicles in a day.

Type of vehicles	Conventional vehicles	Electric vehicles
Scooter	50	2
Bike	150	2
Car	75	3
Auto	10	-
Mini-truck	1	-

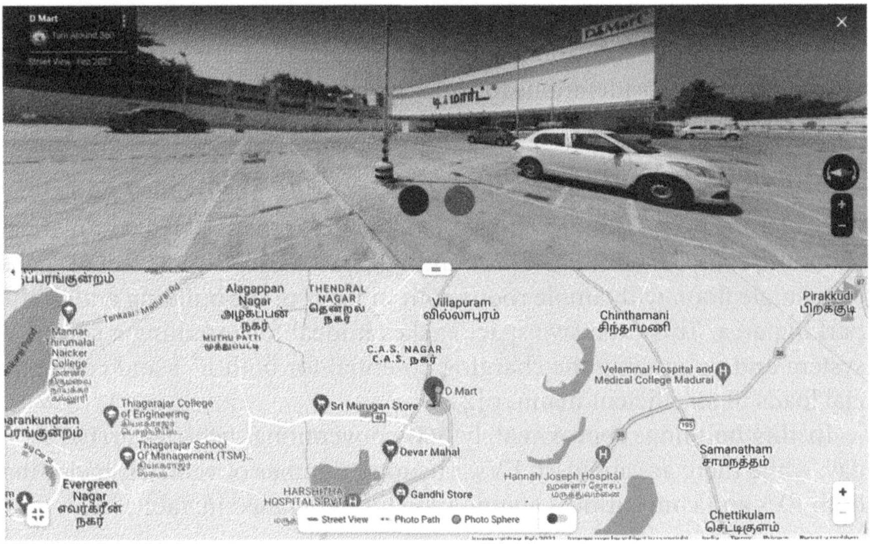

Figure 7.3 Google map of the selected shopping complex.

The shopping complex's location, represented on Google Maps at latitude 9.9252 °N and longitude 78.1198 °E, is illustrated in Figure 7.3.

Furthermore, HOMER software provides the option to retrieve solar global horizontal irradiance levels based on the selected location. This data is dependent on the efficiency of the solar panels. The design of the proposed DC microgrid charging infrastructure includes a depiction of the monthly average solar insolation on a daily basis, along with a clearness index.

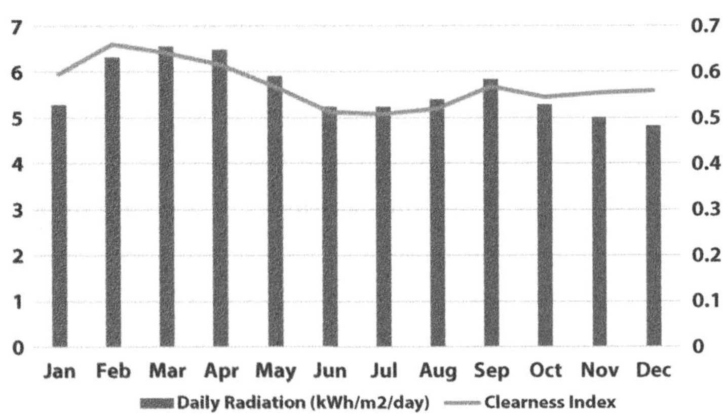

Figure 7.4 Global horizontal insolation level for the chosen site.

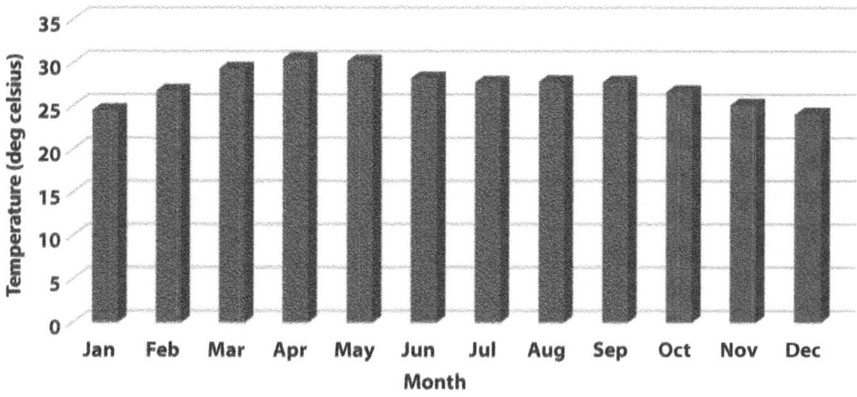

Figure 7.5 Ambient temperature layout for the selected site.

This is illustrated in Figure 7.4. Additionally, the ambient temperature, which is influenced by sunlight exposure on the panels, is obtained for the site location from the NREL database and is shown in Figure 7.5.

7.2.3 Electrical Load Parameters

When considering the charging potential for EVs and commercial loads within the selected shopping complex, it becomes evident that the power demand is significantly higher than the actual supply. To address this issue, the representation of a DC microgrid for the commercial shopping location is crucial. This approach aims to reduce costs and system size and enhance the design efficiency when compared to an AC microgrid. Furthermore, it helps mitigate power fluctuations with fewer converters, thereby improving system efficiency when integrated with solar renewable energy resources. This integration is crucial for reducing greenhouse gas emissions. During power outages and blackouts of the grid-connected system, power consumption is often inefficient. This is the primary reason for implementing a solar design for the selected building as it offers a solution to these problems.

The specific lighting loads for this building are calculated in real time, with load simulations conducted on a daily, monthly, and seasonal basis using software. As a result, it is evident that the highest demand occurs in September, with a load profile of 30.77 kW for the shopping complex. Additionally, the annual demand response is obtained, and the load layout for September 1, 2022 is illustrated in Figure 7.6.

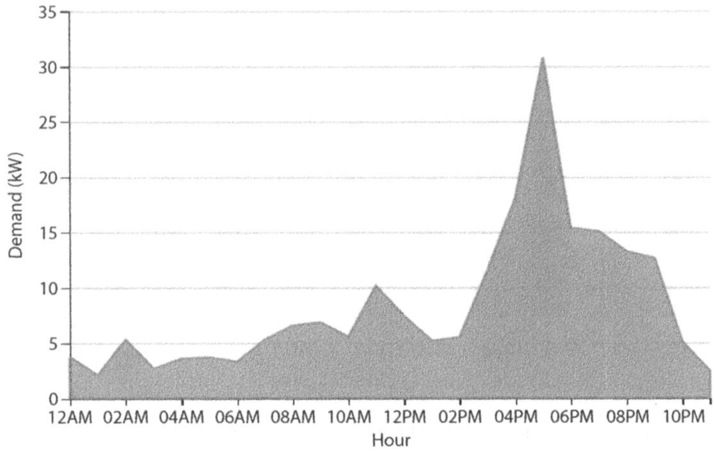

Figure 7.6 Daily profile for the highest demand of the month of September.

The DC microgrid EV charging system powered by solar is designed based on the availability of EVs. To accommodate various EVs, the load structure considers different vehicles, including two-wheelers and four-wheelers, as detailed in Table 7.2. The installation of solar panels on the selected commercial shopping building and the design of a charging infrastructure for electric vehicles are essential due to the large parking area. This initiative aims to support the growth of EVs and plays a crucial role in promoting eco-friendliness. These actions influence market prices and the availability of charging facilities, contributing to a sustainable future when compared to conventional engine vehicles.

In addition, we consider a commercial building equipped with two chargers, each with an output power of 10 kW, totally 11 scaled average sessions per day. The proposed system is designed based on vehicle arrivals for charging, utilizing solar energy production throughout the day. We investigate the starting and running times, which are from 9 A.M. to 10 P.M. The layout of the EV load for the commercial shopping complex is estimated using HOMER software and is displayed in Figure 7.7.

7.2.4 Component Specification Parameters

The implementation of the proposed system for the commercial building necessitates specific components. These include a solar PV system with on-grid extension, primarily driven by environmental concerns, to mitigate power outages. Furthermore, in the selected hot location, solar

Table 7.2 Characteristics of various types of electric vehicles.

Vehicle name	Proportion of EV population	Max EV charging power (kW)	Average charging duration (min)
SLR Dealership Electric Merged Scooter	10	2	120
OLA S1 Hero Electric Photon	18	4	180
Hero Electric Photon	14	1.87	120
Bounce Infinity E1	12	1.9	180
TATA Nexon EV	15	30.2	60
Mahindra E Verito	20	20.74	45
Hyundai Kona Electric	25	39.2	60

energy production is maximized. After utilizing solar energy for both EV and commercial loads, any excess electrical power is stored in batteries for use during power failures. Additionally, a converter is employed to convert AC power from the solar source into DC for EV charging. The proposed microgrid DC system facilitates the charging of EVs and powers

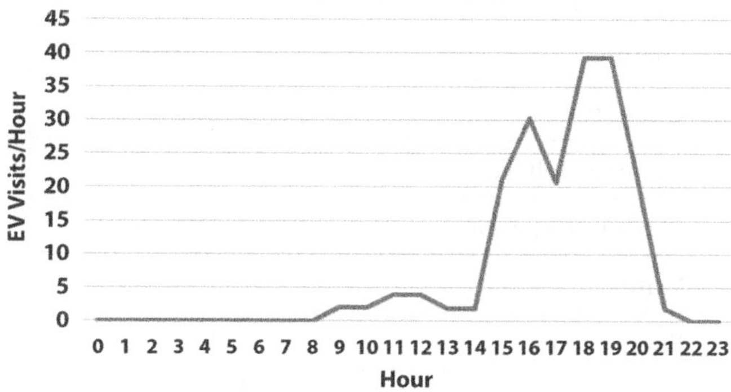

Figure 7.7 Electrical energy supplied to the electric vehicles.

Table 7.3 Equipment specifications and ratings for the proposed microgrid design.

Parameters	**Components**
Canadian Solar PV MaxPower	
Initial cost	₹ 900,000
Operating and maintenance cost	₹ 5,000
Lifetime	25 years
Efficiency	16.94%
Rated capacity	30 kW
Operating temperature	45.00°C
Panel type	Flat plate
Converter–system converter	
Initial cost	₹ 420,000
Operating and maintenance cost	₹ 4,000
Lifetime	15 years
Efficiency	95%
Rated capacity	30 kW
Generic lead acid	
Initial cost	₹ 19,000
Operating and maintenance cost	₹ 4,000
Lifetime	10 years
Quantity	One
Nominal voltage	12 V
Maximum capacity	83.4 Ah
Utility	
Energy charge	₹ 7/kWh
Demand rate	₹ 370/kW
Grid sale limit	10 kW

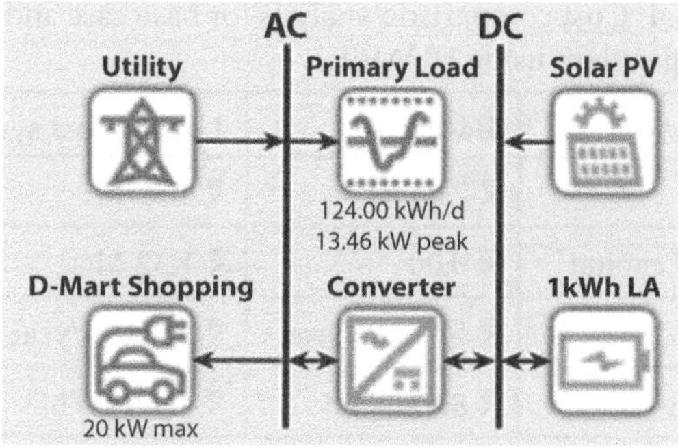

Figure 7.8 Single-line diagram for the solar power DC microgrid.

commercial loads. This comprehensive system combines solar PV, batteries, and converters with grid connections, as illustrated in Figure 7.8. For the design of the charging system, we specify parameters such as capital cost, operating and maintenance expenses, lifetime, efficiency, and capacity ratings. These parameters are detailed in Table 7.3.

7.3 Result Analysis

The configuration of the proposed charging system for the selected shopping complex combines solar PV with a battery on-grid system, optimized using HOMER software. This optimization allows for an evaluation of the system's economic, technical, and environmental aspects, comparing both the base case and the winning case configurations. Moreover, key parameters assessed in the software include NPC, COE, EV load served, and energy production from solar, and these characteristics are described below.

7.3.1 Proposed System Cost Benefits

The economic parameters were studied, considering the capital and operation and maintenance costs of primary and secondary sources. In this context, the on-grid solar PV-battery optimization system for the commercial building undergoes a comprehensive examination through the evaluation of NPC and COE for both the base case and the expected lowest case, anticipating real-time implementation. The base case comparison is made by contrasting the grid-only option in HOMER with the estimated optimal solar system. The analysis reveals that the selected location benefits from

Table 7.4 Cost comparison analysis for base case and winning system using HOMER.

	Base case	Lowest cost system
NPC	₹ 7.44 M	₹ 4.79 M
Initial capital	₹ 0.00	₹ 1.32 M
O&M	₹ 575,762/year	₹ 268,094/year
LCOE	₹ 8.87/kWh	₹ 4.42/kWh

the lowest capital cost and the highest NPC, resulting in a short payback period of just 4.3 years. The correlation between the default base case and the proposed winning system's cost benefits is presented in Table 7.4.

7.3.2 Electrical Analysis

When considering the electrical energy production for the design of solar PV with on-grid interconnection, it becomes evident that the primary source of solar energy is significantly more cost-effective than relying solely on grid consumption. Utility power is utilized when solar energy production is limited, with the additional capability of storing energy in batteries. Furthermore, the selected zone experiences a moderate temperature range of 27.34°C and a solar irradiance level of 5.61 kWh/m^2/day. In addition, the total energy production benefits for the proposed microgrid system designed for EV charging are illustrated in Figure 7.9.

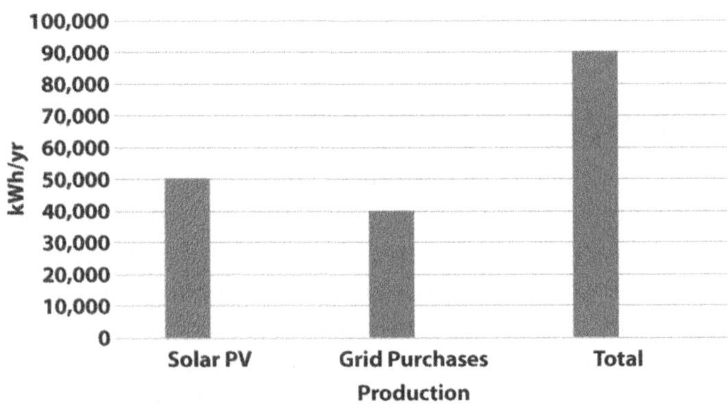

Figure 7.9 Overall energy production comparison.

In addition, the total energy consumption for EVs per year is 18,777 kWh, while the commercial lighting load consumes 26,560 kWh/year. This results in an excess electricity capacity of 6,408 kWh/year, which is stored for later use. During EV charging, a total of 2,643 sessions occurs annually, with a peak power demand of 20 kW and an energy consumption of 7.4 kWh per session. Furthermore, the solar PV system falls into the category of 5.74 kW mean output, producing a total of 138 kWh per day, which is equivalent to 50,244 kWh per year. The system achieves a maximum output of 28 kW during its operational hours, which amounts to 4,376 h per year. The levelized cost is 1.45 Rs/kWh.

7.3.3 Grid and Solar PV Comparison

This study mainly focuses on grid purchases for the proposed system, comparing them with the base case grid power. The analysis in HOMER helps determine the most effective solution that complements the solar PV-generated output and grid power. From this perspective, it is evident that the base case grid purchase is significantly lower than the real-time grid purchase in the winning case, as depicted in Figure 7.10.

When considering the selected location, it becomes evident that it has a high capability for utilizing energy from solar renewable resources. Furthermore, the selected shopping complex offers ample space for solar panel installation, resulting in significantly higher energy utilization compared to relying solely on grid power. As a result, the monthly range of solar production contributes to the total renewable output, averaging at 25 kW, while grid purchases amount to 15 kW. This distribution is illustrated in Figure 7.11.

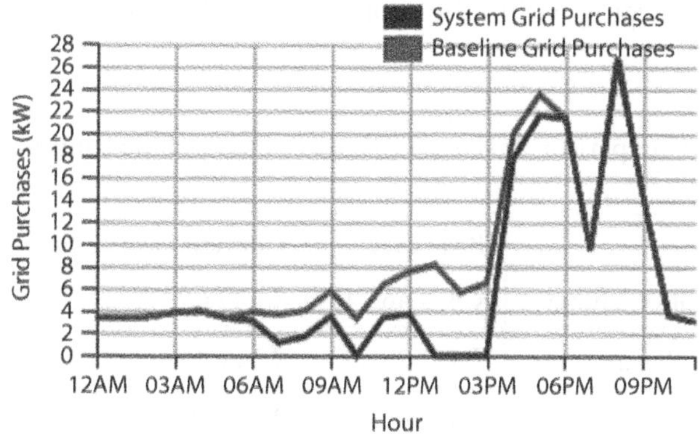

Figure 7.10 Comparison between a base case grid and the proposed case grid level.

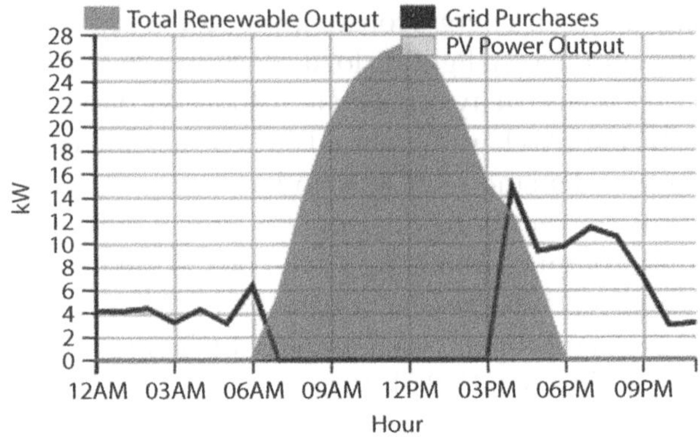

Figure 7.11 Energy production difference between the conventional grid and the solar PV.

7.3.4 Grid Bill Comparison

When considering the utility bill for consumption charges, it is evident that the base case in HOMER incurs significantly higher costs. In contrast, the current case, which aligns with the demand response of the base case, results in lower expenses. The difference between these two cases underscores the cost feasibility of the proposed system, although the savings are still lower than those in the base case of HOMER. Moreover, a comparison of the total annual utility costs for these cases is presented in Table 7.5.

7.3.5 Electric Vehicle State of Charge Analysis

The electric vehicle charging process depends on both the time of day and the state of charge (SOC) of the vehicle. Vehicle arrival is a crucial parameter for EV charging within the shopping complex, helping to avoid noise

Table 7.5 Comparison analysis for the base case and the winning case of grid.

	Base case	**Current case**	**Savings**
Consumption charge	₹ 454,465.40	₹ 149,657.33	₹ 304,808.07
Demand charge	₹ 121,296.93	₹ 111,436.70	₹ 9,860.23
Total	575,762.33	261,094.03	314,668.31

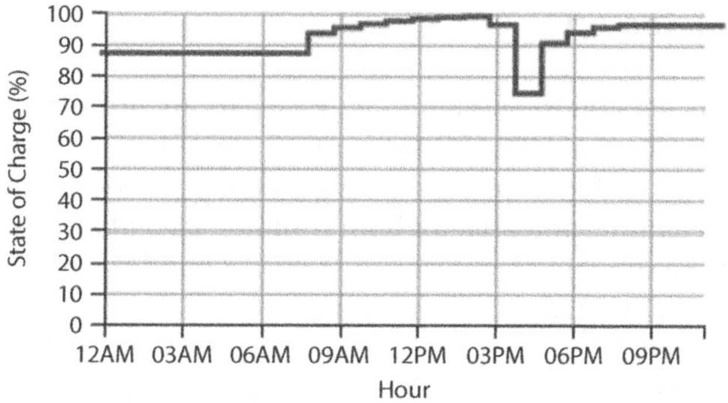

Figure 7.12 Battery state of charge *versus* time in the shopping complex.

and parking issues during charging. The daily charging profile for available vehicles in this location starts from 6 A.M. and continues until 10 P.M. In between this time frame, the charging process is actively running. As a result, the peak SOC occurs around noon, reaching 97% at 3 P.M. The SOC of the battery with respect to time throughout the day is depicted in Figure 7.12.

7.3.6 Emission Analysis

When it comes to delivering power to the load, relying on renewable energy primary sources offers significant benefits, particularly in terms of reducing air pollution. EVs are especially advantageous in this regard as they produce zero emissions. This estimation study focuses on emission determination, utilizing HOMER software to assess carbon dioxide (CO_2), sulfur dioxide (SO_2), and nitrogen oxide (NO). The table below, Table 7.6, illustrates the number of emissions for the selected location with a grid-connected system for EV charging infrastructure.

Table 7.6 Proposed EV microgrid design emission levels.

Quantity	Value (kg/year)
Carbon dioxide	25,379
Sulfur dioxide	110
Nitrogen oxide	53.8

7.4 Conclusion

This study focuses on the design of an EV charging infrastructure within a DC microgrid using HOMER software. The aim is to estimate the optimal solution from both technical and economic perspectives. The study assesses the charging potential for both commercial building lighting loads and EVs in the shopping complex. The selected building for this project is D-Mart, located in Madurai, Tamil Nadu.

The proposed DC microgrid system integrates solar PV with on-grid extension and battery storage to meet peak demand in the commercial complex. Relying solely on battery storage and utility power supply is found to be inefficient for charging EVs and powering commercial loads. Solar PV serves as the primary renewable source, particularly during grid unavailability and power blackouts, with excess energy stored in the battery.

Furthermore, the combination of solar PV and battery storage in a grid-connected system is more effective than a solar PV system with grid-only connection. This optimal solution results in lower NPC and COE compared to the base case grid configuration. Moreover, electrical energy production exceeds grid power utilization for meeting the load demand, leading to reduced emissions due to the use of renewable energy sources. As a result, the proposed DC microgrid design is deemed feasible and reliable in both technical and economic analyses.

References

1. Yap, K.Y., Chin, H.H., Klemeš, J.J., Solar energy-powered battery electric vehicle charging stations: Current development and future prospect review. *Renewable Sustainable Energy Rev.*, 169, 112–862, 2022, https://doi.org/10.1016/j.rser.2022.112862.
2. Sadeghian, O., Oshnoei, A., Mohammadi-Ivatloo, B., Vahidinasab, V., Anvari-Moghaddam, A., A comprehensive review on electric vehicles smart charging: Solutions, strategies, technologies, and challenges. *J. Energy Storage*, 54, 105–241, 2022, https://doi.org/10.1016/j.est.2022.105241.
3. Wu, Y., Wang, Z., Huangfu, Y., Ravey, A., Chrenko, D., Gao, F., Hierarchical operation of electric vehicle charging station in smart grid integration applications—An overview. *Int. J. Electr. Power Energy Syst.*, 139, 108–005, 2022, https://doi.org/10.1016/j.ijepes.2022.108005.
4. AlHammadi, A., Al-Saif, N., Al-Sumaiti, A.S., Marzband, M., Alsumaiti, T., Heydarian-Forushani, E., Techno-economic analysis of hybrid renewable energy systems designed for electric vehicle charging: A case study from the

United Arab Emirates. *Energies*, 15, 18, 6621, 2022, https://doi.org/10.3390/en15186621.

5. Al Wahedi, A. and Bicer, Y., Techno-economic optimization of novel stand-alone renewables-based electric vehicle charging stations in Qatar. *Energy*, 243, 123008, 2022, https://doi.org/10.1016/j.energy.2021.123008.
6. Narasipuram, R.P. and Mopidevi, S., A technological overview & design considerations for developing electric vehicle charging stations. *J. Energy Storage*, 43, 103–225, 2021, https://doi.org/10.1016/j.est.2021.103225.
7. Nandini, K.K., Jayalakshmi, N.S., Jadoun, V.K., An overview of DC micro-grid with DC distribution system for DC loads. *Mater. Today: Proc.*, 51, 635–639, 2021, https://doi.org/10.1016/j.matpr.2021.06.093.
8. AbuElrub, A., Hamed, F., Saadeh, O., Microgrid integrated electric vehicle charging algorithm with photovoltaic generation. *J. Energy Storage*, 32, 101–858, 2020, https://doi.org/10.1016/j.est.2020.101858.
9. Sayed, K., Abo-Khalil, A.G., Alghamdi, A.S., Optimum resilient operation and control DC microgrid based electric vehicles charging station powered by renewable energy sources. *Energies*, 12, 22, 4240, 2019, https://doi.org/10.3390/en12224240.
10. Yang, L. and Ribberink, H., Investigation of the potential to improve DC fast charging station economics by integrating photovoltaic power generation and/or local battery energy storage system. *Energy*, 167, 246–259, 2019, https://doi.org/10.1016/j.energy.2018.10.147.
11. Salameh, T., Ghenai, C., Merabet, A., Alkasrawi, M., Techno-economical optimization of an integrated stand-alone hybrid solar PV tracking and diesel generator power system in Khorfakkan, United Arab Emirates. *Energy*, 190, 116–475, 2020, https://doi.org/10.1016/j.energy.2019.116475.
12. Sharma, A. and Kolhe, M., Techno-economic evaluation of PV based institutional smart micro-grid under energy pricing dynamics. *J. Cleaner Prod.*, 264, 121–486, 2020, https://doi.org/10.1016/j.jclepro.2020.121486.
13. Ekren, O., Canbaz, C.H., Güvel, Ç.B., Sizing of a solar-wind hybrid electric vehicle charging station by using HOMER software. *J. Cleaner Prod.*, 279, 123–615, 2021, https://doi.org/10.1016/j.jclepro.2020.123615.
14. An, Y., Gao, Y., Wu, N., Zhu, J., Li, H., Yang, J., Optimal scheduling of electric vehicle charging operations considering real-time traffic condition and travel distance. *Expert Syst. Appl.*, 213, 118–941, 2022, https://doi.org/10.1016/j.eswa.2022.118941.
15. Taghavifar, H. and Zomorodian, Z.S., Techno-economic viability of on grid micro-hybrid PV/wind/Gen system for an educational building in Iran. *Renewable Sustainable Energy Rev.*, 143, 110–877, 2021, https://doi.org/10.1016/j.rser.2021.110877.
16. Uddin, M.N., Multan Biswas, Md., Nuruddin, S., Techno-economic impacts of floating PV power generation for remote coastal regions. *Sustainable Energy Technol. Assess.*, 51, 101–930, 2022, https://doi.org/10.1016/j.seta.2021.101930.

17. Turkdogan, S., Design and optimization of a solely renewable based hybrid energy system for residential electrical load and fuel cell electric vehicle. *Eng. Sci. Technol. Int. J.*, 24, 2, 397–404, 2021, https://doi.org/10.1016/j.jestch.2020.08.017.
18. Ghotge, R., Snow, Y., Farahani, S., Lukszo, Z., van Wijk, A., Optimized scheduling of EV charging in solar parking lots for local peak reduction under EV demand uncertainty. *Energies*, 13, 5, 1275, 2020, https://doi.org/10.3390/en13051275.
19. Muna, Y.B. and Kuo, C.-C., Feasibility and techno-economic analysis of electric vehicle charging of PV/wind/diesel/battery hybrid energy system with different battery technology. *Energies*, 15, 12, 4364, 2022, https://doi.org/10.3390/en15124364.
20. Schetinger, A.M., Dias, D.H.N., Borba, B.S.M.C., da Silva, G.D.P., Techno-economic feasibility study on electric vehicle and renewable energy integration: A case study. *Energy Storage*, 2, 6, e197, 2020, https://doi.org/10.1002/est2.197.
21. https://www.google.com/search?q=d+mart+shopping+complex+madurai&rlz=1C1UEAD_enIN1002IN1002&oq=d&aqs=chrome.1.69i60j69i59l2j69i57j69i60l4.1486j0j7&sourceid=chrome&ie=UTF-8

8

Design and Simulation of a Rooftop Stand-Alone Photovoltaic Power System for an Academic Institution

Saswati Dey, Nilanjan Mukhopadhyay*, Kuntal Das, Neha Shaw and Rajesh Halder

Department of Electronics and Communication Engineering, Global Institute of Management and Technology, Krishnagar, Nadia, India

Abstract

Being the world's third largest consumer of electricity, India's solar power industry is now an emerging sector and is a part of renewable energy. Thus, renewable energy is the most effective way of shifting India to a non-polluting source of energy to meet future electricity needs. This chapter presents a design and a simulation of a stand-alone photovoltaic system proposed to meet the electrical energy needs of the Electronics and Communication Engineering department of the Global Institute of Management and Technology, India. This system is designed by considering the existing departmental daily electrical load capacity, configuration and installation of solar panel units, sizing of each system component, climate data, and maximum tilt angle of the panels. The simulation of the rooftop photovoltaic system produces a power of 9,290 kWh/year, the required area of which is 35.14 m^2 with 28 units of 250-W_p solar panels and a battery pack of 12 units (48 V, 528 Ah) with a fixed tilt angle of 30° facing south.

***Keywords*:** Stand-alone PV system, solar energy, roof top, PVsyst, off-grid, renewable energy

**Corresponding author*: nilu.opt@gmail.com

P. Sivaraman, C. Sharmeela and P. Sanjeevikumar (eds.) Microgrids for Commercial Systems, (207–222) © 2024 Scrivener Publishing LLC

8.1 Introduction

Over the years, the depletion of fossil fuel reserves led to efforts of identifying an abundant source of energy which is not proportional to the amount of energy being used due to an increase in the earth's population and also to decrease the need for fossil fuels. Solar energy is the new continuous renewable energy source on the Earth's surface without any negative impact on the environment [1–4]. It is expected that the consumption of the world's total energy will become double by the year 2040 as a summation of 1% increase in annual energy consumption [5]. The potential of solar energy can be imagined by the amount of solar radiation received in an hour in two decades. The Indian government is now committed to develop sources of solar energy to bring support for energy security in the future just like Europe and America [6, 7]. If fossil energy supply declines by half and modern renewable energy gets doubled, then the primary energy supply will fall by 30% in 2050 [8].

The photovoltaic (PV) system is becoming more popular and an alternative daily need for large facilities like lighting in highways and transportation through electricity because of production cost reduction and panel efficiency increment [9, 10]. The current status of global solar energy utilization is very poor. The initiative taken by the G-20 pooled countries is a clear indication in switching to renewable sources of energy, preferably solar energy, by replacing other forms of energy dependency [11]. The traditional electricity generation method has many detrimental effects on the environment in terms of CO_2 emission and ozone layer depletion and results in various environmental disasters. The benefit of using solar technology could be visualized as a summation of a single-house contribution to the environment if it is implemented on a large area [12]. Renewable energy is a non-traditional energy source [13]. Renewable energy sources are the only alternative solution for the minimization of environmental pollution and climate difference [14]. Solar energy can reliably compensate for the disadvantages of the popular traditional energy sources [15]. Two forms of environment-friendly solar energy are thermal energy converted through a solar collector and electrical energy converted through photovoltaic cells [16]. PV technology is free from air pollution due to emission of CO_2 and has a wide range of real-life applications [17].

India is full of sustainable energy resources, among which the most popular is solar energy, and is making power generation system more reliable through photovoltaic technology. There are many defining factors in a solar system design like tilt orientation, irradiance, latitude, longitude, and climate. PVsyst simulation is done to detect the efficiency in both fixed and seasonally changed tilt position of the photovoltaic panel [18]. The performance of a solar system has been simulated many times by PVsyst [19, 20]. A study of on-grid solar PV system in India was done [21]. A comparison has been done using PVsyst between generated electrical energy and losses that occurred by PV array in Kathmandu [22]. A solar power plant has been studied using PVsyst based in Tamil Nadu [23]. India, being a developing country, is mostly dependent on fossil fuels for energy supply and now shifting to renewable energy source electricity as an alternative [24]. The calculated factors are load estimation, number of panels, number of batteries, wires, charge controller, inverter, mounting structure, orientation, climate, panel area, and simulation so that the system operates optimally without any failure [25, 26].

The optimization of tilt angle is an important factor in terms of the characteristics of PV power and shading [27, 28]. The cost analysis and optimization of a stand-alone solar PV system were done [29, 30]. Both on grid and stand-alone systems are reviewed [31, 32]. A conclusion has been achieved by analyzing the life cycle cost for diesel and stand-alone PV systems in India where it was found that that the PV system is more suitable [33]. A Nigeria-based off-grid PV system was studied [34]. A stand-alone PV system is implemented in Bhopal, India, for electrification [35].

In this chapter, a case study has been done at the Electronics and Communication Engineering department of Global Institute of Management and Technology, Krishnagar, India, using a trial version of PVsyst 7.2. The main purpose of this chapter is the design and simulation of a stand-alone solar photovoltaic system to fulfil the total electrical requirement for the whole ECE department including classrooms, faculty rooms, and different laboratories considering fluorescent lights, fans, computers, and lab equipment.

The rest of the chapter is divided into the following sections: Section 8.2 deals with the theoretical aspects of the system design, which is followed by the detailed methodology involved in the design in Section 8.3. Finally, a discussion on the obtained results and a conclusion that has been drawn are presented in Section 8.4.

At first step, the analysis is done mathematically by estimating the total load as 18,224 Wh. In the present design, an approximate load has been considered for the purpose of the lab equipment. This chapter presented an introduction followed by system design, calculation, and simulation with results and conclusions considering the site data, daily and monthly irradiation levels, temperature, and wind speed to certify a precise design, where Global Institute of Management and Technology (GIMT)—Krishnagar is selected as the geographical location at latitude 23.38°N, longitude 88.49°E, and altitude of 14 m.

8.2 System Design

In this section, the components associated with the stand-alone PV system design will be introduced. The stand-alone photovoltaic system includes PV modules, charge controller, battery bank, DC–AC inverter, and wiring cables. The block diagram of such a stand-alone PV system is illustrated in Figure 8.1. The design and specification of all the components will be described in the following subsections.

8.2.1 Size of the PV Module

The power of the PV module (P_{PV}) has to be calculated to achieve the daily load energy (E_L) required to run the household appliances. The total daily solar energy (H) is taken as 5 kW/m^2/day. In the design to reach optimum system operation, various system parameters should be considered, such as η_{PV}, T_C, η_C, η_{INV}, and η_B. η_{PV} is the module efficiency, and it has a very small value of approximately 0.2 because of internal series and shunt resistance along with recombination effect in the solar PV module.

T_C is the PV module temperature coefficient and is a vital parameter to be considered for designing as the output depends on its surface temperature. Silicon modules are having a temperature correction factor of approximately -0.5%/°C, so T_C has been considered safely as 0.80 [36].

The values of other system parameters like efficiency of charge controller (η_C) taken as 0.9, the DC–AC inverter efficiency (η_{INV}) taken as 0.9, and the battery efficiency (η_B) taken as 0.8 were also considered.

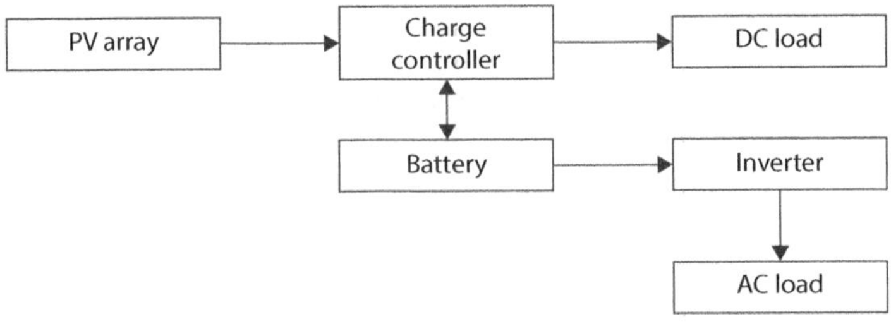

Figure 8.1 Block diagram representing a stand-alone PV system.

Considering all parameters, the calculated PV array total area (A_{PV}) is as follows [37]:

$$A_{PV} = \frac{E_L}{H \times T_C \times \eta_{PV} \times \eta_{INV} \times \eta_C \times \eta_B} \tag{8.1}$$
$$= 35.14 \text{ m}^2$$

The peak power (P_{PV}) of the PV module array is calculated by considering the standard solar irradiation (H_{SC}) as 1,000 W/m² [38];

$$P_{PV} = A_{PV} \times H_{SC} \times \eta_{PV} \tag{8.2}$$
$$= 7{,}000 \text{ W}$$

The peak PV array power is calculated using Equation 8.1 and Equation 8.2. The calculated daily load energy is 18.22 kWh/day, as shown in Table 8.1. The PV module setup model LG 250 S1C-G3 had a peak power of 250 Wp. In total, 28 PV modules of 14 strings times two in series produce the total required electrical energy.

8.2.2 Battery Sizing

Battery is the most important part of a stand-alone PV system in terms of storing electric energy for the night requirement of a cloudy day for the purpose of sizing of storage capacity. The capacity of the storage battery B_C has been calculated in terms of the efficiency of the battery (η_B), efficiency of the inverter (η_{INV}), depth of discharge (DOD), and number of cloudy days (N_C). DOD has been taken as 80% as the batteries are deep-discharge batteries and have a lifetime of nearly 10 years. As it is known, DOD is inversely proportional to battery life. The number of cloudy days N_C is taken as equal to unity for minimizing the cost of the system.

The capacity of the storage battery can be calculated in Wh as follows [39];

$$B_C = \frac{E_L \times N_C}{DOD \times \eta_{INV} \times \eta_B} \tag{8.3}$$
$$= 31631 \text{ Wh}$$

Using Equation 8.3, the storage capacity can be calculated as 31,631 Wh. The required battery capacity will be 658.98 Ah if we consider a 12-V battery. The battery bank consists of three (in parallel) by four (in series) deep-cycle lead-acid battery (Generic Enersol 250) of 200 Ah, at 12 V each. The batteries have their operating temperature fixed at 20°C. For accurate calculation, 48-V DC voltage bus is preferred, which can be obtained by connecting four storage batteries in series in three strings. Therefore, finally, the PV modules are connected in 14 parallel strings, with each string having two modules connected in series.

8.2.3 Charge Controller

To enhance the lifetime of the storage batteries and to prevent overcharging of batteries, charge controllers are used in a stand-alone PV system. For this purpose, a universal MPPT charge controller has been used. To tolerate the maximum charging current of the system, a charge controller with specifications of 48 V and 80 A is placed between the arrays and the batteries in our design. There are many advantages, e.g., high- and low-input voltage disconnection, over-voltage, etc. There is a provision for displaying different parameters like the battery voltage, battery current, etc., in the LCD.

8.2.4 Inverter Sizing

The inverter must be set at a power which is greater than the maximum AC power of the load [30, 37]. Another function of the inverter is that it should protect refrigerators and washing machines against any power surge [40]. The selected inverter specification is 6,000 W. This inverter is suitable for home appliances because it provides excellent performance against different parameters, e.g., battery over-temperature, over-voltage, etc.

Thus, the proposed PV system has 28 PV modules connected in 14 parallel strings, each of which has two modules in series and generates an operating DC voltage of 48 V with a short circuit current of 75 A (the standard conditions are 1,000 W/m^2 solar radiation and 25°C module temperature). To withstand the load of PV modules and also to protect them from wind and storm, a strong mechanical structure is required, which is set with 30° optimal tilt facing south.

System Operation

The average (yearly) PV energy generated is 9,290 kWh/year. The installed system provides a higher average monthly energy than the average monthly demand throughout the year, which indicates proper selection, designing, and optimization of the installed system. Moreover, the estimated annual load energy required for the smooth running of the department is 6,426 kWh, which is much less than the generated energy from the PV system.

8.3 Design Methodology

Several steps need to be followed while designing an off-grid PV system, such as the calculation of the total energy required every day, the estimation of resources of energy, the optimum tilt angle of panels, etc. The design flow chart is shown in Figure 8.2. In the present study, a stand-alone PV system for the Electronics and Communication Engineering department of GIMT has been considered. Simulation of the design was performed in PVsyst 7.2 software.

The steps involved in designing the solar PV system will be elaborated in this section.

Step 1: The total load or energy required to run the department on a daily basis is calculated first. While calculating, the solar irradiance data for the particular location is collected from the meteorological information of the site.

Step 2: Now, depending on the requirement, the capacity of the PV equipment is calculated using basic formulae. It includes the capacity of the solar module, inverter module, MPPT charge controller, battery, etc.

Step 3: Next, all these data are used in the PVsyst simulation software for modeling and simulation of the system considering climate data and daily load consumption.

Step 4: Through the PVsyst software, the best possible tilt angle of the PV panel, at which it receives maximum irradiation and produces enough energy, is fixed.

Step 5: After simulation, the obtained design results are analyzed on the basis of annual energy production, irradiance, and cost. If the analyzed result reflects that enough energy is not produced with the existing system cost, then the process is repeated from step 2 onwards.

8.3.1 Meteorological Information of the Site

Krishnagar is a town and headquarter of Nadia district in the Indian state of West Bengal. GIMT is located 1.8 km away from the Krishnagar Railway station and 3 km from the famous Krishnagar Rajbari of Raja

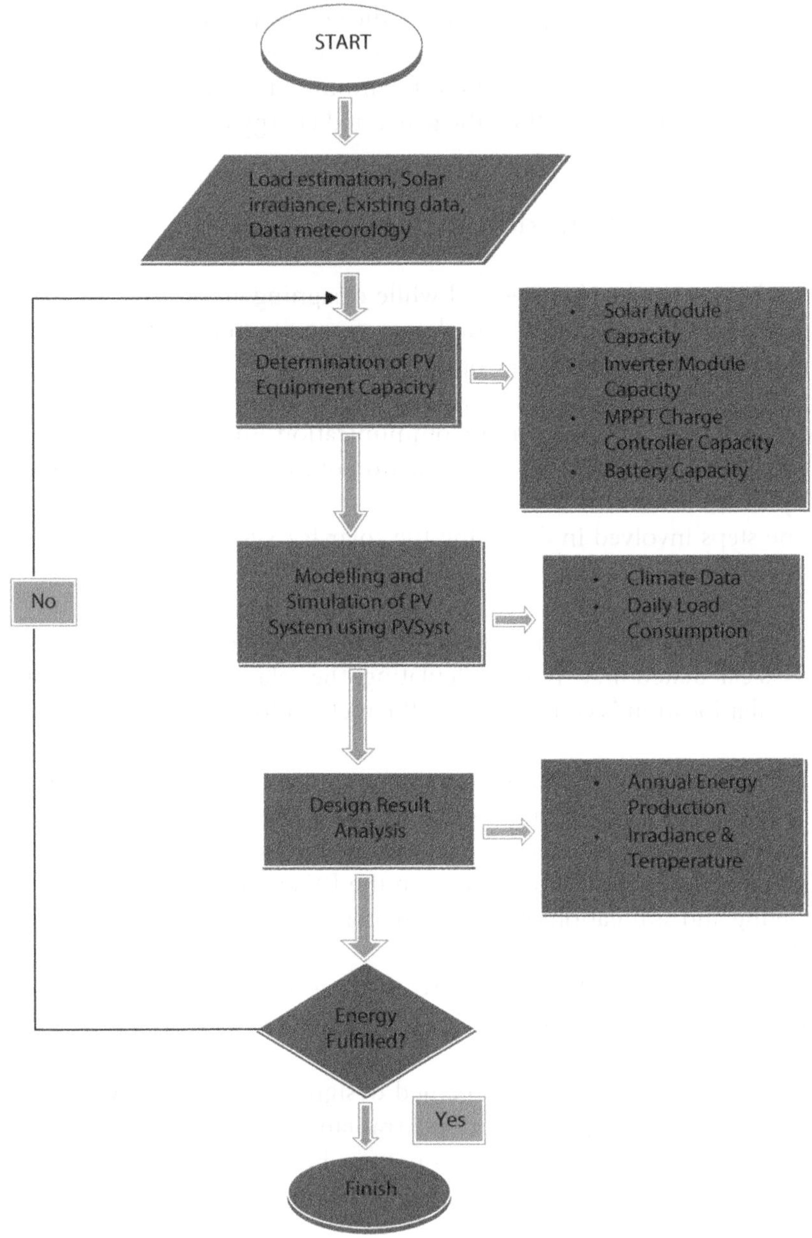

Figure 8.2 Flow chart reflecting the design steps.

Krishna Chandra Ray. Figure 8.3 shows the location of the GIMT campus. However, there are some small buildings spread throughout the college campus along with four main buildings. The grid system is not feasible for providing electricity, whereas the stand-alone PV system can play an important role for supplying electricity. GIMT—Krishnagar is geographically located at latitude 23.38°N, longitude 88.49°E, and altitude of 14 m. Figure 8.4 exhibits the meteorological data of Krishnagar City as obtained from PVsyst software. As illustrated by the table, in the summer months, the selected site has considerably high temperatures along with high irradiation of approximately 180 kWh/m^2, which justifies the implementation of a stand-alone PV system.

The orientation of the PV system should be in a correct direction so that maximum radiation levels can be received throughout the year. As mentioned earlier, Krishnagar is a place with high levels of irradiation and maximum sunshine hours throughout the year. The variation of the levels of irradiation received with the system inclination is shown in Figure 8.5. The best possible tilt angle of the system is close to 30° which is the latitude angle site, and the corresponding azimuth is 0°. This optimization is based on yearly irradiation that yield a transposition factor of 1.06 and loss with respect to optimum of -0.5%.

Figure 8.3 Image illustrating the GIMT campus.

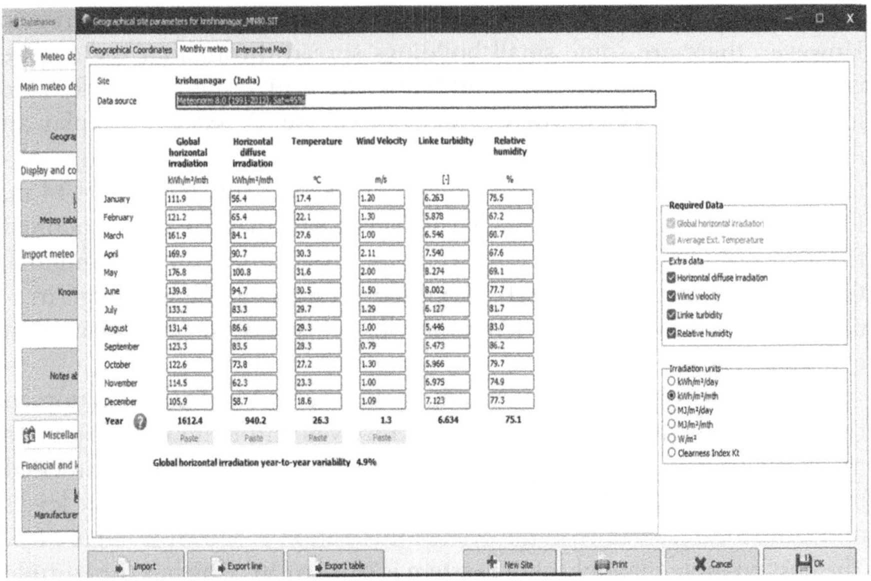

Figure 8.4 Geographical site parameters of Krishnagar over the year.

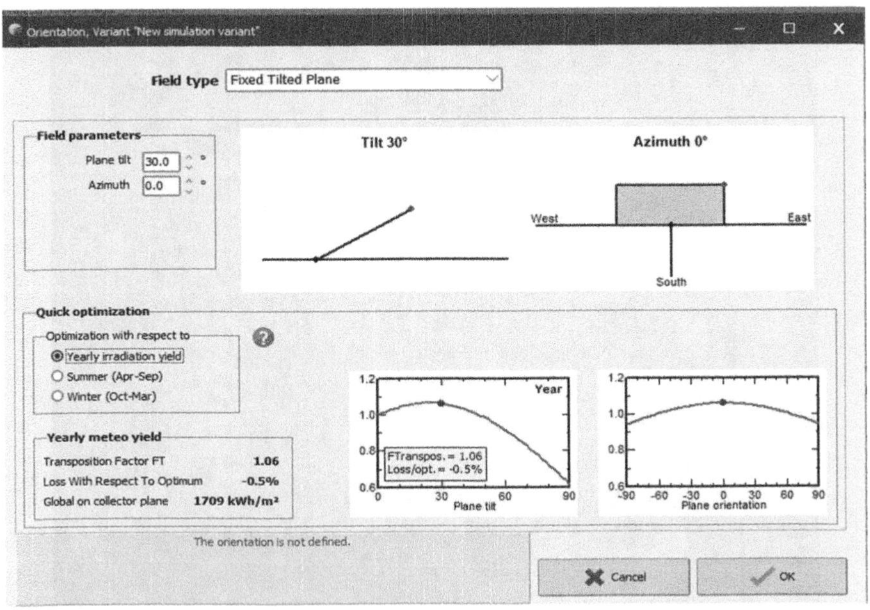

Figure 8.5 Optimum tilt angle of the PV panel.

8.3.2 Daily Load Calculation

After the investigation on solar resource, the next step is to estimate the electric load required to drive the Electronics and Communication Engineering department in GIMT smoothly on daily basis. Here computation of the load distribution has been performed considering the appliances used in different rooms. Table 8.1 shows the detailed load estimation, i.e., total power and energy distribution of the ECE department against the corresponding hourly distribution of loads in a typical day as calculated in PVsyst shown in Figure 8.6. Currently, the department have fluorescent tubes, and these include all the PCs that are used by individual faculty and students in the lab as well. The calculated maximum total daily energy required is 18.220 kWh/day, which will decrease in the winter months.

8.3.3 Cost Analysis

The rooftop PV system aims to provide the energy required to drive the load in the Electronics and Communication Engineering department. According to Table 8.1, the energy required is calculated to be 18.22 kWh/day

Table 8.1 Daily load estimation table for the ECE department.

Rooms in the department	Appliances (including light, fan, PC, lab equipment)	Total power (W)	Daily usage	Energy (Wh/day)
Faculty room including H.O.D.	12	860	7 h	6,020
Communication and Simulation lab	18	1,500	2 h	3,000
Digital Electronics and Microwave Engineering lab	16	1,000	2 h	2,000
Classroom including tutorial rooms	21	1,200	6 h	7,200
	Total	4,560		18,220

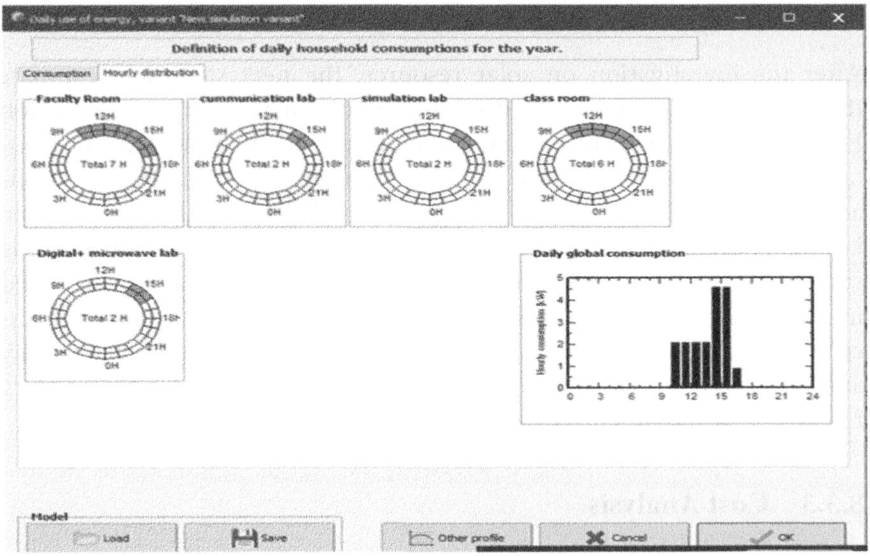

Figure 8.6 Daily energy consumption in the ECE department (room-wise) on an hourly basis.

or 18.22 unit/day and 546.6 unit/month. The cost of electricity according to West Bengal State Electricity Distribution Company Limited is Rs.5.37 per unit, which results into Rs.2,935 (approximate) per month, i.e., Rs.35,222 (approximate) per year. Considering the lifetime of mono-crystalline PV arrays as approximately 20 years, the total electricity bill will be 7,04,458 (approximate), excluding very minimal annual operating and maintenance cost. However, the overall PV system cost is Rs. 4lakhs (approx.). Moreover, the PV system is generating energy of approximately 9,290 kWh, which is greater than the total annual load demand of the department, i.e., 6,426 kWh. The excess energy can be used for other purposes.

8.4 Conclusion

In this communication, a proposal to install a solar-based, stand-alone PV system for the Electronics and Communication Engineering department of Global Institute of Management and Technology (West Bengal, India) has been made through calculations based on the available standards and simulation with the help of PVsyst software. The system requirements obtained through the simulation results have the following specifications:

28 PV modules (of 250 Wp) of 14 strings × two in series, the area needed is 35.14 m^2 with a battery bank consisting of three in parallel × four in series lead-acid battery of 200 Ah, at 12 V each. A charge controller placed between the PV arrays and the batteries has 48 V and 80 A specifications and a 6,000-W inverter with a fixed tilt angle of 30° facing south to supply 18.22 kWh/day. The obtained results suggest that this system provides excess energy and is cost effective as well considering the PV system's lifespan.

References

1. European Photovoltaic Industry Association (EPIA), Global market outlook for photovoltaics 2014-2018, Publication Date:05/2014, Belgium, 2014, http://www.solarpowereurope.org/.
2. Bosshard, P., *An assessment of solar energy conversion technologies and research opportunities*, Standford University, California, US, 2006, Retrieved from Stanford.edu.
3. Pachpande, S.G. and Zope, P.H., Studying the effect of shading on solar panel using MATLAB. *Int. J. Sci. Appl. Inf. Technol.*, 1, 2, 46–51, 2012.
4. Iskandar, H.R., Purwadi, A., Rizqiawan, A., Heryana, N., Prototype development of a low cost data logger and monitoring system for PV application, in: *The 3rd IEEE Conference on Power Engineering and Renewable Energy*, pp. 171–177, 2016.
5. U.S Energy Information Administration, EIA projects world energy consumption will increase 56% by 2040, 2013, July 25, Retrieved from Today in Energy: https://www.eia.gov/todayinenergy/detail.php?id=12251.
6. Breeze, P., *Power generation technologies*, Elsevier, Oxford, ISBN 978-0-7506-6313-7, 2005.
7. Freris, L. and Infield, D., *Renewable energy in power systems*, Wiley, New Jersey, US, 2008, ISBN: 978-0-470-01749-4.
8. Dewan, Energi Nasional Republik Indonesia, Outlook energi Indoensia 2014, 2014, Publication Date:12/2015, https://www.den.go.id/index.php/publikasi/index/EnergyOutlook.
9. Rante, J.C., Patras, A., Rompis, L., Design of a solar micro power plant for home lighting, in: *International Conference on Electrical Engineering and Computer Science*, vol. 17, pp. 453–456, 2018.
10. Isik, A.H., Erden, S., Imeruzz, M., Design of the solar energy system in ITU faculty of electric and electronics, in: *4th International Conference on Renewable Energy Research and Applications*, pp. 22–25, 2015.
11. Richardson, L., Solar electricity vs. fossil fuels: How do they compare?, 2017, May 14, Retrieved from Energsage: https://news.energysage.com/solar-energy-vs-fossil-fuels/.

12. RGS- Rethink your roof. Head to head: Benefits of solar energy vs fossil fuels. Retrieved from RGS: https://rgsenergy.com/home-solar/solar-vs-traditional-power/.
13. Baghaee, H.R., Mirsalim, M., Gharehpetian, G.B., Talebi, H.A., Reliability/cost-based multi-objective Pareto optimal design of stand-alone wind/PV/FC generation microgrid system. *Energy*, 115, 1022–1041, 2016.
14. Kabakian, V., McManus, M.C., Harajli, H., Attributional life cycle assessment of mounted 1.8 kWp monocrystalline photovoltaic system with batteries and comparison with fossil energy production system. *Appl. Energy*, 154, 428–437, 2015.
15. Roohollahi, E., Mehrabian, M., Abdolzadeh, M., Prediction of solar energy gain on 3-D geometries. *Energy Build.*, 62, 315–322, 2013.
16. Shazly, J., Hafez, A., El Shenawy, E., Eteiba, M., Simulation, design and thermal analysis of a solar Stirling engine using MATLAB. *Energy Convers. Manage.*, 79, 626–639, 2014.
17. Ahmad, G., Photovoltaic-powered rural zone family house in Egypt. *Renew. Energy*, 26, 379–390, 2002.
18. Tallab, R. and Malek, A., *Predict system efficiency of 1MWc photovoltaic power plant interconnected to the distribution network using PVSYST software*, IEEE, 2015.
19. Biswas, M.M., Das, K.K., Baqee, I.A., Sadi, A.H., Hossain, M.S., Prospects of renewable energy and energy storage systems in Bangladesh and developing economics. *Glob. J. Res. Eng. J. Gen. Eng.*, 11, 5, 23–31, Version 1.0, 2011.
20. Ali, R.B., Aridhi, E., Mami, A., *Design modelling and simulation of hybrid power system. (Photovoltaics-wind)*, IEEE, 2016.
21. Sharma, Y. and Chandel, S.S., Performance analysis of a 190 kWp grid interactive solar photovoltaic power plant in India. *Energy*, 55, 476–485, 2013.
22. Karki, P., Adhikary, B., Sherpa, K., Comparative study of grid-tied photovoltaic (PV) system in Kathmandu and Berlin using PVsyst, in: *Third IEEE International Conference on Sustainable Energy Technologies*, pp. 196–199, 2012.
23. Kandasamy, C.P., Prabhu, P., Niruba, K., Solar potential assessment using PVSYST software, in: *IEEE International Conference on Green Computing, Communication and Conservation of Energy*, pp. 667–672, 2013.
24. Mancini, T., Heller, P., Butler, B., Osborn, B., Schiel, W., Goldberg, W.G., Associates, V. *et al.*, Dish-Stirling systems: An overview of development and status. *J. Sol. Energy Eng.*, 125, 2, 135, 2003.
25. Farhat, M., Abdul-niby, M., Abdullah, M., A low cost automated weather station for real time local measurements. *Eng. Technol. Appl. Sci. Res.*, 7, 3, 1615–1618, 2017.
26. Adiyasuren, E., Damba, U.O., Tsedensodnom, B., Comparison of power generation from solar panel with various climate condition and selection of best tilt angles in Ulaanbaatar, in: *8th International Forum on Strategic Technology 2013, IFOST 2013 - Proceedings*, vol. 2, pp. 519–521, 2013.

27. Iskandar, H.R., Zainal, Y.B., Sambasri, S., Study and analysis of shading effects on photovoltaic application system, in: *1st International Conference on Industrial Electrical and Electronics (ICIEE) International Board and Program Committee*, pp. 1–8, 2018.
28. Atluri, K., Hananya, S.M., Navothna, B., Performance of rooftop solar PV system with crystalline solar cells, in: *National Power Engineering Conference (NPEC)*, pp. 1–4, 2018.
29. Lu, Y., Wang, S., Shan, K., Design optimization and optimal control of grid-connected and standalone nearly/net zero energy buildings. *Appl. Energy*, 155, 463–477, 2015.
30. Kamalapur, G. and Udaykumar, R., Rural electrification in India and feasibility of photovoltaic solar home systems. *Int. J. Electr. Power Energy Syst.*, 33, 594–599, 2011.
31. Kaundinya, D.P., Balachandra, P., Ravindranath, N., Grid-connected versus stand-alone energy systems for decentralized power—A review of literature. *Renew. Sustain. Energy Rev.*, 13, 2041–2050, 2009.
32. Chakrabarty, S. and Islam, T., Financial viability and eco-efficiency of the solar home systems (SHS) in Bangladesh. *Energy*, 36, 4821–4827, 2011.
33. Kolhe, M., Kolhe, S., Joshi, J., Economic viability of stand-alone solar photovoltaic system in comparison with diesel-powered system for India. *Energy Econ.*, 24, 155–165, 2002.
34. Oko, C., Diemuodeke, E., Omunakwe, N., Nnamdi, E., Design and economic analysis of a photovoltaic system-a case study. *Int. J. Renew. Energy Dev.*, 1, 65–73, 2012.
35. Kumar, A. and Mandapati, R., Designing and life cycle assessment of SPV system for conference hall at Dept. of Energy, MANIT, Bhopal. *Int. J. Wind Renew. Energy*, 1, 79–83, 2012.
36. Agarwal, N. and Kumar, A., Optimization of grid independent hybrid PV–diesel–battery system for power generation in remote villages of Uttar Pradesh, India. *Energy Sustain. Dev.*, 17, 210–219, 2013.
37. Alamsyah, T., Sopian, K., Shahrir, A., Techno economics analysis of a photovoltaic system to provide electricity for a household in Malaysia, in: *Proceedings of the International Symposium on Renewable Energy: Environment Protection & Energy Solution for Sustainable Development*, pp. 387–396, Kuala Lumpur, Malaysia, 2003.
38. Sick, F. and Erge, T., *Photovoltaics in buildings: A design handbook for architects and engineers*, Earthscan, London, UK, 1996.
39. Kai, F.M., Cobb, A., Chua, A.F., Tee, M.H., Ng, B., Gandois, L., Harvey, C., An off-grid PV power system for meteorological and eddy covariance flux station in Kranji, Singapore. *Energy Procedia*, 33, 364–373, 2013.
40. Tarigan, E. and Purba, L., Assessment of PV power generation for household in Surabaya using Solar GIS–PV planner simulation. *Energy Procedia*, 47, 85–93, 2014.

9

Integration of Wind Energy Control with Electric Vehicle

Abhishek Kumar

School of Electronics and Electrical Engineering, Lovely Professional University, Punjab, India

Abstract

The integration of renewable energy sources, such as wind power, into modern power grids has significantly contributed to reducing carbon emissions and enhancing energy sustainability. However, the inherent variability of wind power poses challenges to grid stability, primarily due to frequency deviations caused by fluctuations in power output. This chapter explores the application of proportional–integral–derivative (PID) control as an effective strategy to mitigate frequency deviations in wind power systems. The chapter provides an in-depth analysis of the principles of PID control, its application to wind power systems, and the benefits it offers in terms of stabilizing the grid frequency and ensuring a reliable energy supply.

***Keywords*:** Wind power, frequency deviation, PID control, renewable energy, grid stability, energy sustainability

9.1 Introduction

The integration of renewable energy sources into power generation systems has become an important focus in the search for a sustainable and environmentally friendly energy landscape. Among these resources, wind power has emerged as a pivotal contender due to its potential to harness the natural kinetic energy of the wind and convert it into clean electricity [1]. The transition towards renewable energy sources has gained momentum as societies worldwide recognize the urgent need to reduce greenhouse gas

Email: abkvjti@gmail.com

P. Sivaraman, C. Sharmeela and P. Sanjeevikumar (eds.) Microgrids for Commercial Systems, (223–240)

emissions and mitigate the impacts of climate change. Renewable energy technologies, which include wind power, have emerged as key contributors to this transition. Wind power, in particular, utilizes the natural energy of the wind to produce electricity, offering a clean and sustainable alternative to traditional fossil fuels [2, 3]. This integration of renewable energy into power generation systems represents a crucial step towards achieving energy sustainability and reducing our dependence on finite resources. However, the integration of wind power is not without its challenges, one of the most significant being the inherent variability in power output. This variability gives rise to frequency deviations, which can impact the stability and reliability of power grids. In addressing these challenges, the application of proportional–integral–derivative (PID) control emerges as a promising solution [4, 5].

The integration of wind power brings numerous benefits; it also introduces new complexities to the operation and management of power grids. One of the foremost challenges is the variability in wind power generation due to fluctuations in wind speed and direction. This variability directly impacts the stability of power grids and can lead to frequency deviations. Frequency, measured in hertz (Hz), reflects the balance between power supply and demand within a grid. Any deviation from the nominal frequency can disrupt the synchronization of electrical systems, potentially causing equipment malfunctions, instability, and even power outages. Therefore, managing frequency deviations becomes imperative for ensuring grid stability and a reliable energy supply. To address the challenge of frequency deviation in wind power systems, the application of control strategies becomes crucial. One such strategy is the utilization of PID control [6]. The PID control method is widely employed in engineering and control systems to regulate and maintain desired system variables within specific limits. It operates through three components: proportional, which responds to the current error; integral, which accumulates past errors; and derivative, which predicts future error trends. The combination of these components allows PID control to dynamically adjust control inputs to minimize deviations and maintain setpoints. In the context of wind power systems, PID control offers a systematic and adaptive approach to mitigate frequency deviations. By continually monitoring the frequency and adjusting power outputs in response to variations, PID control aims to stabilize grid frequency and ensure a consistent energy supply. The effectiveness of PID control lies in its ability to respond swiftly to changing conditions and provide accurate adjustments that align with the dynamic nature of wind power generation.

This chapter includes the intricate dynamics of mitigating frequency deviations in wind power systems using PID control. Through a comprehensive exploration of the principles and application of PID control, we aim to highlight its role in enhancing grid stability and the reliable integration of wind power. The subsequent sections of this chapter will provide a deeper understanding of PID control, its relevance to wind power integration, and its impact on frequency regulation. Moreover, real-world case studies and simulations will be presented to showcase the practical implementation and effectiveness of PID control in managing frequency deviations. This chapter focuses on figuring out how PID control and wind power systems interact. We are aware of the crucial role that cutting-edge control techniques play in creating a future powered by sustainable energy. The knowledge learned from this chapter will help us develop a deeper understanding of the challenges and opportunities associated with integrating renewable energy sources, ultimately pointing us in the direction of stronger and more reliable power systems.

9.2 PID Controller

PID control is a widely utilized feedback control technique that plays a crucial role in maintaining desired system behaviors and minimizing deviations from setpoints. Comprising three fundamental components—proportional, integral, and derivative—PID control offers a dynamic and adaptive approach to regulate various processes and systems. Each of these components contributes distinct characteristics to the control mechanism, allowing for the effective management of system variables [7].

9.2.1 Proportional Action

The proportional component, often referred to as the "P" term, is the simplest and most intuitive aspect of PID control. It responds directly to the current error, which is the difference between the desired setpoint and the actual system output. The proportional action generates a control output proportional to this error, aiming to reduce the deviation between the actual value and the desired value. In the context of frequency regulation in wind power systems, the proportional action helps control the immediate response to deviations in grid frequency caused by wind power fluctuations. If the frequency deviates from the desired value, the proportional term generates an output that adjusts the power output of wind turbines to counteract the deviation. While proportional action provides a quick response to deviations, it may lead to oscillations if used alone, as it does not consider past or future trends.

9.2.2 Integral Action

The integral component, represented by the "I" term, addresses the accumulation of past errors over time. It continuously integrates the error signal, ensuring that even small deviations from the setpoint are eventually corrected. This characteristic is particularly useful in scenarios where the proportional action alone may not eliminate steady-state errors. When applied to wind power systems, the integral action is essential for eliminating persistent frequency deviations that may arise due to continuous wind speed variations. It prevents the steady accumulation of frequency deviation errors by gradually adjusting the power output based on the cumulative error over time. This characteristic enhances the system's ability to achieve accurate and consistent frequency regulation.

9.2.3 Derivative Action

The derivative component, denoted by the "D" term, anticipates future trends by analyzing the rate of change of the error signal. It acts as a predictive mechanism, responding to rapid changes in error and aiming to dampen oscillations caused by the proportional component's immediate response. In wind power systems, the derivative action is beneficial in addressing rapid changes in wind speed that may lead to sudden frequency deviations. By considering the rate of change in frequency deviation, the derivative term can apply corrective actions that counteract potential oscillations, promoting smoother and more stable frequency regulation.

The mechanism of PID control is constituted by the proportional, integral, and derivative components, which collectively offer a versatile and adaptive methodology for managing system variables. In the realm of wind power systems, the incorporation of PID control principles holds promising potential for effectively mitigating frequency deviations. This achievement is realized through the amalgamation of the immediate corrective action exhibited by proportional action, the error rectification abilities of integral action, and the forward-looking nature of derivative action. By embracing this holistic approach, grid stability and reliability are significantly enhanced, paving the way for a seamless integration of wind power while upholding grid frequency within specified limits.

PID control extends its applicability across a diverse spectrum of industrial processes and automation domains. In the context of industrial automation, it is notable that nearly 95% of closed-loop frequency PID controllers are employed, as depicted in Figure 9.1. These controllers function in unison to generate a regulated signal. Fundamentally, a PID controller serves as an essential tool for regulating process variables such as temperature, flow, pressure, and speed in various industrial settings.

PID controllers, which combine the principles of proportional, integral, and derivative control loops, stand out as the most precise and reliable controllers. These controllers handle physical processes through the use of feedback loops, offering a tried-and-true method for directing systems toward predetermined target positions or levels. Its vast application in activities like temperature regulation, automation, and various chemical processes serves as an example of its adaptability.

PID control's ability to guarantee that system performance nearly resembles the target set points is what makes it so essential. Through the use of closed-loop control feedback, this is achieved. An example of this occurs during the operation of an engine, where the controller's PID algorithm methodically modifies the engine's power output to quickly and precisely align the measured speed with the target speed, effectively minimizing delays and overshoots. This control strategy involves the application of proportional, integral, and derivative terms to an error value. The error value, denoted as $e(t)$, is essentially the discrepancy between the intended setpoint $r(t)$ and the measured process variable PV=$y(t)$. As a response, the controller regulates a control variable $u(t)$, often manifesting as the adjustment of a control valve, to a new value determined by the weighted sum of these control terms. This orchestrated adjustment contributes to the gradual reduction of the error over time.

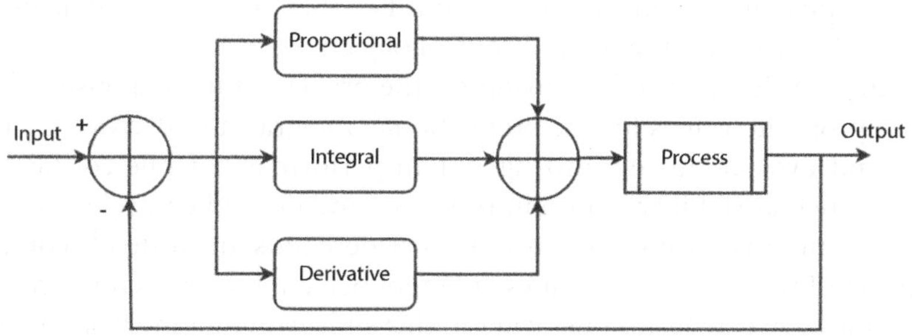

Figure 9.1 Block diagram of the PID controller [8].

9.2.4 PID Controller Design and Tuning

In the world of control systems, the design and tuning of PID controllers represent a critical endeavor to achieve effective and efficient regulation of dynamic processes. The art of crafting a well-suited PID controller and fine-tuning its parameters plays a pivotal role in ensuring system stability, optimizing performance, and mitigating undesirable effects such as overshoot or oscillations. This section dives into the intricacies of PID controller design and tuning, shedding light on key considerations and methodologies that guide these essential tasks.

9.2.5 PID Controller Design

At the heart of PID controller design lie three integral components: proportional, integral, and derivative gains. Each of these elements contributes distinct attributes to the control mechanism, shaping the controller's responsiveness and behavior. The proportional gain (Kp) dictates the magnitude of control action in response to the current error. While a higher Kp yields a more aggressive response, it can also lead to challenges like overshooting or oscillations. The integral gain (Ki) addresses steady-state errors by accumulating past errors over time, seeking to minimize persistent discrepancies. On the other hand, the derivative gain (Kd) offers predictive capabilities, anticipating future error trends and dampening undesirable oscillations for smoother responses. Crucially, the PID controller design entails a delicate equilibrium between these three gains. The proportional gain influences the system's responsiveness, but excessive values can tip the balance toward instability. The integral gain addresses steady-state errors, yet its magnitude can contribute to overshooting issues. Meanwhile, a suitable derivative gain is instrumental in countering oscillations, although it must be balanced to avoid amplifying system noise. Harmonizing these gains is pivotal to crafting a PID controller that effectively regulates the system and achieves the desired control objectives.

Tuning a PID controller constitutes the process of meticulously selecting appropriate values for the proportional, integral, and derivative gains. This endeavor entails optimizing system performance while considering factors such as stability, transient response, and desired control characteristics. Tuning is an intricate process that demands an in-depth comprehension of the system's dynamics, inherent behavior, and intended control outcomes. Diverse techniques cater to this endeavor, ranging from manual tuning to advanced optimization methodologies.

(a) Manual Tuning: This conventional approach involves iteratively adjusting PID gains while observing the system's response. The trial-and-error method helps attain acceptable performance levels, but it might necessitate significant expertise and time investment.
(b) Ziegler–Nichols Method: This is a prevalent technique that employs step–response analysis to determine preliminary PID gains. Parameters like ultimate gain and oscillation period aid in calculating proportional, integral, and derivative gains.
(c)Tuning Rules: The landscape of PID tuning encompasses various established rules and equations. These rules leverage system characteristics to guide the selection of PID gains.
(d) Optimization Techniques: For enhanced precision, advanced optimization algorithms come into play. These techniques, such as genetic algorithms, particle swarm optimization, and model predictive control, systematically fine-tune PID gains. They provide a structured approach to achieve optimal controller performance.

A finely tuned PID controller emerges as the bedrock of accurate control, empowering systems to navigate varying conditions while maintaining stability and desired performance benchmarks.

9.3 Wind Power System Dynamics

Renewable energy is characterized by using wind power systems to harness the kinetic energy of the wind to produce electricity. Because wind power systems are dynamic, there is a subtle interplay between several factors that affect how they behave [9]. This section goes deeply into the complex dynamics of wind power systems, covering aspects such as wind turbine properties, underlying output fluctuations, and resulting frequency deviations within these systems.

9.3.1 Wind Turbine Characteristics

Wind turbines are the cornerstone of wind power systems, transforming wind energy into electrical power. Understanding their characteristics is pivotal in comprehending the overall dynamics of wind power generation.

Wind turbines exhibit an optimal operating range, denoted as the "rated wind speed," where their efficiency is maximized. Below this speed, power generation is limited due to insufficient wind energy, while above it mechanical and aerodynamic constraints curtail power production. This non-linear behavior underscores the necessity of efficient control mechanisms, such as PID control, to manage power output within optimal ranges and maintain grid stability. Wind speed is measured as per Equation 9.1:

$$V_{wt} = V_{wb} + V_{wg} + V_{wc} + V_{wr} \tag{9.1}$$

where V_{wb} is the basic wind, V_{wg} is the gust, V_{wc} is the gradual change wind, and V_{wr} is the stochastic wind.

9.3.2 Wind Power Output Fluctuations

The variability in wind speed and direction introduces inherent fluctuations in wind power output. These fluctuations are influenced by atmospheric conditions, terrain, and seasonal variations. Such inherent variability poses a considerable challenge for grid integration, as the erratic nature of wind power output can lead to sudden changes in grid frequency. This volatility demands precise control strategies to counteract frequency deviations and uphold grid stability. The integration of PID control within wind power systems emerges as a strategic approach to actively manage and mitigate the impact of output fluctuations. The wind turbine power conversion P is given as per Equation 9.2 [6]:

$$P = 0.5\rho\pi R^2 V_{wt} C_p (\lambda, \beta) \tag{9.2}$$

9.3.3 Frequency Deviation in Wind Power Systems

Frequency deviation, arising from the variability in wind power output, is a critical concern in wind power systems. As the wind power output experiences fluctuations, the grid's operating frequency can deviate from its nominal value. This deviation can disrupt the synchronization between interconnected systems, potentially triggering equipment malfunctions and power quality issues. To ensure reliable grid operation, it becomes imperative to regulate frequency deviations effectively.

The complex nature of wind power system dynamics is highlighted by the interaction between wind turbine features, output fluctuations, and frequency aberrations. A thorough control approach that not only maximizes electricity generation but also assures grid stability is required to manage these dynamics. The importance of PID control in solving the frequency deviation challenge within wind power systems will be examined in more detail in the sections that follow. Its potential to improve stability, reduce output variations, and promote a harmonic integration of wind power into conventional energy grids will be underlined.

9.4 PID Control in Frequency Regulation

In the realm of wind power systems, maintaining stable grid frequency is of greatest importance. Wind turbines are inherently affected by variable wind speeds, which can result in fluctuations in the power they generate. PID control plays a crucial role in addressing this challenge. This section delves into the significance of frequency regulation in the context of wind energy and explores how PID control algorithms are adept at mitigating frequency variations. Real-world examples and case studies will illustrate the practical application of PID control in ensuring consistent grid frequency.

9.4.1 PID Control for Output Power Control

Effective power control is fundamental for the efficient operation of wind turbines. Wind conditions can fluctuate rapidly, and turbines need to adapt to maximize power output while avoiding overloading. PID control proves invaluable in this regard. This section discusses the vital role of power control in wind turbines and elucidates how PID control strategies are utilized to optimize power production. Real-world case studies will showcase the versatility and effectiveness of PID control in governing output power.

9.4.2 PID Controller Parameters and Tuning

To harness the benefits of PID control, understanding controller parameters and their tuning is imperative. This section provides a comprehensive overview of the proportional, integral, and derivative components of a PID controller. It delves into the theoretical foundations of PID tuning while addressing the practical considerations in the context of wind power systems. Furthermore, it explores automated tuning methods and tools that aid in achieving optimal PID controller performance.

PID control utilization in wind power systems, covering its role in frequency regulation and power control as well as the essential aspects of controller parameters and tuning. It serves as a valuable resource for those engaged in the field of renewable energy, offering insights into maintaining stability and maximizing power output in wind turbines.

9.4.3 Optimization of PID Parameter

PID controllers are widely employed in process control frameworks. These controllers are initially designed for specific operating conditions, but real-world systems often experience variations in their parameters due to changing operating conditions. As a result, it becomes necessary to periodically reconfigure or replace these controllers to maintain optimal performance and fine-tune their parameters. To address this challenge, a robust approach is introduced in Figure 9.2 based on particle swarm optimization (PSO), which eliminates the need for frequent PID controller retuning. The inspiration for this approach comes from observing swarm intelligence in the animal kingdom, where creatures like birds and fish exhibit complex social interactions. PSO mimics the foraging behavior of these animals, where they simultaneously compete and cooperate to find food. In the context of PSO, each unit in the system represents a possible combination of unknown constraints to be optimized. The process begins with a population of potential solutions, and these particles navigate a multidimensional search space in a randomized manner. As particles move through this space, they constantly adapt their positions based on their individual experiences and the collective behavior of the swarm. The goal is to converge toward the best-fitting function, allowing for the efficient optimization of controller parameters without the need for frequent replacements.

Energy in power sources partially covers power destabilization triggered by load destruction of fluctuation of renewable energy. Frequency deviation within an acceptable range is still difficult to maintain. However, electric vehicles can behave as mobile and great energy storage units using the vehicle-to-grid technique which can isolate and give a solution for load frequency control as presented in Figure 9.2. The presented method increases the reliability of process frequency wait for dynamic operating conditions like intermittent renewable energy resources and constant load instability. The system state constraint and functional circumstances change the scheduled microgrid with an energy source and electric vehicle.

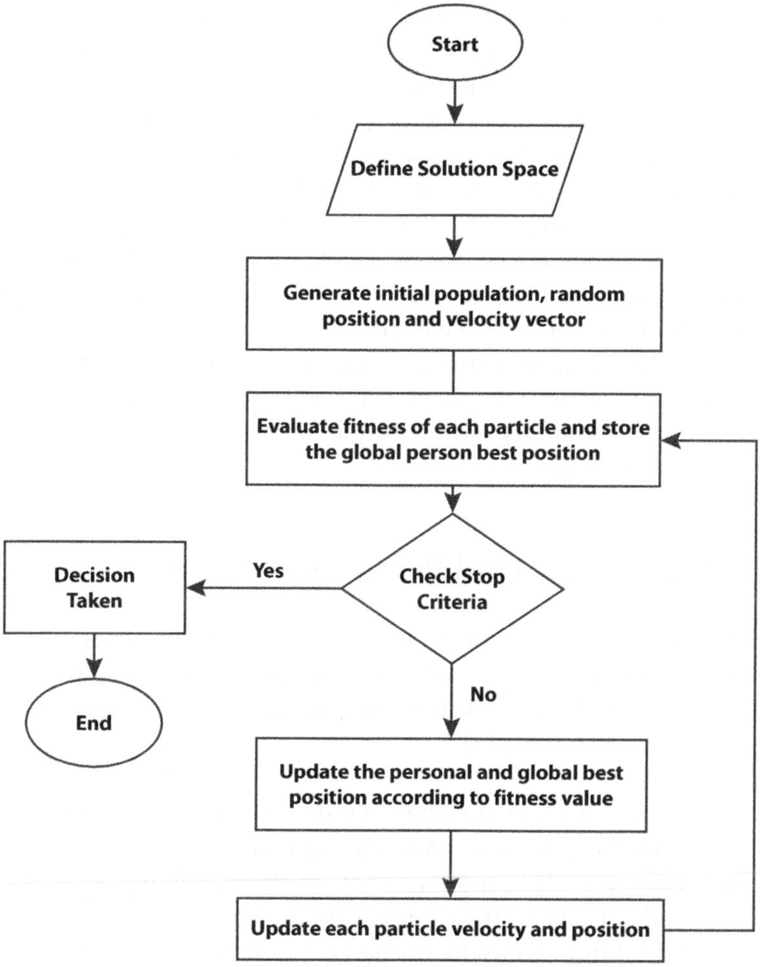

Figure 9.2 Flow chart of the PSO algorithms.

9.4.4 Frequency Deviation With and Without PID Control

This section conducts a thorough comparative analysis between two scenarios: one with the implementation of PID control and another without it. Specifically, the focus is on frequency deviation within a power grid. By contrasting these two approaches, we aim to shed light on the significant impact that PID control has on the stability of the grid. The analysis will encompass real-world data and simulation results to provide valuable insights into the effectiveness of PID control in mitigating frequency deviations. The stability and reliability of a power grid are vital for ensuring uninterrupted energy supply. The fluctuations in grid frequency can lead to disruptions and inefficiencies. The application of PID control can enhance grid stability and bolster its reliability. The use of PID control systems has revolutionized this field. With PID control, the system can swiftly and precisely respond to variations in load and disturbances, maintaining a consistent frequency. However, in scenarios without PID control, addressing frequency deviations often rely on manual interventions and rudimentary control mechanisms. This can lead to slower response times, increased human error, and challenges in adapting to changing conditions. The comparative analysis between frequency deviation with and without PID control underscores the pivotal role of PID algorithms in enhancing the stability and reliability of power systems, minimizing deviations, and optimizing control responses.

9.5 Integrating Wind Power Systems into EV

The integration of wind power systems into electric vehicles (EVs) represents a transformative approach towards sustainable transportation and energy utilization. This innovative integration holds the potential for significant benefits as follows [8]:

(i) Onboard Wind Turbines: Wind turbines can be integrated into the design of EVs, allowing them to harness wind energy while in motion. These compact turbines, strategically placed on the vehicle, capture wind energy and convert it into electricity to charge the vehicle's batteries.

(ii) Regenerative Charging: Wind-powered EVs can employ regenerative charging techniques, where excess energy generated by the onboard wind turbines is stored in the vehicle's batteries. This energy can be used to extend the EV's range or power onboard systems.

(iii) Charging Stations With Wind Power: EV charging stations can be equipped with wind turbines or wind farms to generate clean electricity. EVs can recharge their batteries at these stations with wind-generated power, reducing their carbon footprint.

(iv) Grid Interaction: Wind-powered EVs can interact with the grid, allowing surplus energy from the vehicle's batteries to be fed back into the grid during peak demand or when excess wind power is available. This benefits the grid's stability and reduces strain during high-demand periods.

(v) Sustainable Mobility: Integrating wind power into EVs aligns intending to achieve sustainable and eco-friendly transportation. It reduces greenhouse gas emissions associated with conventional vehicles and promotes clean energy usage.

The integration of wind power systems into EVs represents a promising avenue for sustainable and clean transportation. The primary challenge associated with wind power lies in its intermittent nature. To address this issue, battery energy storage systems (BESSs) [9–12] are employed to mitigate power fluctuations. In Figure 9.3, it is evident that BESS serves the crucial role of compensating for the intermittent power output of the wind turbine. As depicted in Figure 9.2, a doubly fed induction generator is driven by both a wind turbine and a gearbox. The mechanical power output of the wind turbine, influenced by wind speed and blade pitch angle, is detailed within the wind turbine model. Positioned between the DC/AC converter and AC/DC converter of the wind turbine, the BESS facilitates a connection. Additionally, the BESS possesses the capability to charge and discharge as needed, ensuring the smoothing of the wind turbine's output power.

Battery models are typically categorized into three main types: experimental models, electrochemical models, and equivalent circuit models. Among these, the equivalent circuit model stands out as the most suitable

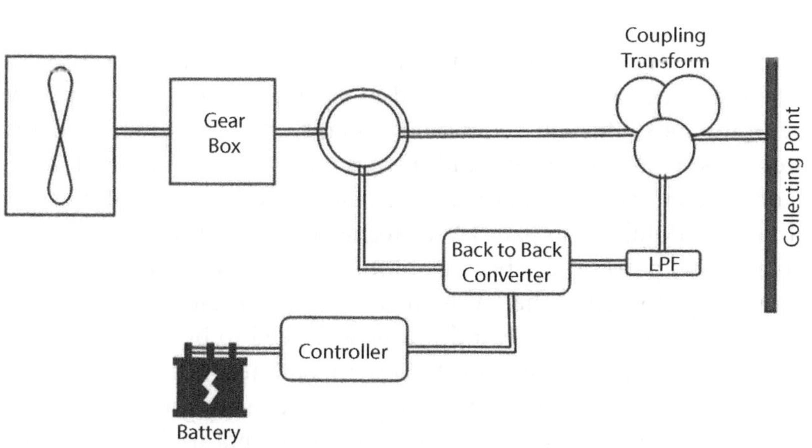

Figure 9.3 Integration of the wind energy system [12].

for dynamic simulations [15]. In dynamic simulations, the controlled voltage source E is determined as Equation 9.3 outlined in [9]:

$$E = E_0 - \frac{Q}{Q - \int i_b dt} + Aexp(-B\int i_b dt) \tag{9.3}$$

where E is no-load voltage (V), E_0 is battery constant voltage (V), K is polarization voltage (V), Q is battery capacity (Ah), A is exponential zone amplitude (V), and B is exponential zone time constant inverse (Ah^{-1}).

Renewable energy is produced by wind power and still increasing as the generation cost is attractive. Wind energy costs can be lower than most other available technologies of electricity generation. Seaward wind control refers to the growth of wind cultivates in streams to generate electricity from the wind. Seaward wind control includes inshore water regions, such as lakes, fjords, and protected seaside regions, using conventional settled base breeze turbine advancements as well as profound water areas using gliding wind turbines, which is unlike the typical usage of the term "seaward" in the maritime industry [13–15]. Close shore wind control is a subtype of seaward wind power. There will be stability issues if more and more wind generators are added to the national grid. Wind turbines that extract cinematic wind energy and turn it into useful mechanical energy are designed to capture as much wind energy as possible for the generation of power by maximizing the number of the speed of the turbine rotor blades.

$$P_w = 0.5\rho AV^3 \tag{9.4}$$

where ρ, A, and V are the density of air, normal area to the speed of the wind, and velocity of wind expressed in kg/m^3, m^2, and m/s, respectively. One of the most efficient generation systems, the wind power generation system was created to harness and convert wind energy into electricity. The use of wind energy to recharge zero-emission electric automobiles is crucial. It is still difficult to plan EV charging and wind power generation to increase wind power usage because of the uncertainties in wind supply and EV mobility. To define the EV load scheduling problem as a multi-stage decision-making problem, wind power will need to be made available in the future. The wind speed affects how much energy is generated by the wind. The power curve, which may be generated using two methods, shows this relationship. The measured wind measurements are first fitted to produce this power curve. The second approach is to use each wind turbine's

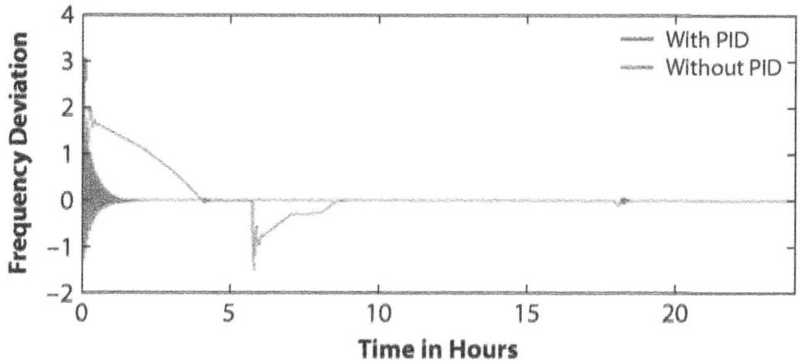

Figure 9.4 Frequency deviations with and without the PID controller.

power curve directly from the manufacturer. The wind's peak velocity can reach 19 m/s, whereas its valley velocity can only reach 2 m/s. Due to the cubic effect, variations in wind speed will make variations in wind power worse. If the large variance of wind power cannot be accurately forecasted, it may result in damage to the electrical grid.

In the present work, electric vehicles and wind as renewable sources are integrated with the existing grid using MATLAB. Electric vehicle output and the output of wind are time-coordinated with the help of a lookup table. The graph shows the frequency deviation of the proposed system when a disturbance of dynamic load in terms of the lookup table is applied. The frequency deviation dropped down to zero because the disturbance was covered by the use of the PID controller. Frequency deviations are suppressed effectively and quickly with the use of a PID controller. The frequency deviation in the proposed system compared with the use of PID controller and without the use of PID controller is shown in Figures 9.4

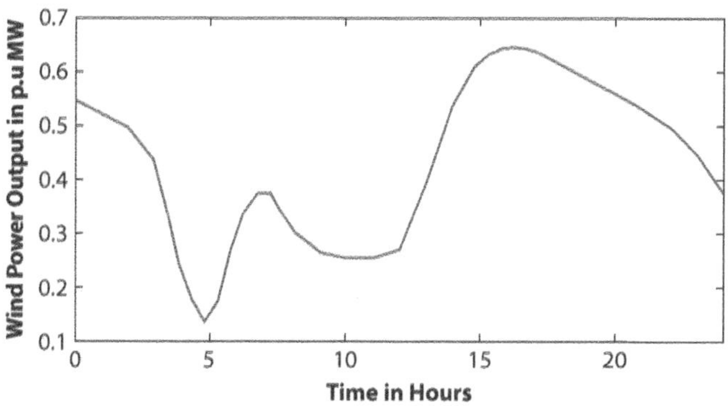

Figure 9.5 Wind output power.

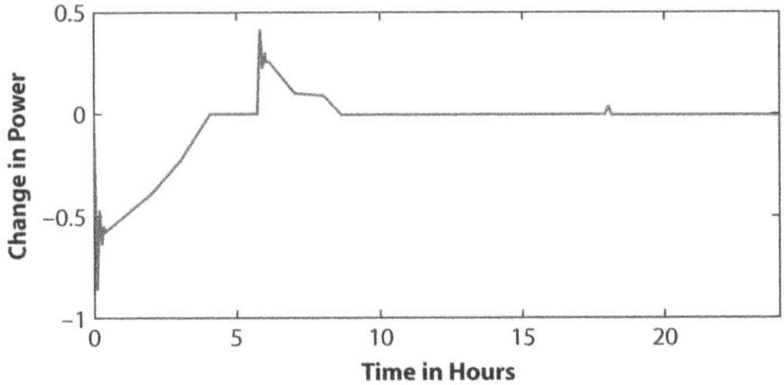

Figure 9.6 Change in power.

and 9.5 With PID control, the deviation is found near the initial hours only and stabilizes. The graph shows the output power of wind corresponding to the speed of the wind for 24 h. Figure 9.5 shows the wind power generation scores with a maximum of 0.65 p.u MW at 17 h and a minimum of 0.11 p.u. MW at 4.8 h. Figure 9.6 shows the output power corresponding to the load disturbance and thus frequency deviation. Power change gradually reaches the zero value by the use of the PID controller.

Figure 9.7 shows the variation in the power output of the electric vehicle. There are two stages of power positive and negative value. The positive and negative stage of power indicates whether the battery is charging (absorbing power from the grid) or discharging (delivering power to the grid)

The utilization of PID control in wind power systems offers benefits and opens new horizons for the future of renewable energy, namely:

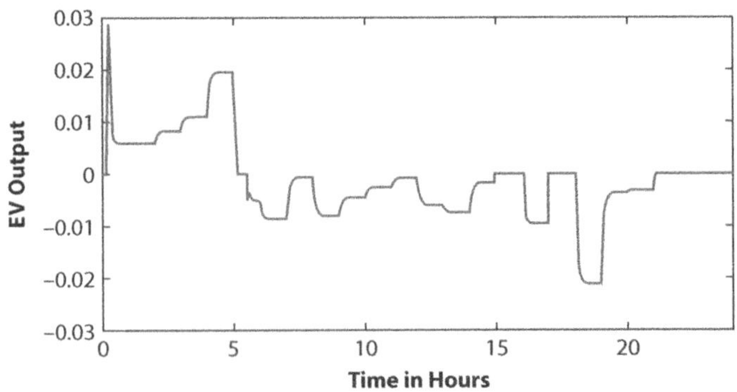

Figure 9.7 EV output power.

(i) Improved Frequency Stability: PID control significantly contributes to enhancing the stability of power grid frequencies. By promptly adjusting control inputs based on real-time feedback, PID control mitigates frequency deviations, ensuring a consistent and reliable power supply.
(ii) Enhanced Grid Resilience: Wind power systems equipped with PID control demonstrate enhanced resilience against dynamic load changes and disturbances. The responsiveness of PID algorithms fortifies grid stability, reducing the susceptibility to disruptions.
(iii) Potential for Hybrid Energy Systems: The versatility of PID control extends to hybrid energy systems. PID algorithms can efficiently manage the integration of multiple energy sources, including wind, solar, and conventional power generation, enabling a seamless transition toward sustainable energy solutions.

Ongoing research and technological advancements continue to refine PID control strategies for wind power systems. Machine learning, advanced control algorithms, and predictive analytics are being harnessed to optimize PID performance. Furthermore, renewable energy sources, like wind power, play an increasingly pivotal role in global energy portfolios. The significance of PID control cannot be overstated. PID control stands as a linchpin for maintaining grid stability and ensuring the successful integration of wind energy. PID control facilitates the achievement of a sustainable and reliable energy supply, aligning with global efforts to reduce carbon emissions and combat climate change. The precise regulation of wind power contributes to a more sustainable and resilient energy future.

9.6 Conclusion

PID control emerges as a cornerstone in realizing the full potential of wind power systems. Its role in ensuring frequency stability, enhancing grid resilience, and paving the way for sustainable energy solutions underscores its significance in shaping the future of renewable energy. As research and innovation progress, PID control will continue to play a pivotal role in achieving a sustainable and reliable energy supply.

References

1. Huang, Q., Jia, Q.S., Guan, X., A review of EV load scheduling with wind power integration. *IFAC-PapersOnLine*, 48, 28, 223–228, 2015.
2. Kothari, D.P. and Nagrath, I.J., *Modern power system analysis*, Tata McGraw-Hill Publishing Company, India, 2003.
3. Ayodele, T.R., Jimoh, A., Munda, J.L., Tehile, A.J., Challenges of grid integration of wind power on power system grid integrity: A review. *Int. J. Renew. Energy Res.*, 2, 4, 618–626, 2012.
4. Johnson, M.A. and Moradi, M.H., *PID control*, pp. 47–107, Springer-Verlag London Limited, London, UK, 2005.
5. Shamseldin, M.A., Adaptive controller with PID, FOPID, and NPID compensators for tracking control of electric–wind vehicle. *J. Robot. Control (JRC)*, 3, 5, 546–565, 2022.
6. Sung, H.C., Tak, M.H., Joo, Y.H., Robust fuzzy controller for mitigating the fluctuation of wind power generator in wind farm. *J. Inst. Control Robot. Syst.*, 19, 1, 34–39, 2013.
7. Lee, D., Sung, H.C., Joo, Y.H., Fuzzy modeling and robust stability analysis of wind farm based on prediction model for wind speed. *J. Inst. Control Robot. Syst.*, 20, 1, 22–28, 2014.
8. Eisenhut, C., Krug, F., Schram, C., Klockl, B., Wind-turbine model for system simulations near cut-in wind speed. *IEEE Trans. Energy Convers.*, 22, 2, 414–420, 2007.
9. Tremblay, O., Dessaint, L.A., Dekkiche, A.I., A generic battery model for the dynamic simulation of hybrid electric vehicles, in: *2007 IEEE Vehicle Power and Propulsion Conference*, 2007, September, pp. 284–289.
10. Shamseldin, M.A., Adaptive controller with PID, FOPID, and NPID compensators for tracking control of electric–wind vehicle. *J. Robot. Control (JRC)*, 3, 5, 546–565, 2022.
11. Pahasa, J. and Ngamroo, I., Coordinated control of wind turbine blade pitch angle and PHEVs using MPCs for load frequency control of microgrid. *IEEE Syst. J.*, 10, 1, 97–105, 2014.
12. Wei, Z., Moon, B.Y., Joo, Y.H., Smooth wind power fluctuation based on battery energy storage system for wind farm. *J. Electr. Eng. Technol.*, 9, 6, 2134–2141, 2014.
13. Slootweg, J.G., Polinder, H., Kling, W.L., Representing wind turbine electrical generating systems in fundamental frequency simulations. *IEEE Trans. Energy Convers.*, 18, 4, 516–524, 2003.
14. Bevrani, H., Habibi, F., Babahajyani, P., Watanabe, M., Mitani, Y., Intelligent frequency control in an AC microgrid: Online PSO-based fuzzy tuning approach. *IEEE Trans. Smart Grid*, 3, 4, 1935–1944, 2012.
15. Chen, H., Cong, T.N., Yang, W., Tan, C., Li, Y., Ding, Y., Progress in electrical energy storage system: A critical review. *Prog. Nat. Sci.*, 19, 3, 291–312, 2009.

10

Interactive Use of D-STATCOM and Storage Resource to Maintain Microgrid Stability for Commercial Systems

Mostafa Eidiani[1]*, Mohammad Kargar[2] and Hossein Zeynal[3]

[1]*Energy Security and Sustainable Energy Institute, Mashhad, Iran*
[2]*Department of Electrical Engineering, Shahrood University of Technology, Shahrood, Iran*
[3]*Dept. of Electrical and Computer Engineering, Buein Zahra Technical University, Buein Zahra, Qazvin, Iran*

Abstract
Microgrids can be used as platforms for the development of renewable energy-based distributed generation (DG) if they have a suitable design and an efficient control system. Microgrids have low inertia because their structure is small and their tolerance for change is low, which makes maintaining voltage and frequency stability difficult, especially in island mode. Unlike power systems, microgrids are primarily powered by inverter-based DG. Although the DG has an efficient control system, the primary energy sources are slow to respond, which makes it impossible for it to maintain the stability of the microgrid on its own. DG, D-STATCOM, and electrical energy storage systems (ESS) are used to develop a multilevel and interactive structure for microgrid frequency and voltage stability. The interactive control system of this microgrid greatly increases its response speed to disturbances. MATLAB has been used to implement and evaluate the proposed structure on a microgrid with two connection modes and an islanding mode. The final goals are as follows:

- The D-STATCOM increases reactive power generation in the microgrid by interacting with ESS. This, in turn, maintains the voltage stability of the microgrid.
- The power control of DGs is used to investigate microgrid voltage stability and frequency in the presence of renewable units.

**Corresponding author*: eidiani@ijesse.net

P. Sivaraman, C. Sharmeela and P. Sanjeevikumar (eds.) Microgrids for Commercial Systems, (241–270) © 2024 Scrivener Publishing LLC

- It is demonstrated that the microgrid performs well in all operating modes.
- It is shown how the control system operates correctly in the event of a severe short circuit in a part of the microgrid and how it maintains the microgrid's stability after a fault.
- As the microgrid transitions from the state of being connected to the main grid to the state of being an island, the control system is shown to maintain voltage and frequency stability.
- D-STATCOM improves power quality indicators by maintaining voltage stability and reducing voltage sag.
- D-STATCOM is recommended for close loads with rapid changes, such as factories and industrial complexes, and places with sensitive electrical loads, such as hospitals.

Keywords: Energy storage systems (ESS), distributed generation (DG), microgrid (MG), renewable energy sources (RES)

10.1 Introduction

By expanding the use of microgrids (MGs) in the power system, traditional power systems can be made more reliable, and both economic and environmental benefits can be realized. In fact, each MG is a combination of energy storage and power generation that can also utilize RES. In addition to increasing the reliability of power, distributed generation units (DG) can improve the operation methods under existing conditions [1].

Typically, microgrid frequency and voltage changes are caused by imbalances between production and consumption or by load disturbances, which can cause the microgrid to become unstable due to its low stability. In island microgrids, which face increasing electricity demand, the situation is more different [2]. To solve its problems, it needs experts who can manage and control the generation of power from various sources, especially wind turbines and solar panels that use RES. As a result, to prevent island microgrids from becoming unstable, experts must be present [3].

Microgrid stability and control are affected by the simultaneous influence of a number of power generators with different dynamic models. However, the benefits of using RES motivate us to overcome these obstacles. Additionally, the microgrid's connection to the distribution network may result in more acute stability problems. MGs are generally operated in two ways: connected to the grid or islanded. In the state connected to the grid, there is a backup power system that can reduce the problem of supplying the load, but in the case of an island state microgrid, no such option exists, so all existing loads must be fed by the microgrid's DG sources. As a result, in order to achieve and maintain the stability of the microgrid and optimal conditions in conditions of power uncertainty, variable power generation of RES, low load due to the lack of rotating storage power in the microgrid, and severe load changes, flexible control strategies are necessary to utilize MG [4, 5].

In island conditions, traditional controllers may not function optimally in all operating modes due to changes in consumer behavior and fluctuating RES production [6].

It uses a multi-level controller system that interacts with energy storage sources and D-STATCOM to solve various microgrid stability problems. D-STATCOM is responsible for maintaining the microgrid's frequency and voltage stability. Inverter-based DG sources with controllable outputs are present in the microgrid, and with the D-STATCOM control system, the voltages of different buses are kept stable by compensating continuously and quickly for the lack of reactive power. However, since D-STATCOM is present in this structure, most of the capacity of DGs is filled by active power.

The primary control system of DGs includes three main levels. The first level controls the lack of voltage and frequency, the second level controls voltage, and the third level controls current [7]. By using local measurements, the proposed controller controls the output power of DG and ensures microgrid stability in two modes: connected to the grid and separated from the grid (islanding) [8]. In this situation, there is no need for a central controller in the microgrid. In addition to creating favorable conditions for the use of distributed renewables, this structure creates the possibility of utilizing these resources despite the uncertainties and inherent changes in their output power due to their high dependence on changing environmental conditions [9]. This type of resource has different production powers depending on weather conditions and other limitations but does not disturb the microgrid's stability [10].

Innovative technologies have been used to reduce climate pollution dramatically over the past few years. Due to the high consumption of electricity coupled with mismanagement of fossil fuel resources, this event has increased global warming and caused environmental pollution. In addition, many of these technologies such as electric vehicles [11], energy storage systems (ESS), DG, and others are present in the power grid, especially in distribution networks. The use of electric cars has reduced the pollution caused by gasoline and diesel cars. Generally, DGs provide electricity from primary energy sources, such as wind, with less pollution [12].

Other sources of DG widely used in power systems include microturbines, fuel cells, and combined heat and power (CHP) systems [13]. Due to the CHP system's efficiency, it has become more popular than other DG sources in the power system due to its simultaneous output of electricity and heat. Along with this type of DG, energy storage devices are used to reduce the effects of uncertainty in the power system. These features have increased the use of these devices in the power system [14].

DGs and energy storages in the power system have caused problems with management and control, especially in the distribution system. This has led to the expansion of the distribution system due to the multiplicity of new network elements. It is possible to solve this challenge by dividing the network into a number of micro-networks that can be controlled and managed more accurately. The IEEE-1547.4-2011 guidelines define a microgrid as a small distribution network where DG and loads can be interconnected and isolated from the upstream network (island mode).

Generally, microgrid loads are classified as sensitive and adjustable. Several factors should be considered when designing a microgrid, including the type of technology, the optimal size and location of the elements of that network, and the proper structure of the microgrid [15].

10.1.1 Microgrid Concept

According to the current technologies, environmental threats, and economic factors, the power system should be exploitable. The current system relies on long distances with very high voltages to transfer large amounts of electrical energy to the consumer. Using microgrids instead of a large distribution network is more economical since energy production, transmission, and management are done locally [16]. A microgrid model is shown in Figure 10.1.

As renewable production is strongly dependent on weather conditions, the production power of this category of equipment cannot withstand the passage of time. As a result, renewable energy can become the main cause of microgrid instability. A microgrid's frequency and voltage change continuously due to the use of renewable resources [17].

As an alternative, it is also possible to use DG and ESS to reduce losses and prevent air pollution as well as increase demand-side reliability. As indicated by the cases mentioned, microgrids have been designed and operated more frequently in recent years.

The microgrid should also operate in island mode during short circuits, blackouts, one-time changes in voltage profiles, and frequency reductions in the main network. Separating the microgrid from the main grid requires

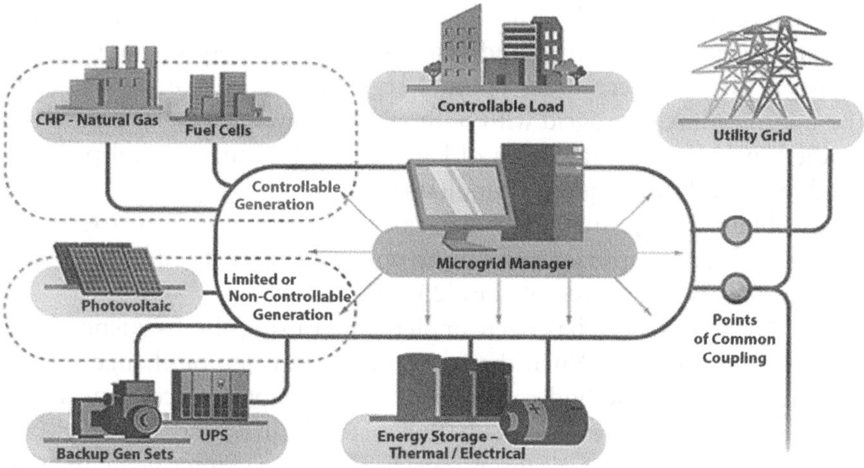

Figure 10.1 Microgrid model.

establishing a balance between production and consumption quickly, which can only be achieved with a suitable control level, to prevent the microgrid from becoming unstable [18].

10.1.2 Review of Past Works

According to [19], the ESS is controlled based on the cost function and the $H\infty$ design in order to reduce the storage cost. The purpose of this paper is to examine the ESS from the point of view of cost reduction, when uncertainty is imposed on the microgrid and energy storage constraints are imposed. The ESS battery control system is designed using mixed quadratic programming in [20]. This paper presents practical plans for reducing power changes despite fluctuations from RES and checking the power system's frequency stability. With the aim of reducing the value of the cost function, the paper applied mixed linear programs to adjust energy charging in hybrid electric cars.

To increase the profitability, we investigate a robust control strategy for ESS control in the microgrid connected to the upstream network in [21]. Despite the net demand or even in the worst conditions, this robust control ensures the highest profit. Several solutions have been presented in [22] for optimizing the energy storage system by dynamically reducing the microgrid control costs. Using DG sources and variable loads at the microgrid level, checking transmission losses, and charging and discharging microgrid storage resources are the proposed processes in this paper.

The microgrid as well as the operation of a smart grid [23] uses a type of centralized control. This study aims to minimize power transmission between MGs and create an optimal regional energy storage system in the microgrid. A microgrid with both wind turbines and solar panels was studied in [24]. A combination of lagging and limiting the real power is used to control the frequency simultaneously.

To obtain the optimal controller values (PI) in an AC microgrid, an online intelligent approach combining optimization methods, fuzzy logic, and particle swarm is presented in [25, 26].

An ARM controller has been implemented using the frequency sampling method in [27]. When the frequency is constant or changes slowly and stepwise, this method can be used with acceptable accuracy. When connecting to a new power system, this method can be used to sample the frequency. A number of solutions have been reviewed to evaluate the performance of the microgrid that is operating as an island and under uncertainty [28]. Its main objective is to reduce fuel costs, improve energy use, and reduce carbon dioxide production by scheduling the resources of DG on an hourly basis. ESS in a microgrid with wind renewable sources was controlled using fuzzy logic in [29] due to the inconsistency of production and the amount of energy stored in it.

In [30], a virtual synchronous generator is used as a mathematical model to reduce the magnitude and frequency fluctuations in microgrids. By incorporating renewable resources into the power grid as much as possible, this proposal offers a suitable solution. It is described in [31] that a multi-objective function is used to formulate the formulation of island MGs. The objectives of this study are (1) fuel consumption cost, (2) voltage stability index, and (3) total voltage instability. For determining voltage stability, the old NRS method [32] and its improved method [33] are the best suggestions. The hybrid optimization method combined with a genetic algorithm was used to achieve these objectives.

Following is a description of the proposed method for this chapter.

10.2 The Proposed Structure

The control system is designed in a decentralized manner. As a result, it can guarantee voltage and frequency stability without expensive or complex communications lines and a central controller between the microgrid components. D-STATCOM is also proposed as a control system for maintaining frequency and voltage stability in microgrids. By continuously and quickly compensating for the lack of reactive power, it prevents voltage instability in the microgrid.

10.2.1 Primary Controller

This part of the system involves a voltage source control for DGs which can adjust the power received from DGs by measuring the voltage, current, and output power locally in order to maintain the time and production of the microgrid and the frequency and voltage near nominal values. Voltage-based DGs act as slack buses in power grids, so island mode is necessary [1]. Figure 10.2 shows the compact system of the primary control system.

10.2.1.1 Drop Controller

In this control system, the governor drop of synchronous generators is used to adjust the reactive and active power of DG outputs based on network frequency and voltage drop [1]. DG is shared by this controller [34]. Figure 10.3 shows the block diagram of a drop controller.

Shown as follows is the relationship between reactive and active power transmission:

$$P = \frac{U^2}{Z}\cos\theta - \frac{UV}{Z}\cos(\delta + \theta) \tag{10.1}$$

$$Q = \frac{U^2}{Z}\sin\theta - \frac{UV}{Z}\sin(\delta + \theta) \tag{10.2}$$

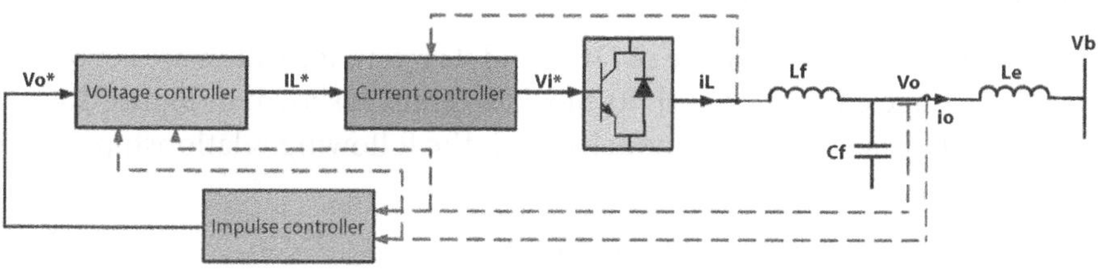

Figure 10.2 Compact system of the primary control system.

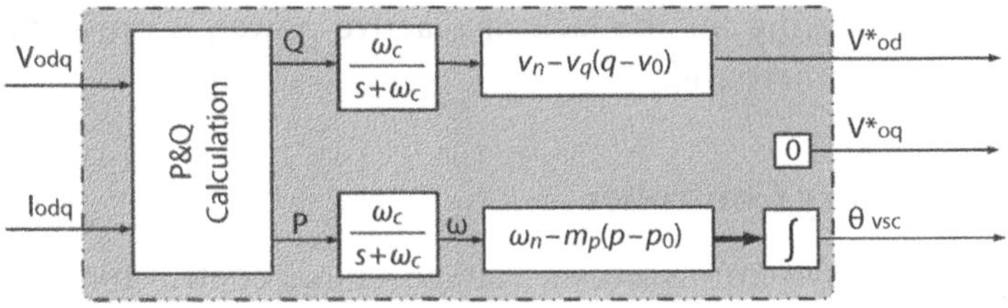

Figure 10.3 Drop controller block diagram.

With placement $Ze^{j\theta} = R + jX$ have:

$$P = \frac{U(R(U - V\cos\delta) + XV\sin\delta)}{R^2 + X^2} \tag{10.3}$$

$$Q = \frac{U(X(U - V\cos\delta) - RV\sin\delta)}{R^2 + X^2} \tag{10.4}$$

As a result of the small values of line resistance and power angle, the real power is influenced by the power factor and frequency and the reactive power by the voltage. As a result, the equations can be rewritten as follows:

$$\delta = \frac{XP}{UV} \tag{10.5}$$

$$U - V = \frac{XQ}{U} \tag{10.6}$$

DGs share power loss control [34].

10.2.1.2 Voltage Controller

It controls the output voltage of DGs, and its reference values are set by the previous stage controller [34]. Its block diagram is shown in Figure 10.4.

The voltage controller is modeled with the following relationships:

$$\dot{\varphi}_d = v_{od}^* - v_{od} \tag{10.7}$$

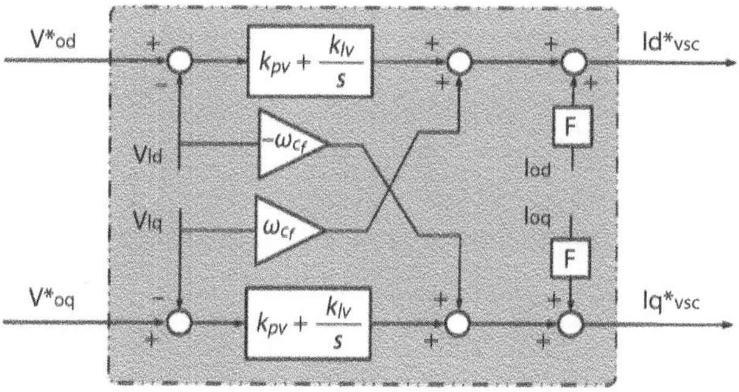

Figure 10.4 Voltage controller's block diagram.

$$\dot{\varphi}_q = v^*_{oq} - v_{oq} \tag{10.8}$$

In these equations, the state variables with φ_d and φ_q have been expressed. The state space form of this PI controller is expressed as follows:

$$[\dot{\varphi}_{dq}] = [0][\varphi_{dq}] + B_{v1}\left[v^*_{odq}\right] + B_{v2}\begin{bmatrix} i_{ldq} \\ v_{odq} \\ i_{odq} \end{bmatrix} \tag{10.9}$$

wherein

$$i^*_{ld} = k_{iv}.\varphi_d - k_{pv}\left(v_{od} - v^*_{od}\right) - \omega_n.C_{f.}v_{oq} + F.i_{od} \tag{10.10}$$

$$i^*_{lq} = k_{iv}.\varphi_q - k_{pv}\left(v_{oq} - v^*_{oq}\right) + \omega_n.C_f.v_{od} + F.i_{oq} \tag{10.11}$$

10.2.1.3 Current Controller

This part is used for current control [35]. The current controller structure is shown in Figure 10.5.

The equations of this controller are as follows:

$$\dot{\gamma}_d = i^*_{ld} - i_{ld} \tag{10.12}$$

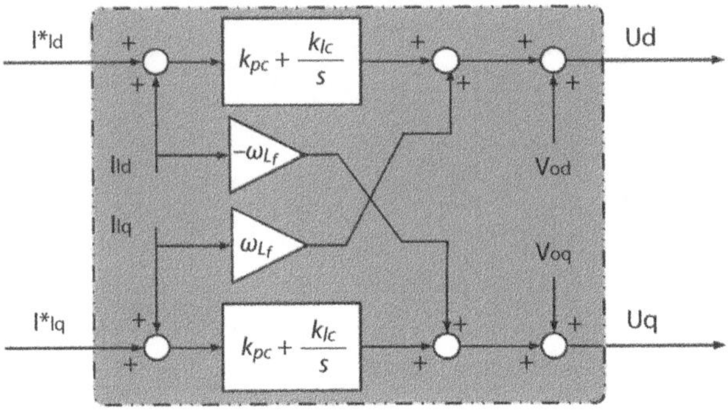

Figure 10.5 Current controller's block diagram.

$$\dot{\gamma}_q = i^*_{lq} - i_{lq} \tag{10.13}$$

The state space variables for the present controller, γ_d and γ_q, are taken into account in the equations above. The state space form can be implemented as follows:

$$[\dot{\gamma}_{dq}] = [0][\gamma_{dq}] + B_{c1}\left[i^*_{ldq}\right] + B_{c2}\begin{bmatrix} i_{ldq} \\ v_{odq} \\ i_{odq} \end{bmatrix} \tag{10.14}$$

$$v^*_{iq} = k_{ic}.\gamma_q - k_{pc}\left(i_{lq} - i^*_{lq}\right) + \omega_n.L_f.i_{ld} \tag{10.15}$$

$$v^*_{id} = k_{ic}.\gamma_d - k_{pc}\left(i_{ld} - i^*_{ld}\right) + \omega_n.L_f.i_{lq} \tag{10.16}$$

Figure 10.6 illustrates the DG control system's final block diagram.

10.2.2 Secondary Control

Since DGs do not have a high response speed, a low response speed on the primary side causes a drastic drop in DC bus voltage. For this reason, the inverter must increase the modulation factor [36] in order to return this level of DC link voltage drop, and there was no change in the delivered voltage.

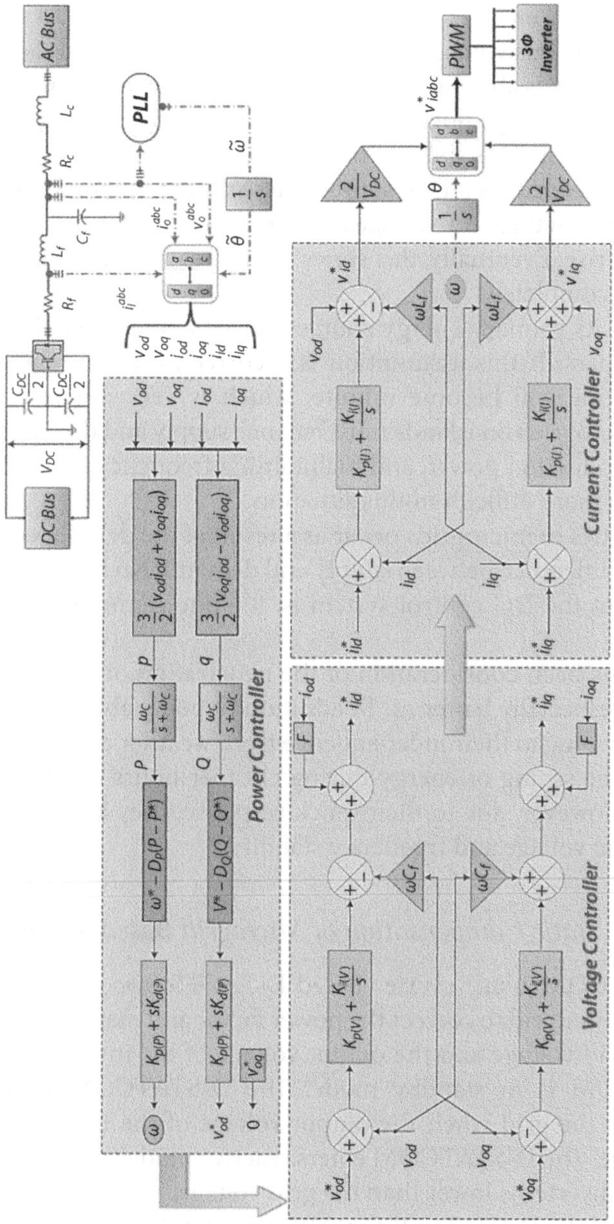

Figure 10.6 Architecture of the distributed generation control system.

If the primary energy source does not have the capacity to supply power to keep the DC bus voltage constant, the modulation factor is increased to the maximum possible value. If the maximum allowed factor is exceeded, frequency and voltage fluctuations increase, and the microgrid protection relays remove the DG sources from the circuit if the maximum allowed factor is exceeded. As DG sources depart, the microgrid's conditions become more acute, and voltage and frequency deviations from nominal values also rise. Eventually, this series of events will cause the microgrid to shut down completely.

DG and its primary energy source are considered ideal in many studies and papers. If this assumption is incorrect, then the energy source's response time will become infinite, which is very different from what occurs. Microgrids on islands must balance supply and demand using only internal production power, and maintaining frequency and voltage stability is much more difficult in this situation.

Instabilities in microgrids occur as a result of the delay in compensating the power difference between supply and demand. No matter how accurate and efficient the DG control system is, it alone cannot ensure microgrid stability.

There has been consideration of the penetration of ESS energy storage resources, especially batteries. In addition to being able to be used in different hours due to their independence from weather conditions, they also allow for the saving of energy during cheaper hours due to their lack of demand. However, due to their quick response, they are very suitable for maintaining voltage and frequency stability.

10.2.2.1 Static Compensation of Microgrid Based on Inverter

A D-STATCOM is an inverter-based D-FACTS that is placed in parallel with the power grid to correct the power factor and maintain voltage levels.

If the grid voltage and the output voltage of the inverter are equal, the D-STATCOM is in standby mode. The D-STATCOM injects reactive power into the grid when the output voltage of the inverter exceeds the grid voltage. The D-STATCOM enters reactive mode when the output voltage of the inverter is lower than the grid voltage.

To model the D-STATCOM in ABC coordinates, ignore the harmonics produced by the inverter.

$$\frac{d}{dt}\begin{bmatrix} i_a \\ i_b \\ i_c \end{bmatrix} = \begin{bmatrix} -\frac{R}{L} & 0 & 0 \\ 0 & -\frac{R}{L} & 0 \\ 0 & 0 & -\frac{R}{L} \end{bmatrix} \times \begin{bmatrix} i_a \\ i_b \\ i_c \end{bmatrix} + \frac{1}{L}\begin{bmatrix} (V_{sa} - V_{ia}) \\ (V_{sb} - V_{ib}) \\ (V_{sc} - V_{ic}) \end{bmatrix} \quad (10.17)$$

In this equation, the currents i_a, i_b and i_c are the output phases of the inverter. For the purpose of setting the network frequency using the Park transformation, the equation above can be expressed as follows:

$$\frac{d}{dt}\begin{bmatrix} i_d \\ i_q \end{bmatrix} = -\frac{1}{L}\left(\begin{bmatrix} R & -L\omega \\ L\omega & R \end{bmatrix} \times \begin{bmatrix} i_d \\ i_q \end{bmatrix} + \begin{bmatrix} V_{id} - V_{sd} \\ V_{iq} - V_{sq} \end{bmatrix}\right) \quad (10.18)$$

In this equation, ω represents the grid angular velocity, and DC currents and voltages are controlled by a proportional–integral (PI) controllers. An inverter's output voltage controls the reactive power, and it can also be expressed as follows:

$$V_{id} = \omega L i_q + V_{sd} - L u_d \quad (10.19)$$

$$V_{iq} = -\omega L i_d - L u_q \quad (10.20)$$

We can derive Eq. (10.18) from the above equation:

$$\frac{di_d}{dt} = -\frac{R}{L} i_d + u_{id} \quad (10.21)$$

$$\frac{di_q}{dt} = -\frac{R}{L} i_q + u_{iq} \quad (10.22)$$

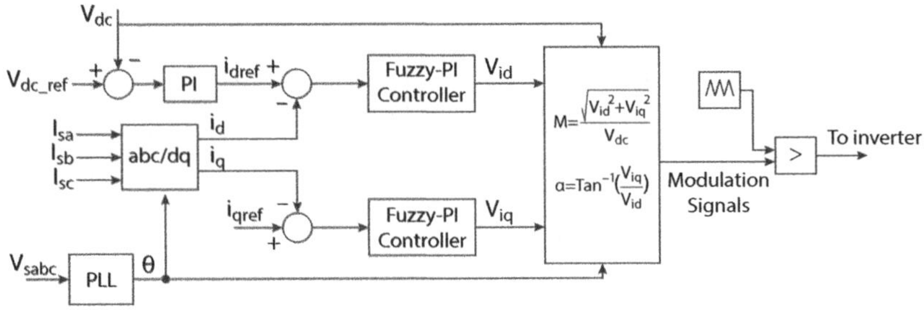

Figure 10.7 Block diagram of the compensator control section.

In the equation above, u_{id} and u_{iq} are controlled signs. These signals are modeled for the PI controller as follows:

$$u_{id} = \left(K_p + \frac{K_i}{S} \right)(i_{dref} - i_d) \tag{10.23}$$

$$u_{iq} = \left(K_p + \frac{K_i}{S} \right)(i_{qref} - i_q) \tag{10.24}$$

Lastly, Figure 10.7 shows the complete block diagram of the system.

The DC current and voltage are controlled by three PI controllers for vectors d and q. It is important to note that d and q are identical vector current controllers. In the D-STATCOM architecture, the phase locked loop is also used to produce the dq conversion and simultaneously measure the output voltage of the inverter and the AC network voltage.

10.3 Simulation

10.3.1 Commercial LV Distribution Network

The interactive control structure is implemented on the microgrid, whose illustration of a single line of a test network is shown in Figure 10.8, and the network data is presented in Tables 10.1 to 10.3. The selected microgrid is of the AC type and is being operated in Denmark [37]. The simulation of various parts of the microgrid has been done using MATLAB software. The nominal values of this microgrid are 50 Hz and 380 V. In the high voltage side of the network, the radial commercial network is connected to a 300-kVA, 20-kV/380-V three-phase transformer. By connecting the

Table 10.1 Data of available resources in the microgrid.

Name	DG1	DG1	**ESS**	**WTG**
Type	Diesel generator	Fuel cell	**ESS**	Wind turbine
Rated power (kVA)	**100**	**50**	**100**	**21**

Table 10.2 Microgrid consumer data.

Bus no.	**S (kVA)**	**P (kW)**	**Q (kVAr)**
1	23.3	20.5	12.1
2	8.84	16.5	5.84
3	23.5	19	11.9
4	8.84	16.2	4.1
5	12.75	8.1	4.1
6	21.64	19.8	7.8
7	12.55	8.1	3.9
8	21.64	19.9	8.1
Total load	134	120.1	57.84

Table 10.3 Conditions of the microgrid lines.

From bus	**To bus**	**Resistance (Ω)**	**Reactance (Ω)**
1	2	0.1501	0.0775
2	3	0.1879	0.0969
3	4	0.3755	0.1937
4	5	0.5637	0.1905
5	6	0.1501	0.0775
6	7	0.3381	0.1747
7	8	0.1501	0.0775

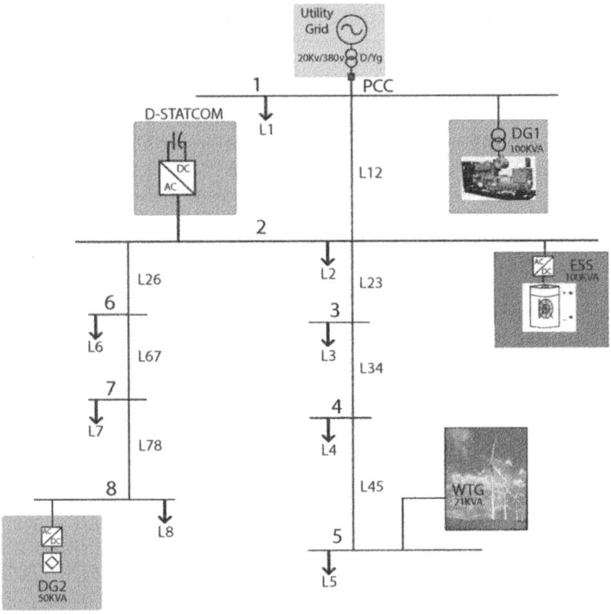

Figure 10.8 An illustration of a single line of a test network.

transformer to a ΔYg, it can handle single-phase loads in either the primary or secondary phase. A T–T arrangement is used for the earth, in which the potential earth of the customer and the electricity supplier is separated at the transformer side. Commercial customers use more energy between 8 a.m. and 12 a.m. The feeder supplies energy to all customers radially and along the transmission line. The point of connection circuit (PCC) breaker switches between connected and island mode. Due to this fact, the microgrid can be both connected to and isolated from the upstream grid.

The microgrid consists of three DG systems, each of which is connected to a microgrid bus. In bus 1, DG1 is a diesel generator, while in bus 8, DG2 is a fuel cell, both of which can be controlled. Wind energy is a renewable resource in bus 5, and the speed of the wind determines how much it produces.

The bus 2 of this microgrid contains an energy storage source of 100 kW (as much as the production power of the largest DG).

A static compensator has been used to control the microgrid voltage at 380 V and its frequency at 50 Hz. D-STATCOM is connected to the power grid by three single-phase transformers. The components in the D-STATCOM are a voltage source inverter with a switching speed of 1.68 kHz, a capacitive damping filter connected to the output of the inverter with a coil size of 800 μH and a capacitor of 300 μF, a capacitance of 1,000 microfarads for the DC source of the inverter and response to the required reactive power and the voltage regulator that stabilizes the voltage of bus 2.

For the purpose of determining the efficiency and performance of the designed system, various conditions on the microgrid have been investigated in light of the presented control structure as well as the necessity for coordination between DG, ESS, and D-STATCOM. The permissible range for voltage and frequency fluctuations is considered in this study.

10.3.2 Test No. 1: Changing the State of the Microgrid from Connected to Island

For this purpose, we assume that the microgrid is in a normal state of operation. According to Figure 10.9, the power grid connection is disconnected at the moment $t = 3$ s, and the microgrid changes state. This action takes place after changing the state of the PCC breaker as shown in the figure.

Initially, the power grid provides approximately 65 kW of the microgrid's consumption load, and then a power shortage of 65 kW is expected after the microgrid is disconnected from the power grid. We assume that wind speed is less than the lower limit of the source, i.e. 5 m/s, for the renewable source to make it more difficult for the microgrid. The wind turbine has no output in this scenario, so the control system will have to work harder (see Figure 10.10).

In this situation, it is necessary for DGs to increase the output power. Microgrids will be disrupted if the generated power is less than the required value or if it takes time to increase the power from the resources. The frequency and voltage values of the microgrid will go out of range, and the microgrid will shut down.

If the primary controller is used alone, after changing the state of the microgrid, the first part of the controller, which is responsible for checking the drop rate, detects a lack of production amount by decreasing the voltage

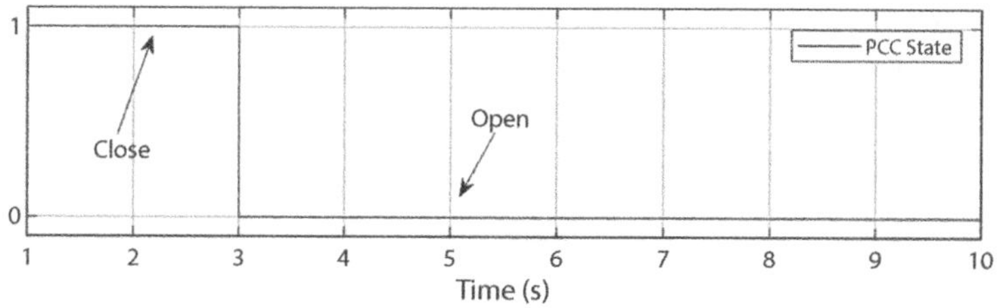

Figure 10.9 Point of connection circuit breaker status change.

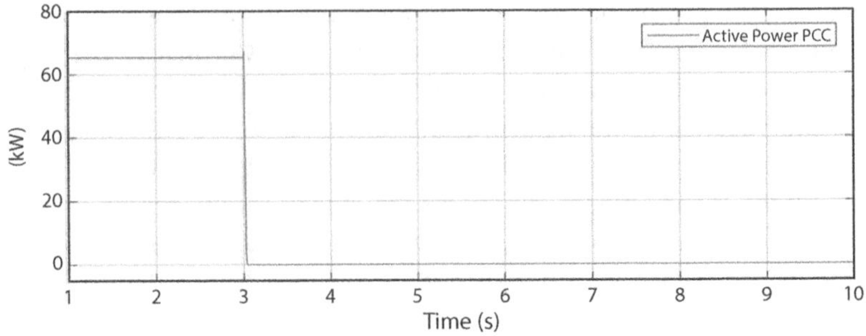

Figure 10.10 Total power received by the microgrid from the power system.

and frequency and then sends an order to the DG sources to increase production. However, since bringing power to the circuit takes a while, in this situation, despite sending the correct control command from the control system, there has been a disturbance in the microgrid due to the low speed of increasing the power of DGs, and the voltages and frequencies are out of range. The microgrid was unstable because the available sources could not quickly compensate for the lost power (see Figure 10.11).

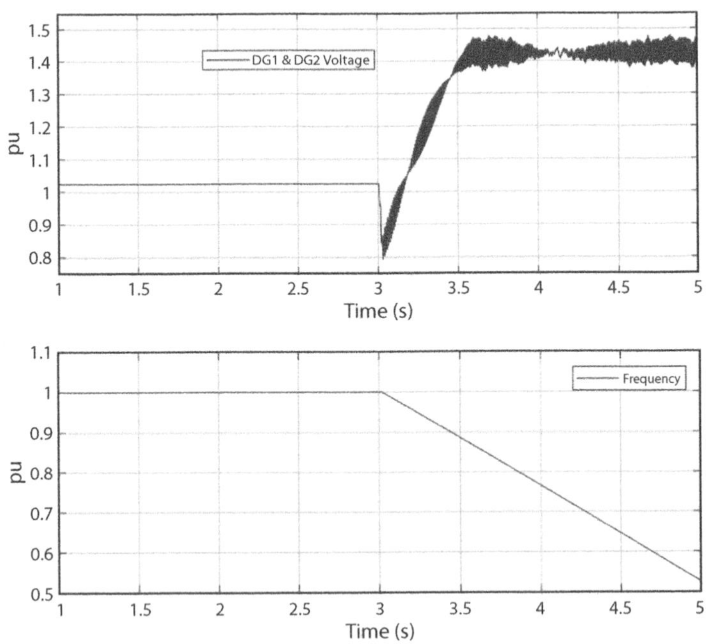

Figure 10.11 Getting out of the allowed range of voltage and frequency.

In the situation where the secondary control is also applied on the microgrid to maintain the frequency and voltage stability of the microgrid with the presence of the storage source, the required load is rapidly injected into the microgrid and prevents the voltage and frequency values from going outside their allowed range. Since ESS sources have high speed, but low work continuity, DG sources increase their production level to allow the reserve source to be fully withdrawn (see Figures 10.12, 10.13, and 10.14).

In Figures 10.12, 10.13, and 10.14, the microgrid controller's design has been shown to perform well in different operating modes (connected, transient, and islanded) and ensure microgrid stability.

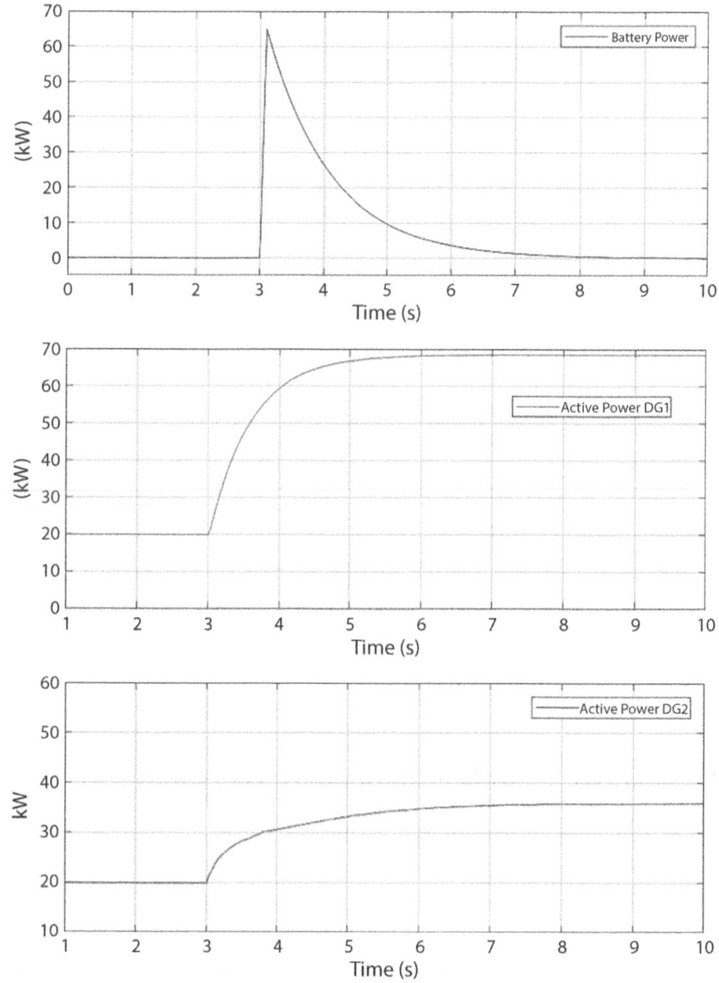

Figure 10.12 Power injected by the storage source and distributed generation sources.

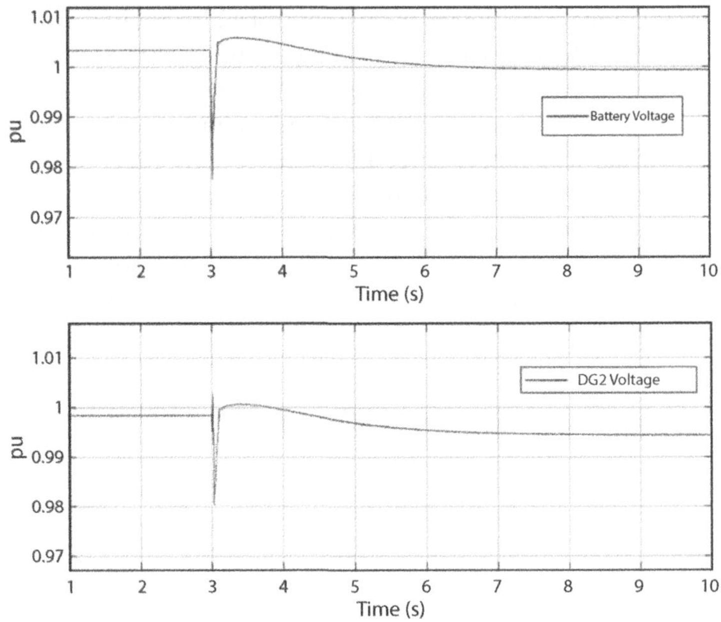

Figure 10.13 Voltage fluctuations of distributed generation and storage source.

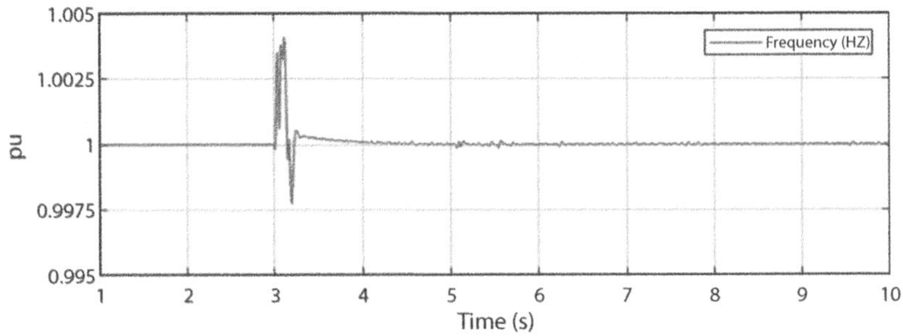

Figure 10.14 Microgrid frequency range.

10.3.3 Test No. 2: Adding Demand to the Microgrid

This test is designed to investigate the behavior of the designed control system under large and consecutive demand changes. For this scenario, we assume that the microgrid is in an island state and investigate the load changes in a multistage process. According to Figure 10.15, the load changes occur sequentially and in short intervals. Microgrid load changes are applied at time $t = 5$ s with a value of 5 kW, at time $t = 7$ s with a value of

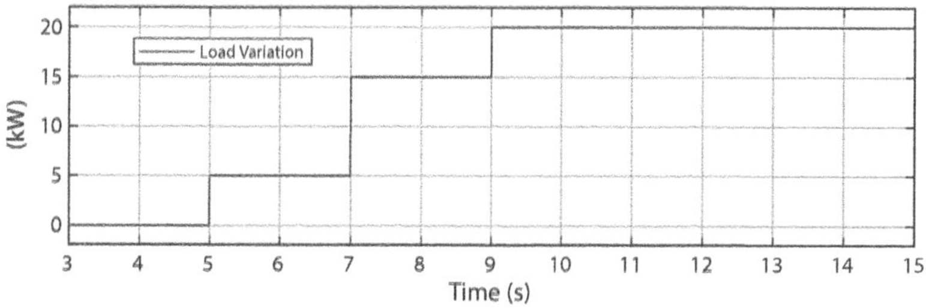

Figure 10.15 Step changes of demand in the islanded microgrid.

10 kW, and at time $t = 9$ s with a value of 5 KW. In the microgrid, the gradual increase in load has upset the balance between supply and demand.

Microgrids have low latency, so power changes cause deviations in frequency and voltage. If a timely and appropriate response is not given, the microgrid's stability will be compromised. According to Figure 10.16, the storage source and static compensator quickly compensate for any power loss resulting from the appropriate speed in controlling and responding to loads.

A RES's production power depends on climate changes that are beyond its control. In this test, the wind speed is at the maximum allowed value, i.e., 15 m/s; therefore, the production power of the wind unit is at its nominal value (21 kW). Figure 10.17 shows that the production power of the renewable source is not affected by the constant wind blowing and does not increase with demand. In contrast, each DG source adjusts its production power to balance the microgrid's production and consumption.

During load increases, the frequency and voltage values have deviated from their nominal, as shown in Figures 10.18 and 10.19. The microgrid remained stable despite these deviations due to the timely response of the control unit and maintaining a balance between production and consumption.

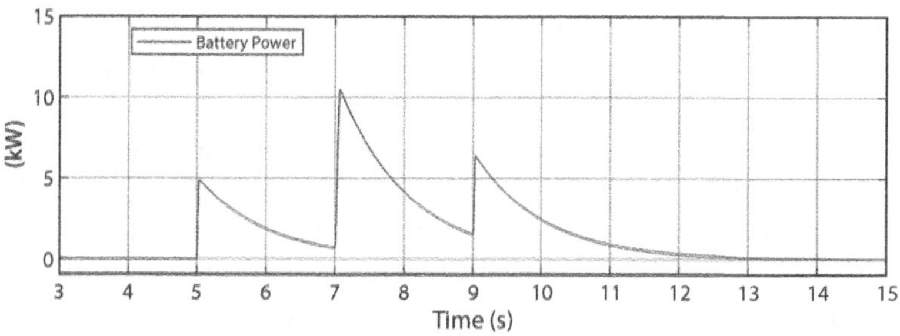

Figure 10.16 Compatibility of the stored power.

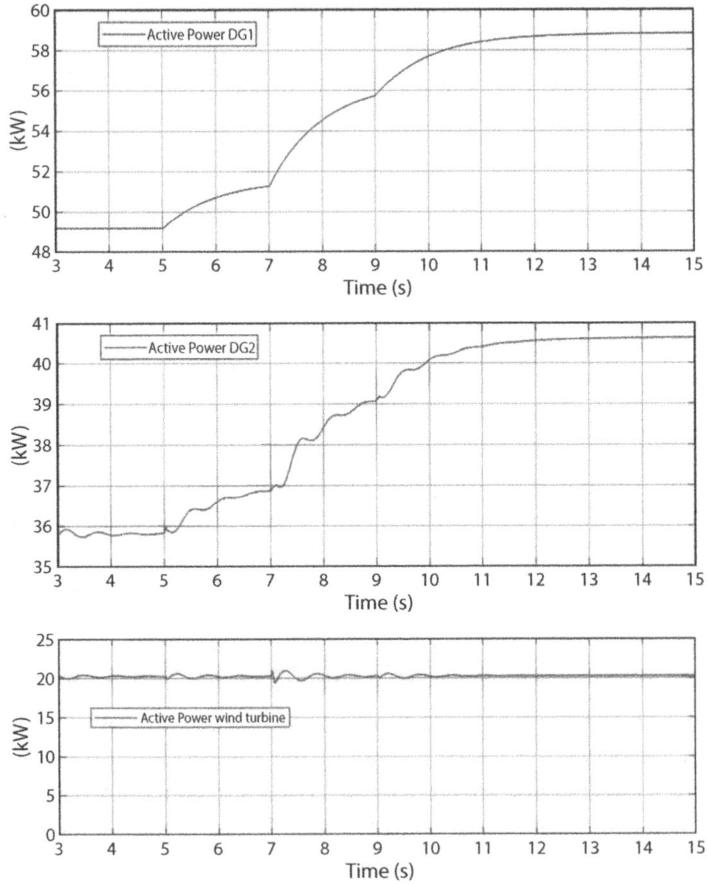

Figure 10.17 Changes in the power of distributed generations.

10.3.4 Test No. 3: Changes in the Production of Renewable Resources

During this test, the control system is tested to see how it reacts when the renewable source is lost. Due to this condition, an error in the estimation of the system state occurs, causing the power balance in the microgrid to be disturbed and then become unstable. Microgrid stability is constantly at risk because the production rate of renewable units cannot be predicted. As such, in order to check the performance of the designed control system under the worst possible conditions, this test was conducted in the conditions that result in the maximum possible change in this source's output.

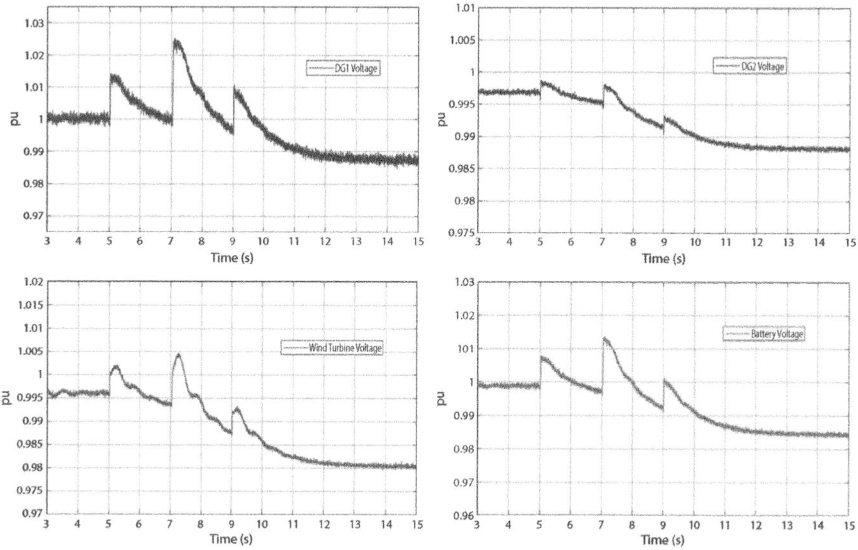

Figure 10.18 Voltage energy storage source and distributed generations.

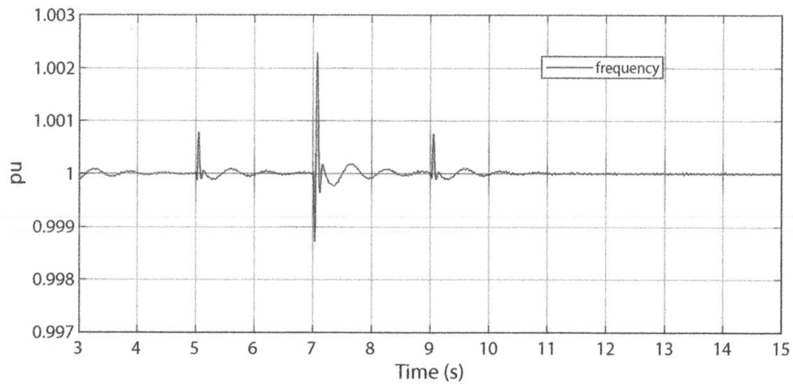

Figure 10.19 Microgrid frequency.

When the wind speed exceeds the upper limit of the turbine, renewable wind sources are equipped with storm protection systems. This unit produces power when the wind speed is within its permissible range (5–15 m/s), and when the wind speed exceeds the permissible limit, the rotor is locked and the output power is reduced once from maximum to zero in order to prevent damage to the turbine. Figure 10.20 illustrates how the production power of the renewable unit changed.

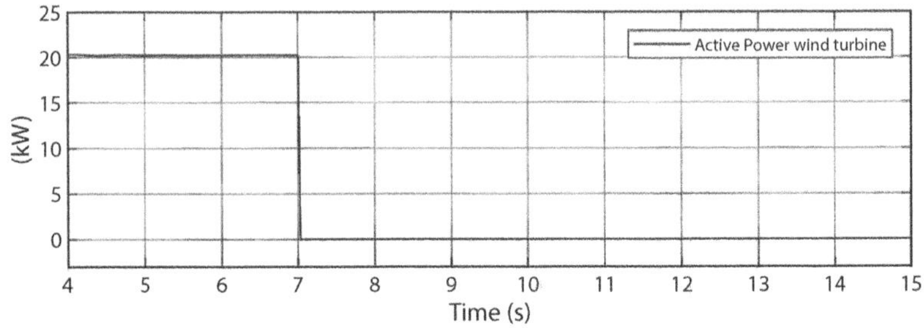

Figure 10.20 Changes in renewable source power.

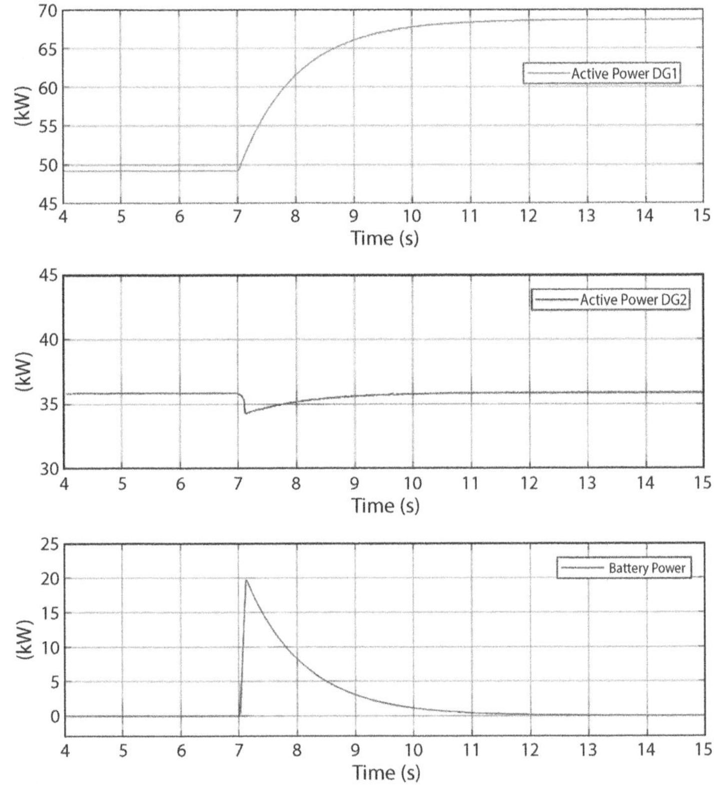

Figure 10.21 The injected power of the storage source and distributed generation sources after the exit of the renewable source.

As the microgrid operates in island mode, power compensation and production increase should be achieved internally. When the renewable source is removed, the lack of power is compensated by the interaction of the static compensator, the storage unit, and then the gradual production of distributed sources. Figure 10.21 shows that, with the increase in production capacity of the storage source, the lack of power will be compensated at the right time, and with the gradual arrival of DG resources, the storage source's capacity will decrease.

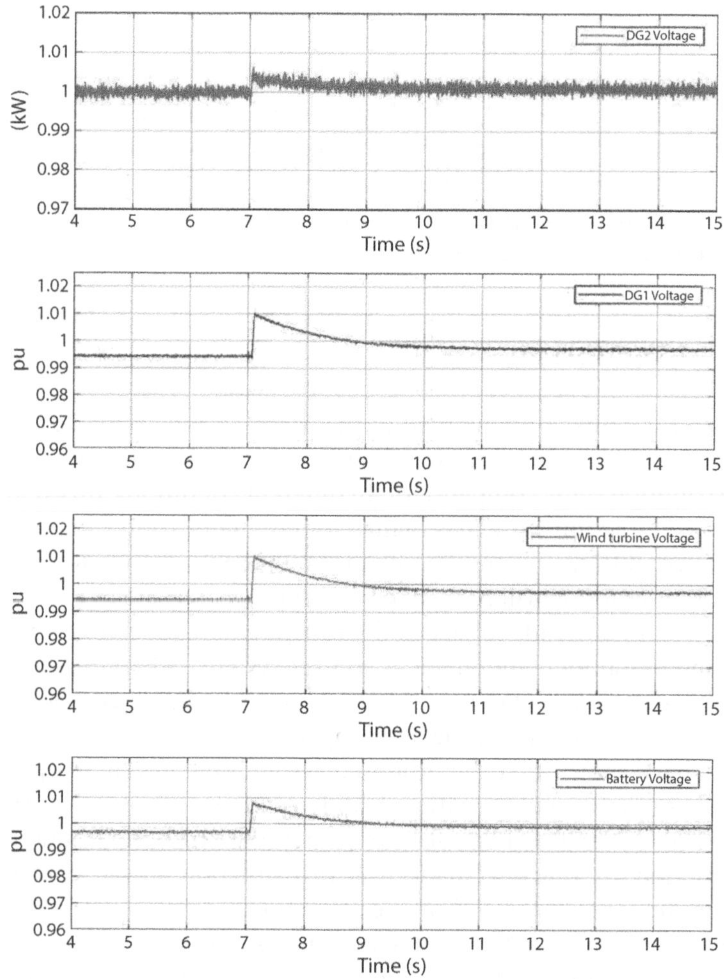

Figure 10.22 Voltage fluctuations of distributed generation and storage source.

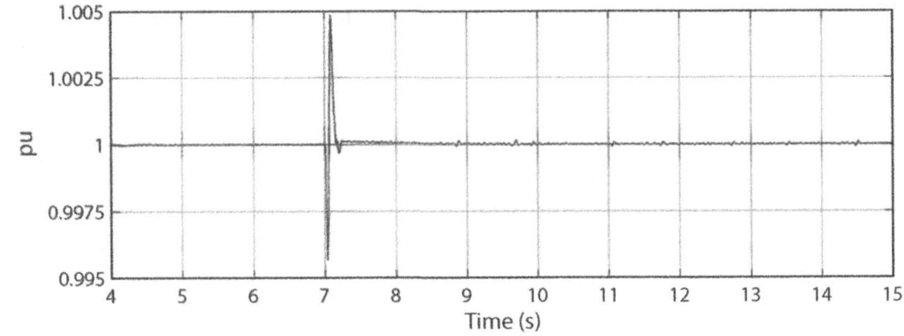

Figure 10.23 Microgrid frequency.

DG sources and energy storage sources are shown in Figure 10.22, and the output of the renewable unit in the microgrid is shown in Figure 10.23. Thus, the proposed control system and its interaction with different components of the microgrid have been able to ensure the stability of the microgrid despite voltage and frequency changes within the allowed range. By using the proposed structure, the issue of uncertainty of renewable resources can be addressed.

10.4 Conclusion

Microgrid stability is maintained in the islanded and connected states by using the control system derived from combining the storage source and the distribution static compensator. In different conditions, this system can work in conjunction with DGs to maintain frequency and voltage stability. In these conditions, on the one hand, by interacting DGs with the control unit, microgrids will be able to respond to disturbances more quickly, and on the other hand, voltage and frequency stability can be ensured by designing an optimal and efficient control system. Microgrids are of low latency, so D-STATCOM next to DGs increases the speed of reactions while allowing more use of the lines' capacity.

Tests have shown that the proposed system operates in all operational conditions (connected, transitional, and island) as well as passes transient states with minimal voltage and frequency fluctuations. Production and consumption are quickly balanced in this situation. Power consumption changed dramatically in a short period, and this system handled it well. The short time interval between load changes did not cause any problems with the microgrid's control system. This structure compensated for the power loss resulting from the growing use of renewable resources in the power system. The proposed model was found to be effective in the tests conducted.

References

1. Eidiani, M. and Kargar, M., Frequency and voltage stability of the microgrid with the penetration of renewable sources. *2022 9th Iranian Conference on Renewable Energy & Distributed Generation (ICREDG)*, pp. 1–6, 2022, doi: 10.1109/ICREDG54199.2022.9804542.
2. Eidiani, M., Zeynal, H., Ghavami, A., Zakaria, Z., Comparative analysis of mono-facial and bifacial photovoltaic modules for practical grid-connected solar power plant using PVsyst. *2022 IEEE International Conference on Power and Energy (PECon)*, pp. 499–504, 2022, doi: 10.1109/PECon54459.2022.9988872.
3. Ghardashi, G., Gandomkar, M., Majidi, S., Eidiani, M., Dadfar, S., Accuracy and speed improvement of microgrid islanding detection based on PV using frequency-reactive power feedback method. *2022 International Conference on Protection and Automation of Power Systems (IPAPS)*, pp. 1–8, 2022, doi: 10.1109/IPAPS55380.2022.9763190.
4. Nikam, V. and Kalkhambkar, V., A review on control strategies for microgrids with distributed energy resources, energy storage systems, and electric vehicles. *Int. Trans. Electr. Energy Syst.*, 31, 1, e12607, 2021.
5. Zeynal, H., Jiazhen, Y., Azzopardi, B., Eidiani, M., Impact of electric vehicle's integration into the economic VAr dispatch algorithm. *2014 IEEE Innovative Smart Grid Technologies, ISGT Asia, Conference-Asia*, Kuala Lumpur, Malaysia, May 20-23, 2014, pp. 780–785, 2014, https://doi.org/10.1109/ISGT-Asia.2014.6873892.
6. Eidiani, M. and Ghavami, A., New network design for simultaneous use of electric vehicles, photovoltaic generators, wind farms and energy storage. *2022 9th Iranian Conference on Renewable Energy & Distributed Generation (ICREDG)*, pp. 1–5, 2022, doi: 10.1109/ICREDG54199.2022.9804534.
7. Eidiani, M., Integration of renewable energy sources, in: *Handbook of Smart Energy Systems*, M. Fathi, E. Zio, P.M. Pardalos (Eds.), Springer, Cham, 2022, https://doi.org/10.1007/978-3-030-72322-4_41-1.
8. Eidiani, M., Applying optimization techniques to develop a renewable energy supply map, in: *Handbook of Smart Energy Systems*, M. Fathi, E. Zio, P.M. Pardalos (Eds.), Springer, Cham, 2022, https://doi.org/10.1007/978-3-030-72322-4_61-1.
9. Eidiani, M., Zeynal, H., Zakaria, Z., Shaaban, M., Analysis of optimization methods applied for renewable energy integration. *2023 IEEE 3rd*

International Conference in Power Engineering Applications (ICPEA), 1570861924, 6-7 March 2023.

10. Eidiani, M., Zeynal, H., Zakaria, Z., Shaaban, M., A comprehensive study on the renewable energy integration using DIgSILENT. *2023 IEEE 3rd International Conference in Power Engineering Applications (ICPEA)*, 1570860205, 6-7 March 2023.
11. Zeynal, H., Jiazhen, Y., Azzopardi, B., Eidiani, M., Flexible economic load dispatch integrating electric vehicles. *Proceedings of the 2014 IEEE 8th International Power Engineering and Optimization Conference, PEOCO2014*, Lankawi, Malaysia, 24-25 March 2014, pp. 520–525, 2014, https://doi.org/10.1109/PEOCO.2014.6814484.
12. Vergara, P. *et al.*, Energy management in microgrids, in: *Microgrids Design and Implementation*, pp. 195–216, Springer, Cham, 2019.
13. Elkazaz, M., Mark, S., David, T., Energy management system for hybrid PV-wind-battery microgrid using convex programming, model predictive and rolling horizon predictive control with experimental validation. *Int. J. Electr. Power Energy Syst.*, 115, 105483, 2020.
14. Sufyan, M., Rahim, N., Aman, M., Tan, C., Raihan, S., Sizing and applications of battery energy storage technologies in smart grid system: A review. *J. Renew. Sustain. Energy*, 11, 1, 014105, 2019.
15. Kabalci, E. and Kabalci, Y., Introduction to smart grid architecture, in: *Smart Grids and Their Communication Systems*, pp. 3–45, Springer, Singapore, 2019.
16. Marnay, C. *et al.*, Japan's pivot to resilience: How two microgrids fared after the 2011 earthquake. *IEEE Power Energy Mag.*, 13, 3, 44–57, 2015.
17. Hassan, L.H. *et al.*, Current state of neural networks applications in power system monitoring and control. *Int. J. Electr. Power Energy Syst.*, 51, 134–144, 2013.
18. Olivares, D.E. *et al.*, Trends in microgrid control. *IEEE Trans. Smart Grid*, 5, 4, 1905–1919, 2014.
19. Chiu, W.Y., Hongjian, S., Vincent, P.H., Demand-side energy storage system management in smart grid. *IEEE Third International Conference on Smart Grid Communications (SmartGridComm)*, IEEE, 2012.
20. Jiang, Q., Yuzhong, G., Haijiao, W., A battery energy storage system dual-layer control strategy for mitigating wind farm fluctuations. *IEEE Trans. Power Syst.*, 28, 3, 3263–3273, 2013.
21. Choi, J. *et al.*, Robust control of a microgrid energy storage system using various approaches. *IEEE Trans. Smart Grid*, 10, 3, 2702–2712, 2018.
22. Xu, Y. and Xinwei, S., Optimal control based energy management of multiple energy storage systems in a microgrid. *IEEE Access*, 6, 32925–32934, 2018.
23. Ouammi, A., Hanane, D., Roberto, S., Optimal control of power flows and energy local storages in a network of microgrids modeled as a system of systems. *IEEE Trans. Control Syst. Technol.*, 23, 1, 128–138, 2014.

24. Eidiani, M., Asghari Shahdehi, N., Zeynal, H., Improving dynamic response of wind turbine driven DFIG with novel approach. *Proceedings-2011 IEEE Student Conference on Research and Development, SCOReD 2011*, Putrajaya, Malysia, 19th-20th Dec. 2011, pp. 386–390, p. 6148770, 2011, https://doi.org/10.1109/SCOReD.2011.6148770.
25. Zeynal, H., Hui, L.X., Jiazhen, Y., Eidiani, M., Azzopardi, B., Improving lagrangian relaxation unit commitment with cuckoo search algorithm, PECON 2014. *2014 IEEE International Power and Energy Conference*, pp. 77–82, 2014, https://doi.org/10.1109/PECON.2014.7062417.
26. Eidiani, M. and Zeynal, H., New approach using structure-based modeling for the simulation of real power/frequency dynamics in deregulated power systems. *Turk. J. Electr. Eng. Comput. Sci.*, 22, 1130–1146, 2014, https://doi.org/10.3906/elk-1208-90.
27. Li, Z. and Penghao, W., A high-precision and wide-range method of frequency measurement in the new energy power grid connection. *2017 Chinese Automation Congress (CAC)*, IEEE, 2017.
28. Zeynal, H. and Eidiani, M., Hydrothermal scheduling flexibility enhancement with pumped-storage units. *The 22nd Iranian Conference on Electrical Engineering, ICEE 2014*, May 20-22, 2014, Shahid Beheshti University, pp. 820–825, 2014, https://doi.org/10.1109/IranianCEE.2014.6999649.
29. Jayapriya, M. *et al.*, Implementation of fuzzy based frequency stabilization control strategy in raspberry pi for a wind powered microgrid. *Procedia Comput. Sci.*, 115, 151–158, 2017.
30. Hirase, Y. *et al.*, A novel control approach for virtual synchronous generators to suppress frequency and voltage fluctuations in microgrids. *Appl. Energy*, 210, 699–710, 2018.
31. Zeynal, H., Zadeh, A.K., Nor, K.M., Eidiani, M., Locational marginal price (LMP) assessment using hybrid active and reactive cost minimization. *Int. Rev. Electr. Eng.*, 5, 5, 2413–2418, Oct. 2010.
32. Eidiani, M., Zeynal, H., Zadeh, A.K., Mansoorzadeh, S., Nor, K.M., Voltage stability assessment: An approach with expanded newton raphson-sydel. *5th International Power Engineering and Optimization Conference, PEOCO 2011*, Shah Alam, Selangor, Malaysia, pp. 31–35, 2011, 5970424, https://doi.org/10.1109/PEOCO.2011.5970424.
33. Eidiani, M., A reliable and efficient holomorphic approach to evaluate dynamic available transfer capability. *Int. Trans. Electr. Energy Syst.*, 31, 11, 1–14, e13031, 2021, https://doi.org/10.1002/2050-7038.13031.
34. Palizban, O. and Kimmo, K., Hierarchical control structure in microgrids with distributed generation: Island and grid-connected mode. *Renew. Sustain. Energy Rev.*, 44, 797–813, 2015.
35. Prodanovic, M. and Green, T.C., High-quality power generation through distributed control of a power park microgrid. *IEEE Trans. Ind. Electron.*, 53, 5, 1471–1482, Oct. 2006, doi: 10.1109/TIE.2006.882019.

36. Eidiani, M., Ashkhane, Y., Khederzadeh, M., Reactive power compensation in order to improve static voltage stability in a network with wind generation. *International Conference on Computer and Electrical Engineering, ICCEE 2009*, Dubai, UAE, December 28-30, vol. 1, pp. 47–50, 2009, https://doi.org/10.1109/ICCEE.2009.239.
37. Liu, Z. *et al.*, A multiagent system-based protection and control scheme for distribution system with distributed-generation integration. *IEEE Trans. Power Deliv.*, 32, 1, 536–545, 2016.

11

Power System Studies for Microgrids

Aditi Chatterjee* **and Festus Okinyi Maklago**

Department of Electrical Engineering, C. V. Global University, Bhubaneswar, India

Abstract

Renewable sources of energy have been recognized as alternative sources of energy for meeting power demands. A microgrid is an integrated network of distributed generation (DG) plants, loads, and energy storage devices. The microgrid can operate in either standalone or grid-connected mode. In this study, a solar photovoltaic-based microgrid is designed to operate in both grid-tied and islanded mode. The power generation capacity of the microgrid is decided taking into account the load profile of a school campus in rural area. A load flow analysis and a transient analysis are performed for the microgrid model using the Electrical Transient Analysis Program (ETAP) software. The load flow analysis is essential to identify the buses violating the voltage limits. A short circuit analysis is performed to obtain the bus voltage and bus current during both symmetrical and unsymmetrical faults occurring in the microgrid. The short circuit analysis report can be utilized to design the microgrid equipment for its protection and reliable operation.

Keywords: Solar photovoltaics, distributed generation, load flow analysis, short circuit analysis, transient analysis, ETAP

11.1 Introduction

In present-day scenario, renewable-energy-based power generation has gained a tremendous interest. Among the renewable sources, the major contributors in power generation mix are solar, wind, hydro, and geothermal. It has been presented in the International Energy Outlook (IEO) report that solar is the highest contributor, followed by wind [1].

**Corresponding author*: aditichatterjee.ee@gmail.com

P. Sivaraman, C. Sharmeela and P. Sanjeevikumar (eds.) Microgrids for Commercial Systems, (271–306) © 2024 Scrivener Publishing LLC

The microgrid concept assumes a cluster of loads and micro sources operating as a single unit with control capability, which means that it can disconnect from the traditional grid and operate autonomously. The microgrid is an integrated network of distributed generation (DG) plants, loads, and energy storage devices. These distributed generation units can be based on either renewable or non-renewable energy sources. It provides backup for the grid in case of emergencies and enhances local reliability. Renewable-energy-based distributed generation plants are mostly reliant on wind, solar, hydro, biomass, geothermal, and fuel cell energy. The non-renewable-energy-based distributed generation plants are based on gas turbines, micro-turbine, and reciprocating engines. Renewable-energy-based distributed generators are in focus for more than a decade because they are environmentally friendly. They help in reducing the environmental pollution due to the toxic gases released during the combustion of fossil fuels like coal and petroleum. Thereby, they also aid in saving the depleting fossil fuel reserves [2–5].

These MGs are generally operated in parallel with the utility grid and can seamlessly transfer to the islanded mode of operation in case of transient situations like faults. The stable operation in islanded mode is a generation–load balance problem and sometimes involves load shedding of non-critical loads in case of inadequate generation. So, the realization of future smart grids will require solutions to a wide range of technical issues like power and energy balance, power quality, hosting capacity, peak planning, the optimal location of distributed energy resources (DERs), and protection of the system in both grid-connected and islanded modes [6]. A microgrid is prone to capricious and undesirable faults. Protection of the microgrid is a perplexing issue due to the bidirectional power flow, inconstant current, and intermittent nature of the renewable energy source. Hence, load flow analysis, short circuit analysis, and voltage stability analysis are crucial to ensure the apposite design of the microgrid equipment for its unswerving operation [7–20].

The Electrical Transient Analysis Program software or ETAP was developed in 1983. It is a powerful analytical tool for the simulation, control, monitoring, and automation of a power system. It is a user-friendly platform and very convenient to use. Creation of a single line diagram of the system model, renewable energy modules, load flow analysis, generator load sharing, cable sizing analysis, transformer MVA sizing, short circuit analysis, transient analysis, protection coordination analysis, arc flash analysis, motor starting analysis, harmonic analysis, ground grid analysis, etc., are some of the attractive and interesting features of ETAP.

When distributed generators (DG) are embedded in a power distribution system, there is an alteration in the power flow regime. As a consequence, the fault current changes, and the coordination between the protection devices is disturbed. In [7], the authors introduced a new protection coordination which they validated by simulating a real 11-kV feeder using the ETAP package. In [8], a new algorithm for the optimal setting of overcurrent relays with variation in fault levels which is a consequence of variable wind speed was presented. The proposed algorithm was validated on a modified IEEE-30 bus system with integrated wind farms using the ETAP software. The authors in [9] have highlighted the significance of System Integrity and Protection Schemes (SIPS) for attaining a secure and reliable grid operation during power system contingencies. They have specified three major SIPS and the shortcomings associated with it and also proposed a methodology to rectify the consequences. The results were validated using ETAP. In [10], an optimal and single protection scheme suitable for all operating modes of a microgrid along with every type of phase faults in the system, i.e., three-phase symmetrical, line-to-line, and double-line-to-ground faults, was proposed. The proposed scheme was validated on the modified IEEE 9-bus 4-DG microgrid in the ETAP platform. A smart protection scheme (SPS) is proposed using micro-phasor measurement units (μPMUs) to obtain the continuous, rapid, and synchronized phasor measurement data for a microgrid system in [12]. A detailed model of an MG including a 10-bus connected distribution system with integrated DG is simulated using the ETAP software. In [14], the authors proposed a transient analysis of distributed generation AC microgrid using ETAP. They focused on transient phenomenon like a large load step and the loss of distributed generators which are mainly affecting the stability of the generation available altogether by the distributed generations.

In [15], ETAP was used to perform a harmonic analysis of an AC–DC hybrid microgrid. In [16], the authors proposed an optimal power flow optimization model for minimizing power loss and harmonics emanating from RES with emphasis on reducing their impact on the locational marginal prices. The efficiency of the model was tested on the modified IEEE six-bus and 24-bus systems using ETAP. In [17], the author demonstrated the design of an electrical system based on load flow analysis and short circuit analysis using ETAP and the selection of electrical equipment parameters based on the short circuit analysis report. In [18], a 1,240-MW combined cycle power plant was designed using ETAP and load flow; voltage stability and short circuit analyses were performed. In [19], a photovoltaic-based micro-grid with an electrical energy storage system (EESS) was designed utilizing ETAP. The microgrid performance was examined at different climate and load conditions. In [20], the authors introduced ETAP for designing a microgrid with wind and photovoltaic generators along with energy storage. The author was able to provide a simulation analysis of a micro-grid system under different operating conditions, including the transient analysis of a micro-grid system under an external power network fault and the recovery operation in a grid-connected state.

In this chapter, a solar photovoltaic (SPV)-based microgrid is designed to operate in both islanded and grid-tied mode. The power generation capacity of a microgrid is decided taking into account the load profile of a school campus. The margin for power generation is reserved at 40% to compensate for the intermittent nature of irradiance. A load flow analysis and a transient analysis were performed for the microgrid model using the ETAP software. The load flow analysis is essential to identify the buses violating the voltage limits. Under-voltage and over-voltage alerts are set to identify the weak buses. A short circuit analysis was performed to obtain the bus voltage and bus current during both symmetrical and unsymmetrical faults. The load flow report and short circuit report are essential to design the power system equipment like the circuit breaker, transformer, cables, etc. Some other transient operating conditions like loss of a DG unit and sudden increase in load demands were also simulated in ETAP. From the comparative result analysis, it is observed that, in islanded mode, the voltage drop is large when there is a sudden increase in load demand. As a consequence, the inverters get overloaded when one of the DG unit gets disconnected.

This chapter is structured in sections as follows: Microgrid design in islanded mode is demonstrated in Section 11.2. Harmonic load flow analysis in islanded mode is described in Section 11.3. Section 11.4 presents the transient analysis of microgrid in islanded mode. Design and load flow analysis of microgrid in grid-connected mode is depicted in Section 11.5. Section 11.5.3 gives a description of the harmonic load flow analysis in grid tied mode. In Section 11.6, transient analysis like faults conditions and loss of PV generators is explained. A comparative analysis of a microgrid operating in both islanded and grid-tied mode is presented in Section 11.7. Finally, the paper is concluded in Section 11.8.

11.2 Description of a Microgrid Model Operating in Islanded Mode

The block diagram of the microgrid system functioning in islanded mode is shown in Figure 11.1. It consists of three SPV generators along with a

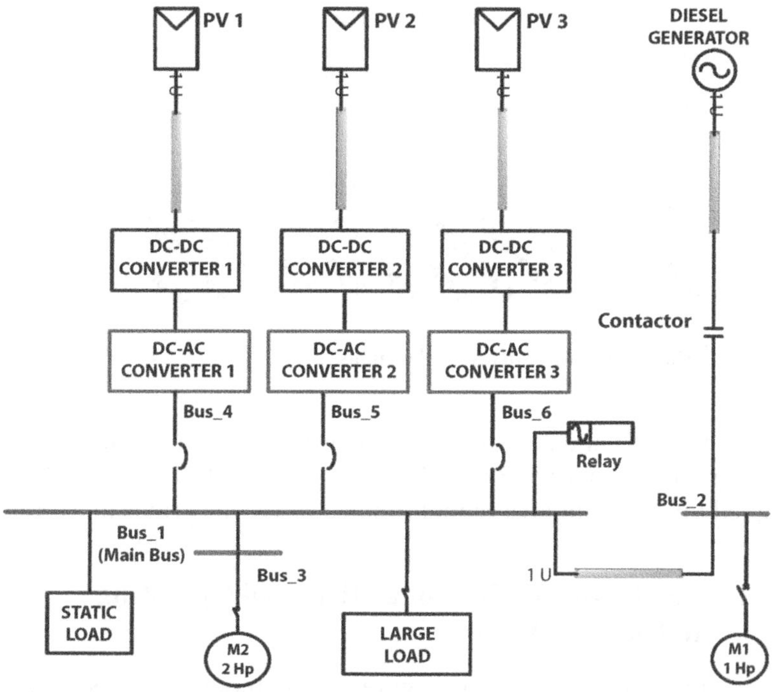

Figure 11.1 Schematic diagram of a microgrid operating in islanded mode.

backup diesel generator with a generation capacity of 6 kW. The SPV generators are connected to the main bus *via* two converters connected back to back. Converter 1, converter 2, and converter 3 are dc–dc boost converters, and then there are three dc-ac converters. The PV arrays are connected on a single three-phase AC main bus. The output is connected on a single 415-V, three-phase AC bus from which both single-phase as well as three-phase loads are connected. The backup generator will be turned on by the relay once the voltages at the 415-V-main bus go below the minimum level. The circuit breakers in the system will protect the system from impairment caused by overcurrent, overloads, and short circuits.

The generation capacity of PV3 is 23.09 kW, and the solar panels PV1 and PV2 have an equal generation capacity of 11.58 kW. The generation capacity is decided considering the load profile of a school campus depicted in Figure 11.2. There are three types of load in the system: one static load and two motor loads. The rating of the static load, motor 1 load, and motor 2 load is 20.22, 1.2, and 3.81 KVA respectively. There is a large load present in the system of 30-KVA rating which is switched on during the peak demand

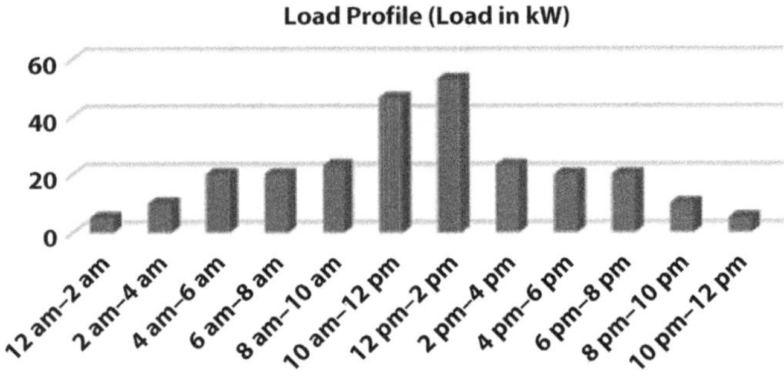

Figure 11.2 Daily load profile of a school campus.

period. Inverter 1 and inverter 2 are having the same rating, i.e., 11.6 KVA, and the rating of inverter 3 is 20.79 KVA. A standalone microgrid can be designed for a remote-area school campus where the grid is inaccessible.

11.2.1 Load Flow Analysis of a Microgrid Operating in Islanded Mode

The system depicted in section Figure 11.1 is modeled and simulated in ETAP. Load flow studies were performed under two diverse operating scenarios. In the first case, the large load of 30 kVA is not allied, and in the second case, the large load is allied along with all other loads. The load flow gives the nominal power flow through the bus bars and the loads during steady state. The significance of the load flow studies is to identify the buses which are crossing the voltage limits for both over-voltage and under-voltage. The under-voltage and over-voltage limits are depicted in Table 11.1.

11.2.1.1 Operating Scenario 1

The load flow analysis is performed with the static load, and all motor loads are on. The large load is not allied. It is turned on only during the peak demand periods. All the PV panels are generating power at their

Table 11.1 Voltage alert levels.

Bus voltage	Marginal	Critical
Under-voltage	95%	90%
Over-voltage	105%	110%

maximum capacity, and the diesel generator is off. The simulation model in the ETAP platform is shown in Figure 11.3. The load flow report is depicted in Table 11.2. There are a total of six buses in the load flow report. Bus 1 is the main bus, bus 2 is the auxiliary bus connected to 1-HP motor load, and bus 3 is connected to 3-HP motor load. The load flow report generated by

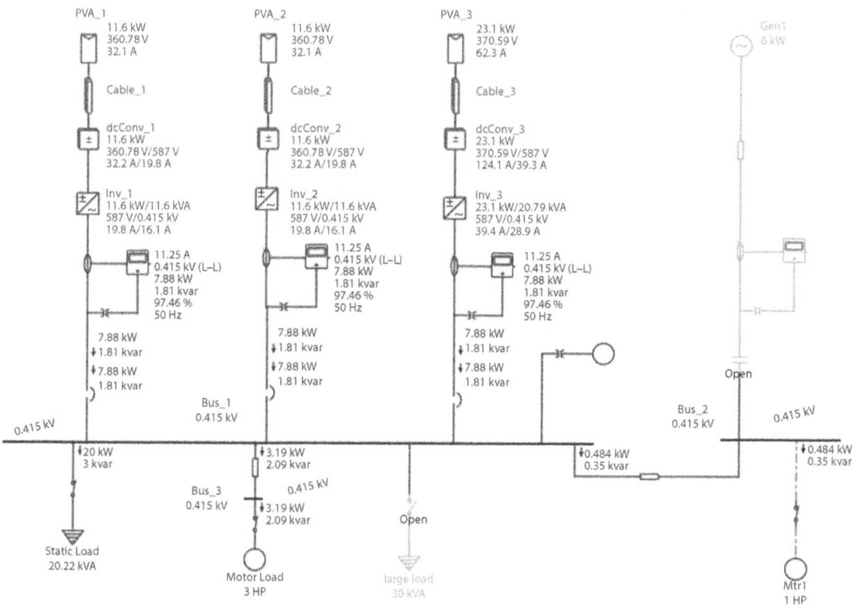

Figure 11.3 Load flow model in electrical transient analysis program for the islanded mode during operating scenario 1.

Table 11.2 Load flow report operating scenario 1 in islanded mode.

Bus_ID	Nominal KV	Volt (kV)	kW loading	kvar loading	Amp loading
Bus_1	0.415	0.415	23.65	5.44	33.76
Bus_2	0.415	0.415	0.484	0.35	0.832
Bus_3	0.415	0.415	3.19	2.09	5.301
Bus_4	0.415	0.415	7.88	1.81	11.25
Bus_5	0.415	0.415	7.88	1.81	11.25
Bus_6	0.415	0.415	7.88	1.81	11.25

ETAP considers inverter 1, inverter 2, and inverter 3 connected to dummy bus 4, bus 5, and bus 6 respectively. It can be observed from the load flow report in Table 11.2 that the voltage at all buses is stable at 415 V. There is no under-voltage or over-voltage situation. There is also no overloading of converters.

11.2.1.2 Operating Scenario 2

In the second scenario, the load flow analysis was performed with the large 30-kVA load connected to the main bus. This is to verify whether the bus voltage levels are maintained within the limits when there is a sudden increase in load demand during peak load hours in between 12 pm and 2 pm. The ETAP model during load flow is shown in Figure 11.4, and the load flow report is presented in Table 11.3.

From the load flow report in Table 11.3, it can be noticed that the voltages at all buses are constant at 415 V, so the backup generator is not put into service. However, both inverter 1 and inverter 2 get overloaded as indicated in Figure 11.4.

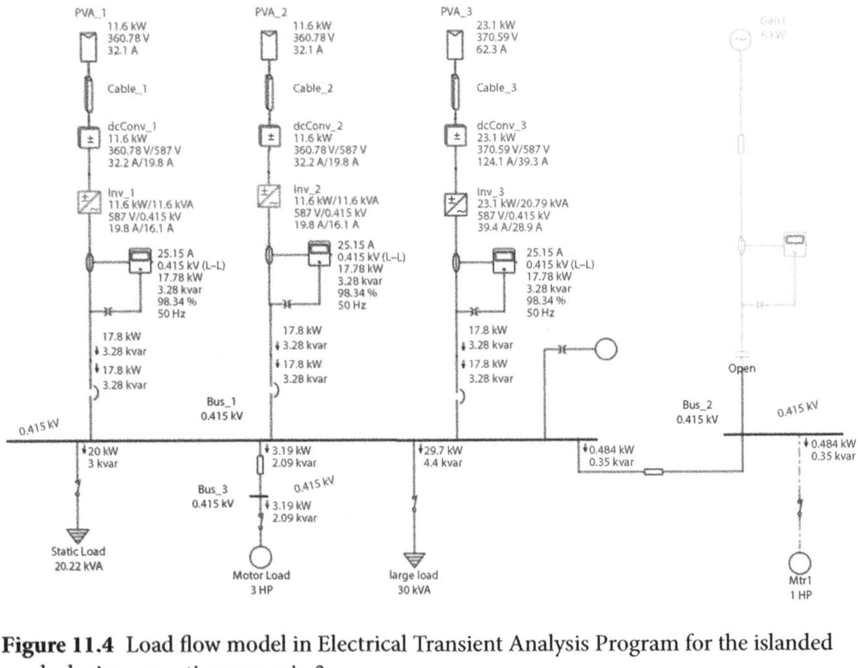

Figure 11.4 Load flow model in Electrical Transient Analysis Program for the islanded mode during operating scenario 2.

Table 11.3 Load flow report operating scenario 2 in islanded mode.

Bus_ID	Nominal KV	Volt (kV)	kW loading	kvar loading	Amp loading
Bus_1	0.415	0.415	53.33	9.83	75.44
Bus_2	0.415	0.415	0.484	0.35	0.832
Bus_3	0.415	0.415	3.19	2.09	5.301
Bus_4	0.415	0.415	17.88	3.28	25.15
Bus_5	0.415	0.415	17.88	3.28	25.15
Bus_6	0.415	0.415	17.88	3.28	25.15

11.3 Harmonic Load Flow Analysis in Islanded Mode

Harmonic contamination is a major concern in microgrid systems because of the usage of power converters. The power converters employed for power conversion are non-linear in nature and introduce harmonics into the system. Harmonics introduce distortions in bus voltages and line currents and thereby deteriorate the power quality [3, 4]. There are various indexes to evaluate the extent of harmonic contamination in a system, out of which total harmonic distortion (THD) is a substantial index. THD is a harmonic index denoting the content of harmonics in a distorted waveform. It is defined as the ratio of the root mean square (RMS) value of the harmonic components to the RMS value of fundamental component and is given by (11.1):

$$THD = \frac{\sqrt{\sum_{n=2}^{\infty} V_n^2}}{V_1} \tag{11.1}$$

where V_n is the RMS value of the nth harmonic component, and V_1 is the RMS value of the fundamental component of the signal. Similarly, the THD value for current can be evaluated by (11.2):

$$THD = \frac{\sqrt{\sum_{n=2}^{\infty} I_n^2}}{I_1} \tag{11.2}$$

The harmonics have a detrimental effect on the performance of an electrical power system. It causes overheating of the microgrid equipment, increased losses, increase in current, poor power factor, etc. The THD should be below 5% as per the standards prescribed by IEEE. Hence, harmonics analysis is mandatory before setting up a microgrid. In this study, harmonic load flow analysis is performed for the standalone microgrid system designed in ETAP with the large load switched on during peak demand period. The harmonic analysis is accomplished as per the IEEE 519 standards in ETAP. The harmonic load flow report generated by ETAP is presented. The voltage harmonics bus information is given in Table 11.4. It can be observed from Table 11.4 that the THD does not exceed the limit, but the individual harmonic distortion of each bus surpasses the limit as indicated by the # symbol in the ETAP report. The bus voltage waveform is shown in Figure 11.5. The voltage harmonic spectrum for each bus is demonstrated in Figure 11.6. The voltage harmonics are within limits. The current harmonic information is given in Table 11.5. It can be seen that the current THD limits are not violated.

Table 11.4 Voltage harmonics bus information in islanded mode.

Bus		Voltage distortion								
ID	kV	Fund. %	RMS	ASUM	THD %	TIF	TIHD %	TSHD %	THDG %	THDS %
# Bus_1	0.415	100.00	100.03	103.46	2.31	14.77	0.00	0.00	2.31	2.31
# Bus_2	0.415	99.99	100.01	103.45	2.31	14.76	0.00	0.00	2.31	2.31
# Bus_3	0.415	99.99	100.01	103.45	2.31	14.77	0.00	0.00	2.31	2.31
# Bus_4	0.415	100.00	100.03	103.46	2.31	14.77	0.00	0.00	2.31	2.31
# Bus_5	0.415	100.00	100.03	103.46	2.31	14.77	0.00	0.00	2.31	2.31
# Bus_6	0.415	100.00	100.03	103.46	2.31	14.77	0.00	0.00	2.31	2.31

* Indicates THD (Total harmonic distortion) exceeds the limit.
# Indicates IHD (Individual harmonic distortion) exceeds the limit.

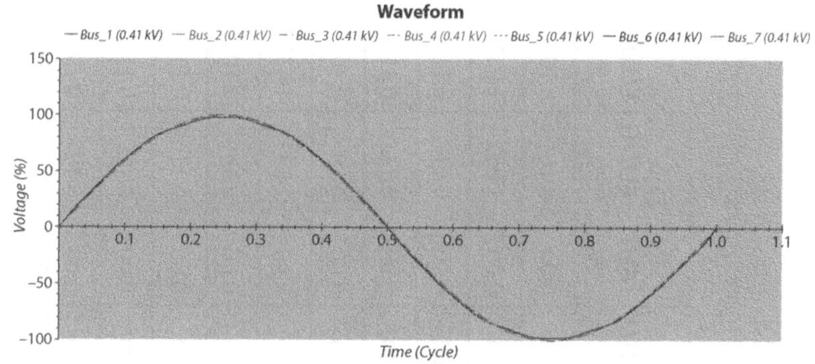

Figure 11.5 Voltage waveform in islanded mode.

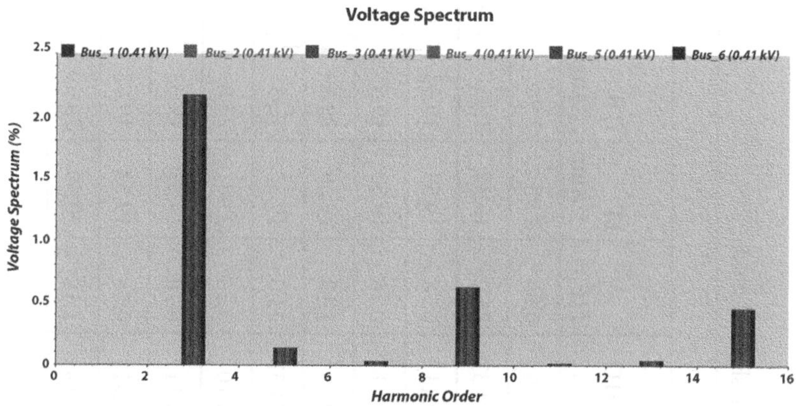

Figure 11.6 Voltage harmonic spectrum in islanded mode.

11.4 Transient Analysis of a Microgrid System in Islanded Mode

The microgrid system is susceptible to various transient conditions such as faults, sudden increase in peak demand, loss of DG units, etc. The most common types of faults occurring in a power system are the unsymmetrical faults which are line-to-ground (LG) faults, line-to-line (LL) faults, and double-line-to-ground (LLG) faults. A three-phase fault is a symmetrical fault where all the three phases carry the same fault current. Faults are unpredictable and undesirable; hence, the design of the circuit breaker is crucial for any power system. The different fault conditions can be effortlessly simulated, and a short circuit analysis report recording the voltage

Table 11.5 Current harmonics bus information in islanded mode.

Bus		Current distortion											
From bus ID	To bus ID	Fund. Amp	RMS Amp	ASUM Amp	THD %	TIF	IT Amp	ITB Amp	ITR Amp	TIHD %	TSHD %	THDG %	THDS %
Bus_1	Bus_3	5.30	5.30	5.31	0.10	0.56	2.98	2.98	0.00	0.00	0.00	0.10	0.10
	Bus_2	0.83	0.84	0.97	14.37	23.05	19.37	1.33	19.32	0.00	0.00	14.37	14.37
	Bus_4	25.15	25.15	25.19	0.11	0.78	19.53	19.53	0.00	0.00	0.00	0.11	0.11
	Bus_5	25.15	25.15	25.19	0.11	0.78	19.53	19.53	0.00	0.00	0.00	0.11	0.11
	Bus_6	25.15	25.15	25.22	0.19	1.18	29.59	29.59	0.00	0.00	0.00	0.19	0.19
Bus_2	Bus_1	0.83	0.84	0.97	14.37	23.05	19.37	1.33	19.33	0.00	0.00	14.37	14.37
Bus_3	Bus_1	5.30	5.30	5.31	0.10	0.56	2.98	2.98	0.00	0.00	0.00	0.10	0.10
Bus_4	Bus_1	25.15	25.15	25.19	0.11	0.78	19.53	19.53	0.00	0.00	0.00	0.11	0.11
Bus_5	Bus_1	25.15	25.15	25.19	0.11	0.78	19.53	19.53	0.00	0.00	0.00	0.11	0.11
Bus_6	Bus_1	25.15	25.15	25.22	0.19	1.18	29.59	29.59	0.00	0.00	0.00	0.19	0.19

magnitude at each bus and the fault current in each line can be generated in ETAP.

The short circuit calculation report can be utilized for designing switchgear, bus bar, transformers, cable sizing, and the selection of a circuit breaker. The methodology to calculate the equipment rating is based on IEC 60909 standards and is described in [17].

11.4.1 Fault at Main Bus

The main bus in the microgrid model is bus 1 as shown in Figure 11.3. Different faults are created at the main bus, and the phase voltages at different buses and the fault currents from the buses are tabulated. A line-to-ground fault is created at bus 1. The voltage magnitude and angles during fault is tabulated in Table 11.6. The fault currents in the three phases are tabulated in Table 11.7. The voltage profile and current profile during LG fault is presented in Figures 11.7 and 11.8 respectively. Similar short circuit reports can also be generated for other unsymmetrical faults such as line-to-line (LL) and double-line-to-ground (LLG) faults.

11.4.2 Three-Phase Fault at Main Bus

A three-phase fault is created at the main bus. The voltage magnitude and the symmetrical current (in kA) during fault is tabulated in Table 11.8. The severe drops in voltage levels due to the fault is highlighted in Table 11.8. The voltage profile during the three-phase fault is shown in Figure 11.9.

Table 11.6 Bus voltage during line-to-ground fault.

		% voltage at from bus					
		V_a		V_b		V_c	
From bus	**To bus**	**Mag**	**Ang**	**Mag**	**Ang**	**Mag**	**Ang**
Bus 1	Total	0.00	0.0	105.39	-123.0	114.25	120.2
Bus 3	Bus 1	0.04	-46.5	105.40	-123.0	114.28	120.2
Bus 2	Bus 1	0.36	-50.5	105.47	-123.0	114.31	120.2
Inv 3	Bus 6	100.00	0.0	100.00	-120.0	100.00	120.0
Inv 2	Bus 5	100.00	0.0	100.00	-120.0	100.00	120.0
Inv 1	Bus 4	100.00	0.0	100.00	-120.0	100.00	120.0

Table 11.7 Bus current during line-to-ground fault.

		Current at from bus (kA)								
From bus	To bus	I_a		I_b		I_c		Sequence current (kA)		
		Mag.	Ang.	Mag.	Ang.	Mag.	Ang.	I1	I2	I0
Bus 1	Total	0.121	-13.6	0.000	0.0	0.000	0.0	0.040	0.040	0.040
Bus 3	Bus1	0.015	-53.3	0.006	149.2	0.010	114.3	0.009	0.006	0.000
Bus 2	Bus 1	0.020	-53.7	0.005	-115.6	0.004	140.7	0.008	0.006	0.007
Inv 3	Bus 6	0.046	0.0	0.003	13.7	0.006	-58.6	0.012	0.016	0.018
Inv 2	Bus 5	0.025	0.0	0.002	13.7	0.004	-58.6	0.007	0.009	0.010
Inv 1	Bus 4	0.025	0.0	0.002	13.7	0.004	-58.6	0.007	0.009	0.010

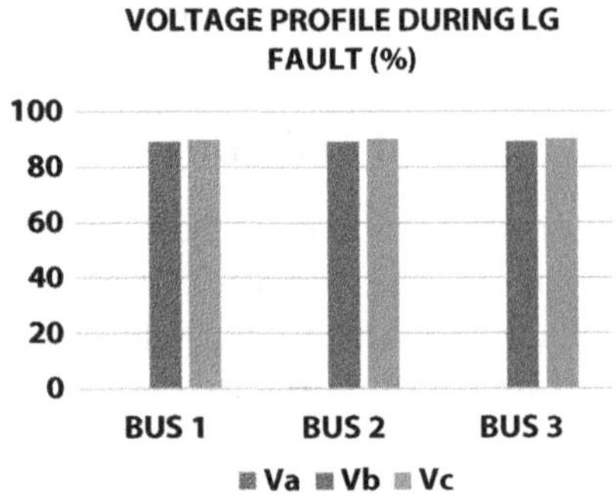

Figure 11.7 Voltage profile during line-to-ground fault.

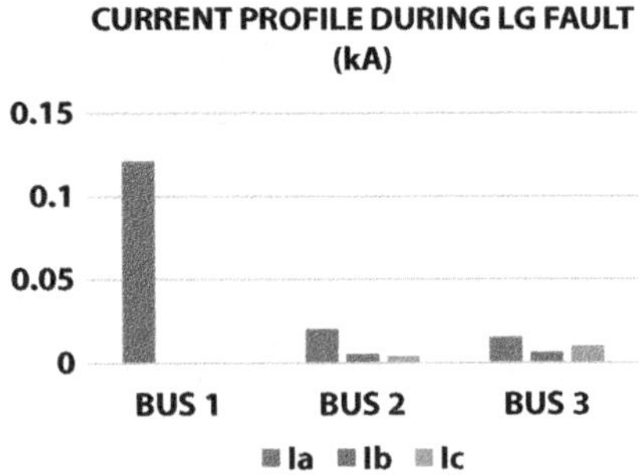

Figure 11.8 Current profile during line-to-ground fault.

Table 11.8 Bus voltage and current during the three-phase fault.

Contribution		Three-phase fault	
From bus	**To bus**	**%V from bus**	**kA symm. rms**
Bus 1	Total	0.00	0.150
Bus 3	Bus 1	**0.09**	0.033
Bus 2	Bus 1	**0.56**	0.031
Inv 3	Bus 6	100.00	0.046
Inv 2	Bus 5	100.00	0.025
Inv 1	Bus 4	100.00	0.025

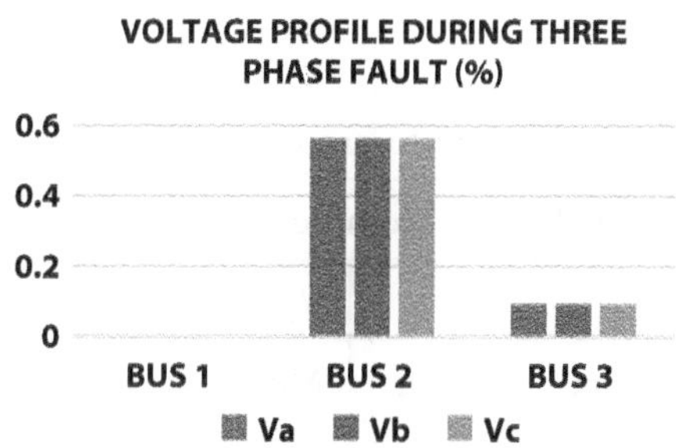

Figure 11.9 Voltage profile during the three-phase fault.

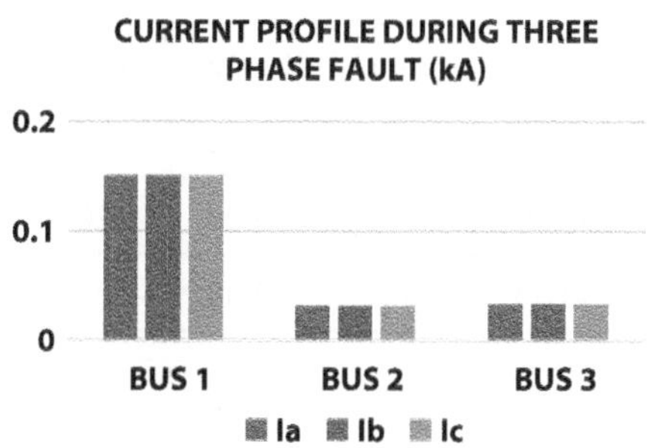

Figure 11.10 Current profile during the three-phase fault.

The current profile during the three-phase fault is illustrated in Figure 11.10. It can be observed from the short circuit analysis report that the severity of three-phase faults is more than the unsymmetrical faults.

11.4.3 Fault at Bus 3 Connected to Motor Load

A line-to-ground fault is created at bus 3, that is the bus where a 3-HP motor load is connected as interpreted from Figure 11.3. This scenario describes the case where there is fault at the load side, and accordingly, the breaker can be designed by considering the short circuit analysis report. The bus voltage during the LG fault is tabulated in Table 11.9, and the current during fault is tabulated in Table 11.10. The voltage profile and the current profile during LG fault at bus 3 is illustrated in Figures 11.11 and 11.12, respectively.

Table 11.9 Bus voltage during line-to-ground fault at bus 3.

		% voltage at from bus					
		V_a		**V_b**		**V_c**	
From bus	**To bus**	**Mag**	**Ang**	**Mag**	**Ang**	**Mag**	**Ang**
Bus 3	Total	0.00	0.0	105.38	-123.0	114.27	120.2
Bus 1	Bus 3	0.28	-1.6	105.38	-123.0	114.24	120.2
Motor load	Bus 3	100.00	0.0	100.00	-120.0	100.00	120.0
Bus 2	Bus 1	0.59	-29.5	105.46	-123.0	114.30	120.2
Inv 3	Bus 6	100.00	0.0	100.00	-120.0	100.00	120.0
Inv 2	Bus 5	100.00	0.0	100.00	-120.0	100.00	120.0
Inv 1	Bus 4	100.00	0.0	100.00	-120.0	100.00	120.0

Table 11.10 Bus current during LG fault at bus 3.

From bus	To bus	Current at from bus (kA)						Sequence current (kA)		
		I_a		I_b		I_c				
		Mag	Ang	Mag	Ang	Mag	Ang	I_1	I_2	I_0
Bus 3	Total	0.120	-13.6	0.000	0.0	0.000	0.0	0.040	0.040	0.040
Bus 1	Bus 3	0.109	-8.5	0.006	-30.8	0.010	-65.8	0.033	0.036	0.040
Motor load	Bus 3	0.015	-53.4	0.006	149.2	0.010	114.2	0.009	0.006	0.000
Bus 2	Bus 1	0.020	-53.7	0.005	-115.6	0.004	140.8	0.008	0.006	0.007
Inv 3	Bus 6	0.045	0.0	0.003	13.6	0.006	-58.6	0.012	0.016	0.017
Inv 2	Bus 5	0.025	0.0	0.002	13.6	0.004	-58.6	0.007	0.009	0.010
Inv 1	Bus 4	0.025	0.0	0.002	13.6	0.004	-58.6	0.007	0.009	0.010

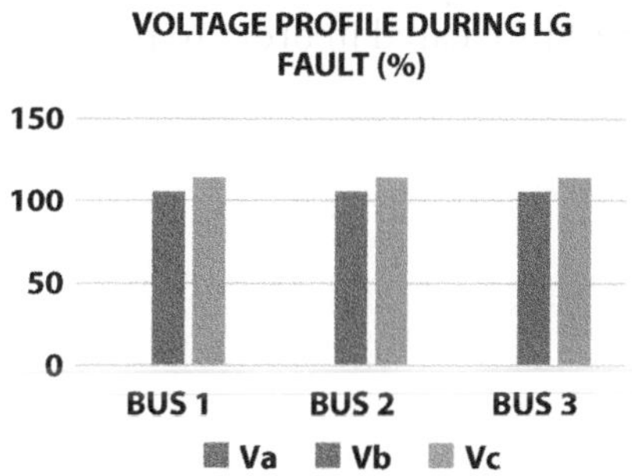

Figure 11.11 Voltage profile during line-to-ground fault at bus 3.

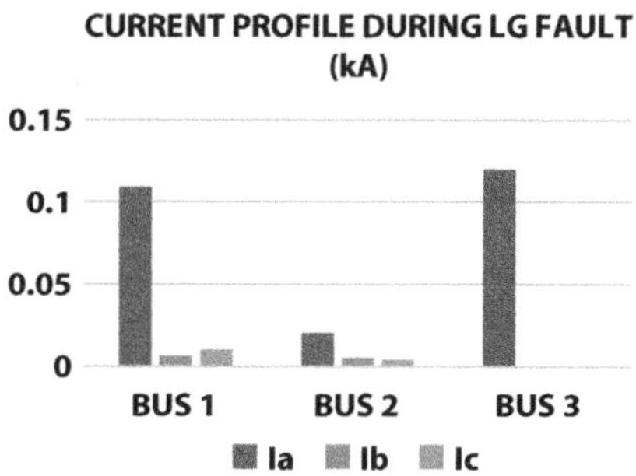

Figure 11.12 Current profile during line-to-ground fault at bus 3.

11.4.4 Loss of One PV Generator in Islanded Mode Operation

In a microgrid, there are operating scenarios where the DG may be lost or the power generation may cease from the DG plant. Such a case is being illustrated here where the PV1 unit generating 11.6 kW of power is put out of service, and the diesel generator is turned on and load flow analysis is performed to identify the weak buses. The load flow analysis model in ETAP with PV 1 unit put out of service is shown in Figure 11.13. The load flow report generated is shown in Table 11.11.

From the load flow analysis model illustrated in Figure 11.13, it can be observed that PV1 unit is out of service and the inverter 2 is overloaded, even though the backup diesel generator is brought into service. From the load flow report, it can be perceived that the voltage at all the buses is stable, and no alert level is being raised but inverter is overloaded.

11.4.5 Critical Clearing Time and Critical Clearing Angle

Critical clearing time (CCT) and critical clearing angle (CCA) are significant factors for transient analysis. The CCT is the maximum time interval by which the fault must be cleared in order to preserve the system stability.

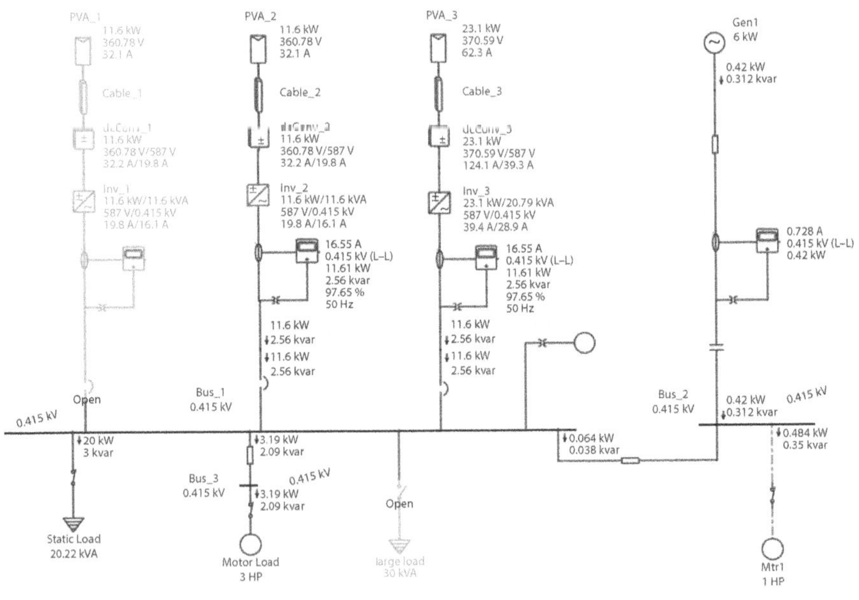

Figure 11.13 Load flow analysis model in electrical transient analysis program with PV1 put out of service in the islanded mode.

Table 11.11 Load flow report during loss of PV 1 unit in islanded mode.

Bus_ID	Nominal KV	Type	Volt (kV)	kW loading	kvar loading	Amp loading
Bus_1	0.415	Load	0.415	23.65	5.12	33.09
Bus_2	0.415	Load	0.415	0.484	0.35	0.832
Bus_3	0.415	Load	0.415	3.19	2.09	5.301
Bus_5	0.415	Swing	0.415	11.61	2.56	16.55
Bus_6	0.415	Swing	0.415	11.61	2.56	16.55
Bus_7	0.415	Swing	0.415	0.42	0.312	0.728

The critical clearing angle is the maximum permissible change in the power angle before clearing the fault without loss of synchronism. Both CCT and CCA impact the microgrid transient performance adversely, especially in islanded mode [21].

The conventional formula to calculate the CCT is given by (11.3):

$$t_{cr} = \sqrt{\frac{4H}{2\pi f}(\delta_{cr} - \delta_0)} \tag{11.3}$$

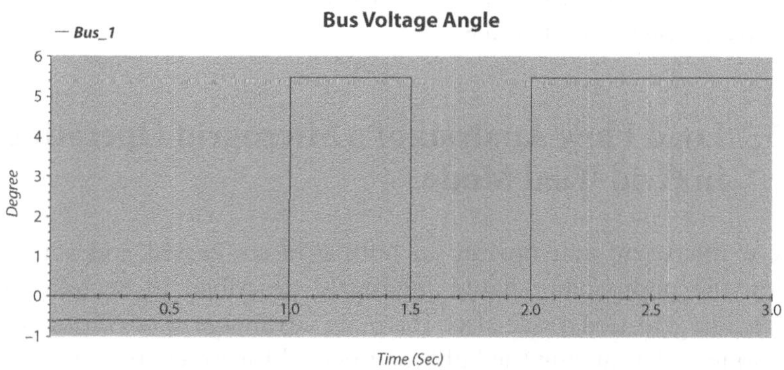

Figure 11.14 Critical clearing time and critical clearing angle for the three-phase fault created at the main bus.

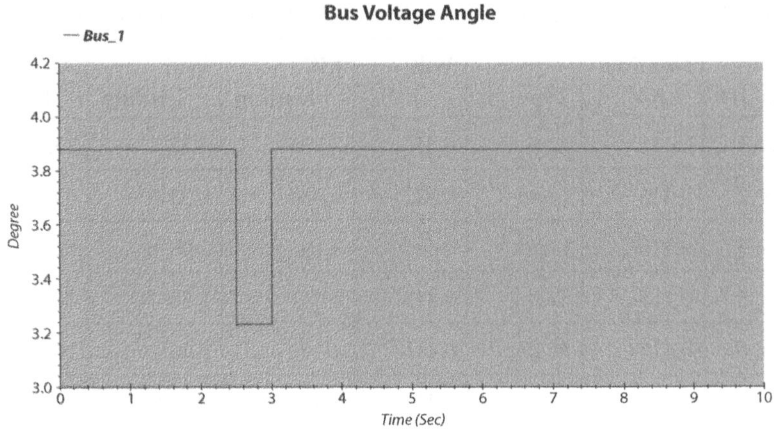

Figure 11.15 Critical clearing time and critical clearing angle for line-to-ground fault created at the main bus.

where H is the inertia constant of the synchronous generator, f is the rated frequency of the system, δ_{cr} is the critical clearing angle, and δ_0 is the initial power angle.

In a microgrid, the DG plants affect the system stability since the DG plants are the generation units apart from the synchronous generators. In Figure 11.14, it can be clearly observed that the three-phase fault is created at 1.00 s, and it is cleared at 1.5 s. The critical clearing time is found to be 0.5 s. The critical clearing angle is found to be 5.4°.

Similarly, in Figure 11.15, it can be observed that the LG fault is created at main bus at 2.5 s and cleared at 3 s. The critical clearing time is 0.5 s, and the critical clearing angle is 3.88°.

11.5 Load Flow Analysis of a Microgrid Operating in Grid-Tied Mode

Since a microgrid can operate in both grid-connected and standalone modes, the photovoltaic-based microgrid described in Section 11.2 is operated in grid-tied mode also. The main advantage of operating in grid-tied mode is that during the light load period the excess power generated by the DG plant can be injected into the grid. During peak load periods or during a sudden increase in load demands when the inverters get overloaded, power can be drawn from the grid. The power generation capacity of the microgrid is decided by taking into account the same load profile of

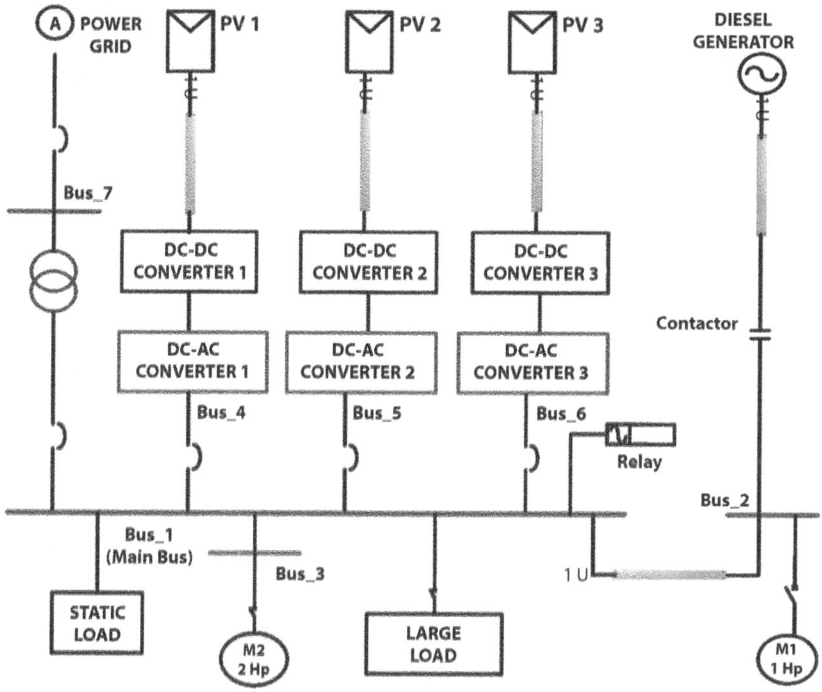

Figure 11.16 Block diagram of grid-tied microgrid.

the school campus as shown in Figure 11.2. The grid-connected model of the microgrid is shown in Figure 11.16. The components of the microgrid are the same as that of the system described in Figure 11.1, but now it is tied to a three-phase 11-kV grid. The grid is tied to 415-V bus through a 6-MVA step down transformer. The parameters of the PV panels, the backup diesel generator, loads, and converters are the same.

Next, load flow analysis was performed again under two operating scenarios, and the load flow report generated was demonstrated to identify the feeble buses.

11.5.1 Operating Scenario 1

The load flow analysis was performed with the static load and all motor load connected but with the large load not connected. All the PV panels are generating power at their maximum capacity, and the diesel generator was turned off. The simulation model in the ETAP platform is shown in Figure 11.17. The load flow report is depicted in Table 11.12. In this case, in addition to six buses, an extra bus 7 is included in the load flow report, which is the 11-kV bus connected to the grid. From the load flow report generated, it can be noted that all the buses are stable, and no alert level was generated.

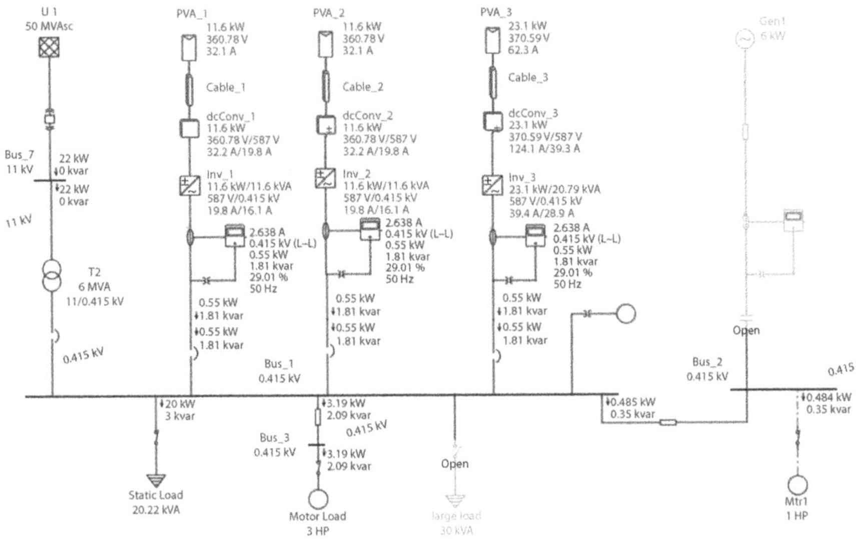

Figure 11.17 Load flow model in electrical transient analysis program for grid-tied mode during operating scenario 1.

Table 11.12 Load flow table in grid-tied mode during operating scenario 1.

Bus_ID	Nominal KV	Type	Volt (kV)	kW loading	kvar loading	Amp loading
Bus_1	0.415	Load	0.415	23.65	5.44	33.76
Bus_2	0.415	Load	0.415	0.484	0.35	0.832
Bus_3	0.415	Load	0.415	3.19	2.09	5.301
Bus_4	0.415	Swing	0.415	0.55	1.81	2.639
Bus_5	0.415	Swing	0.415	0.55	1.81	2.639
Bus_6	0.415	Swing	0.415	0.55	1.81	2.639
Bus_7	11	Gen	11	22	0.0001	1.155

11.5.2 Operating Scenario 2

During this situation, the 30-kVA large load is connected along with the static loads and motor loads in the grid-integrated mode. This depicts the peak demand periods of the school campus. The ETAP model during load flow is shown in Figure 11.18, where it can be observed that the backup

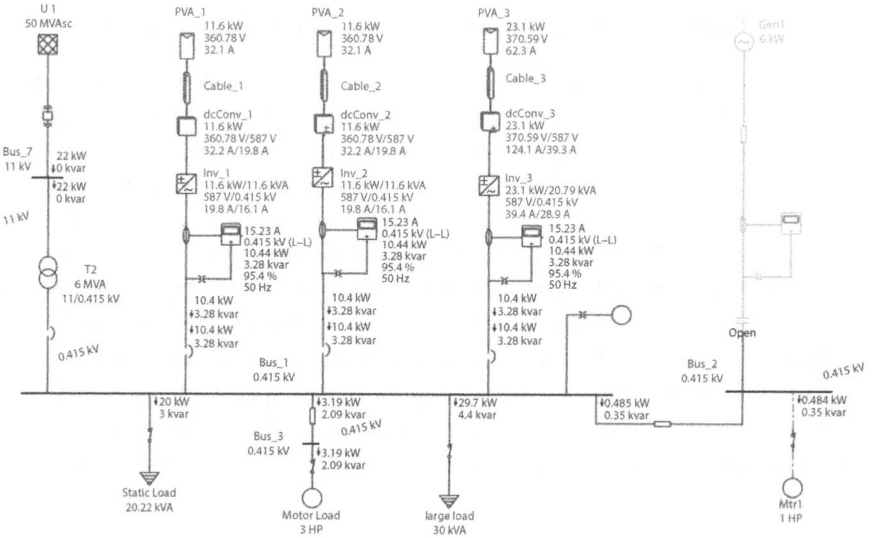

Figure 11.18 Load flow model in electrical transient analysis program for grid-tied mode during operating scenario 2.

Table 11.13 Load flow table in grid-tied mode during operating scenario 2.

Bus_ID	Nominal KV	Type	Volt (kV)	kW loading	kvar loading	Amp loading
Bus_1	0.415	Load	0.415	53.33	9.84	75.44
Bus_2	0.415	Load	0.415	0.484	0.35	0.832
Bus_3	0.415	Load	0.415	3.19	2.09	5.301
Bus_4	0.415	Swing	0.415	10.44	3.28	15.23
Bus_5	0.415	Swing	0.415	10.44	3.28	15.23
Bus_6	0.415	Swing	0.415	10.44	3.28	15.23
Bus_7	11	Gen	11	22	0.0001	1.155

generator was also not put into service. The load flow report is presented in Table 11.13. It can be observed from the generated load flow report that during the peak demand period also the bus voltages are within limits. The inverters are also not overloaded as can be witnessed from Figure 11.18. This is one of the major advantages of operating the microgrid in grid-tied mode.

11.5.3 Harmonic Load Flow Analysis in Grid-Tied Mode

Harmonic load flow analysis was performed for the grid-integrated mode operation of the microgrid system designed in ETAP. During harmonic load flow the large load switched on during the peak demand period as it was executed for a standalone mode. The harmonic load flow report generated by ETAP are presented. The voltage harmonics percentage in each bus is given in Table 11.14. It can be observed from the table that both the THD levels and the IHD levels are within limits. The voltage waveform in the grid-tied mode is shown in Figure 11.19. The 11-kV bus, i.e., bus 7, voltage and the other bus' voltage waveforms are almost sinusoidal as the distortion is very minimal. The voltage harmonic spectrum for each bus is demonstrated in Figure 11.20. The current harmonics information in each bus is shown in Table 11.15. The current THD and IHD values are also not violated.

Table 11.14 Voltage harmonics bus information in grid-tied mode.

Bus		Voltage distortion								
ID	kV	Fund. %	RMS	ASUM	THD %	TIF	TIHD %	TSHD %	THDG %	THDS %
Bus_1	0.415	100.00	100.00	100.01	0	0.42	0.00	0.00	0.00	0.00
Bus_2	0.415	99.99	99.99	99.99	0.01	0.42	0.00	0.00	0.00	0.00
Bus_3	0.415	99.99	99.99	99.99	0	0.42	0.00	0.00	0.00	0.00
Bus_4	0.415	100.00	100.00	100.01	0	0.42	0.00	0.00	0.00	0.00
Bus_5	0.415	100.00	100.00	100.01	0	0.42	0.00	0.00	0.00	0.00
Bus_6	0.415	100.00	100.00	100.01	0	0.42	0.00	0.00	0.00	0.00
Bus_7	11.000	100.00	100.00	100.00	0	0.42	0.00	0.00	0.00	0.00

* Indicates THD (Total harmonic distortion) exceeds the limit.
# Indicates IHD (Individual harmonic distortion) exceeds the limit.

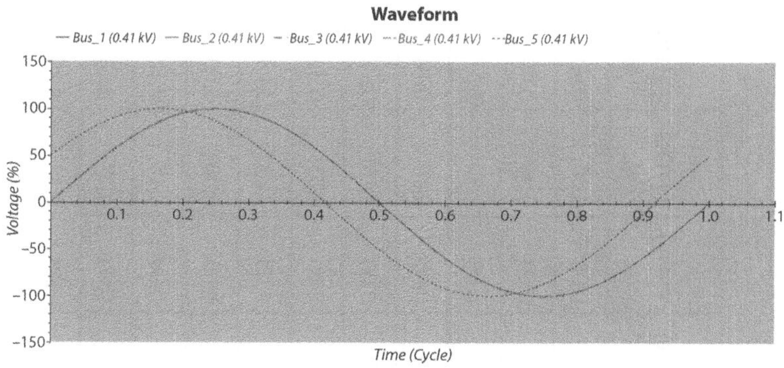

Figure 11.19 Voltage waveform in grid-tied mode.

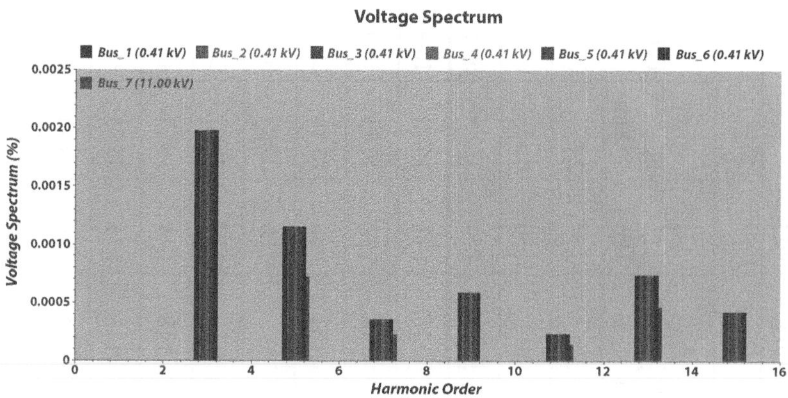

Figure 11.20 Voltage harmonic spectrum in grid-tied mode.

11.6 Transient Analysis of Microgrid System in Grid-Tied Mode

Similar to Section 11.4, a short circuit analysis is performed on the grid-connected microgrid model using ETAP. A short circuit analysis during LG fault and symmetrical three-phase fault is performed, and a short circuit report is generated. This report is an essential tool for designing microgrid equipment.

Table 11.15 Current harmonics bus information in grid-tied mode.

Bus		Current distortion											
From bus ID	To bus ID	Fund. Amp	RMS Amp	ASUM Amp	THD %	TIF	IT Amp	ITB Amp	ITR Amp	TIHD %	TSHD %	THDG %	THDS %
Bus_1	Bus_3	5.30	5.30	5.30	0	0.50	2.65	0.00	0.00	0.00	0.00	0.00	0.00
	Bus_2	0.83	0.84	0.97	0	0.50	0.42	0.42	0.02	0.00	0.00	0.00	0.00
	Bus_7	30.61	30.62	31.65	2.57	4.34	132.97	59.13	119.10	0.00	0.00	2.57	2.57
	Bus_4	15.23	15.23	15.27	0.18	1.11	16.93	16.93	0.00	0.00	0.00	0.18	0.18
	Bus_5	15.23	15.23	15.27	0.18	1.11	16.93	16.93	0.00	0.00	0.00	0.18	0.18
	Bus_6	15.23	15.23	15.30	0.33	1.85	28.14	28.14	0.00	0.00	0.00	0.33	0.33
Bus_2	Bus_1	0.83	0.83	0.83	0	0.50	0.42	0.42	0.02	0.00	0.00	14.37	14.37
Bus_3	Bus_1	5.30	5.30	5.30	0	0.50	2.65	2.65	0.00	0.00	0.00	0.00	0.00
Bus_4	Bus_1	15.23	15.23	15.27	0.18	1.11	16.93	16.93	0.00	0.00	0.00	0.18	0.18
Bus_5	Bus_1	15.23	15.23	15.27	0.18	1.11	16.93	16.93	0.00	0.00	0.00	0.18	0.18
Bus_6	Bus_1	15.23	15.23	15.30	0.33	1.85	28.14	28.14	0.00	0.00	0.00	0.33	0.33
Bus_7	Bus_1	1.15	1.15	1.16	0.35	1.93	2.23	2.23	0.00	0.00	0.00	0.35	0.35

11.6.1 Fault at the Main Bus

A line-to-ground fault is created at bus 1. The voltage magnitude and angles during fault is tabulated in Table 11.16. The fault currents in the three phases are tabulated in Table 11.17. The voltage profile and the current profile during LG fault are presented in Figures 11.21 and 11.22, respectively.

Table 11.16 Bus voltage during line-to-ground fault.

		% voltage at from bus					
From bus	To bus	V_a		V_b		V_c	
		Mag	Ang	Mag	Ang	Mag	Ang
Bus 1	Total	0.00	0.0	88.96	-104.2	89.71	104.1
Bus 3	Bus 1	0.06	-48.4	88.96	-104.2	89.75	104.1
Bus 2	Bus 1	0.40	-48.1	89.01	-104.3	89.93	104.1
Bus 7	Bus 1	62.16	52.5	100.01	-90.0	63.31	126.8
Inv 3	Bus 6	100.00	0.0	100.00	-120.0	100.00	120.0
Inv 2	Bus 5	100.00	0.0	100.00	-120.0	100.00	120.0
Inv 1	Bus 4	100.00	0.0	100.00	-120.0	100.00	120.0

Table 11.17 Bus current during line-to-ground fault.

		Current at from bus (kA)								
From bus	To bus	I_a		I_b		I_c		Sequence current (kA)		
		Mag	Ang	Mag	Ang	Mag	Ang	I_1	I_2	I_0
Bus 1	Total	56.102	-84.2	0.000	0.0	0.000	0.0	18.701	18.701	18.701
Bus 3	Bus 1	0.022	-55.3	0.010	159.9	0.015	103.1	0.014	0.008	0.000
Bus 2	Bus 1	0.022	-49.6	0.007	-170.8	0.012	110.3	0.013	0.007	0.002
Bus 7	Bus 1	56.055	-84.4	0.043	-2.4	0.043	-36.6	18.676	18.682	18.697
Inv 3	Bus 6	0.046	0.0	0.013	-179.3	0.013	-179.2	0.019	0.019	0.007
Inv 2	Bus 5	0.025	0.0	0.007	-179.3	0.007	-179.2	0.011	0.011	0.004
Inv 1	Bus 4	0.025	0.0	0.007	-179.3	0.007	-179.2	0.011	0.011	0.004

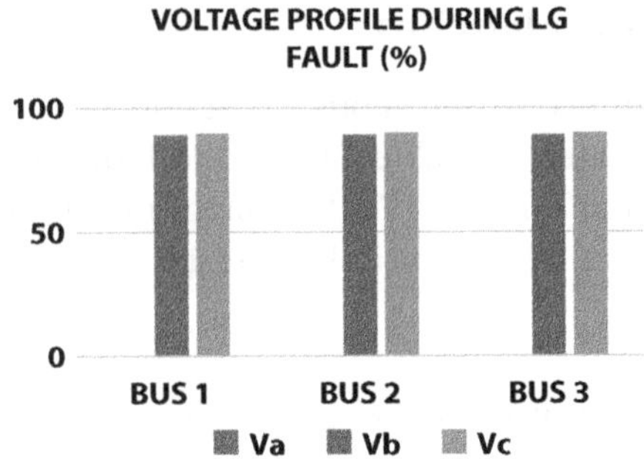

Figure 11.21 Voltage profile during line-to-ground fault at bus 1.

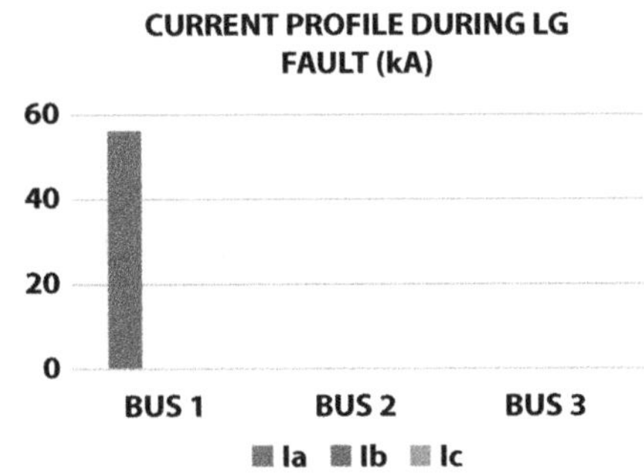

Figure 11.22 Current profile during line-to-ground fault at bus 1.

11.6.2 Three-Phase Fault at the Main Bus

A three-phase fault is created at the main bus. The voltage magnitude and the symmetrical current (in kA) during fault are tabulated in Table 11.18. The drop in voltage levels are also highlighted. The voltage profile during the three-phase fault is shown in Figure 11.23. The current profile during the three-phase fault is illustrated in Figure 11.24. It can be observed from the short circuit analysis report that even in the grid-tied mode there is a severe under-voltage during the three-phase symmetrical fault.

11.6.3 Fault at Bus 3 Connected to 3-HP Motor Load

A line-to-ground fault is created at bus 3, that is, at the motor load side. The voltage magnitude and angles during this fault are tabulated in Table 11.19,

Table 11.18 Bus voltage during three-phase fault at the main bus.

Contribution		Three-phase fault	
From bus	To bus	%V from bus	kA Symm. rms
Bus 1	Total	0.00	43.798
Bus 3	Bus 1	**0.09**	0.033
Bus 2	Bus 1	**0.56**	0.031
Bus 7	Bus 1	34.14	43.740
Inv 3	Bus 6	100.00	0.046
Inv 2	Bus 5	100.00	0.025
Inv 1	Bus 4	100.00	0.025

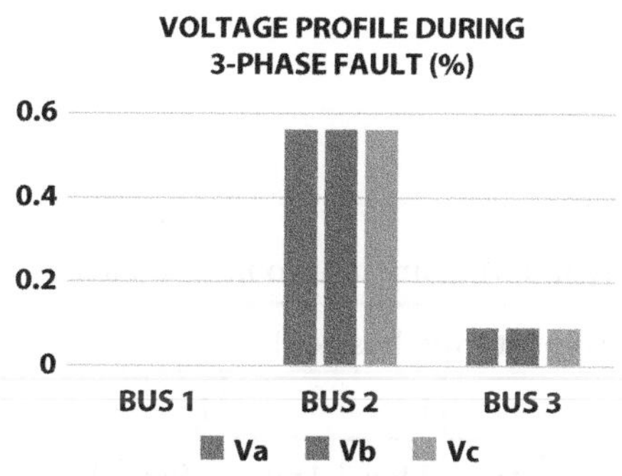

Figure 11.23 Voltage profile during three phase fault at the main bus.

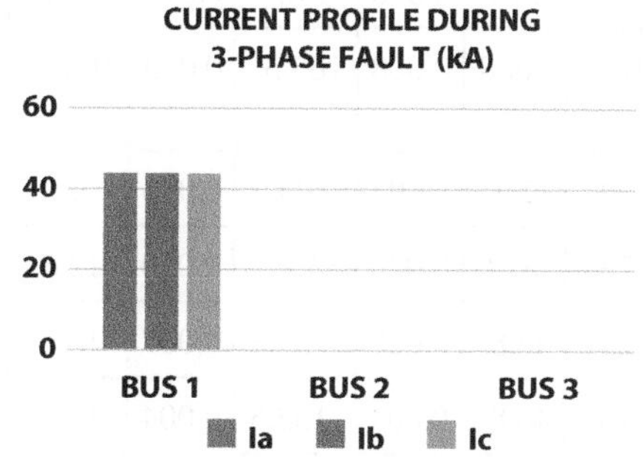

Figure 11.24 Current profile during three-phase fault at the main bus.

Table 11.19 Bus voltage during line-to-ground fault at bus 3.

		% voltage at from bus					
		V_a		V_b		V_c	
From bus	**To bus**	**Mag**	**Ang**	**Mag**	**Ang**	**Mag**	**Ang**
Bus 3	Total	0.00	0.0	85.74	-117.9	105.40	112.4
Bus 1	Bus 3	74.68	-30.6	85.73	-117.9	105.37	112.4
Motor load	Bus 3	100.00	0.0	100.00	-120.0	100.00	120.0
Bus 2	Bus 1	74.86	-30.5	85.80	-117.9	105.45	112.4
Bus 7	Bus 1	76.16	24.2	100.00	-90.0	97.76	135.3
Inv 3	Bus 6	100.00	0.0	100.00	-120.0	100.00	120.0
Inv 2	Bus 5	100.00	0.0	100.00	-120.0	100.00	120.0
Inv 1	Bus 4	100.00	0.0	100.00	-120.0	100.00	120.0

Table 11.20 Bus current during line-to-ground fault at bus 3.

		Current at from bus (kA)						Sequence current (kA)		
		I_a		I_b		I_c				
From bus	**To bus**	**Mag**	**Ang**	**Mag**	**Ang**	**Mag**	**Ang**	**I1**	**I2**	**I0**
Bus 3	Total	29.272	-37.5	0.000	0.0	0.000	0.0	9.757	9.757	9.757
Bus 1	Bus 3	29.253	-37.5	0.008	-14.7	0.013	-71.4	9.745	9.751	9.797
Motor load	Bus 3	0.019	-49.8	0.008	165.3	0.013	108.6	0.012	0.007	0.000
Bus 2	Bus 1	0.012	-2.9	0.004	-124.0	0.006	157.1	0.007	0.004	0.001
Bus 7	Bus 1	29.239	-37.6	0.023	30.2	0.020	-12.6	9.738	9.745	9.755
Inv 3	Bus 6	0.024	46.8	0.007	-132.5	0.007	-132.4	0.010	0.010	0.003
Inv 2	Bus 5	0.013	46.8	0.004	-132.5	0.004	-132.4	0.006	0.006	0.002
Inv 1	Bus 4	0.013	46.8	0.004	-132.5	0.004	-132.4	0.006	0.006	0.002

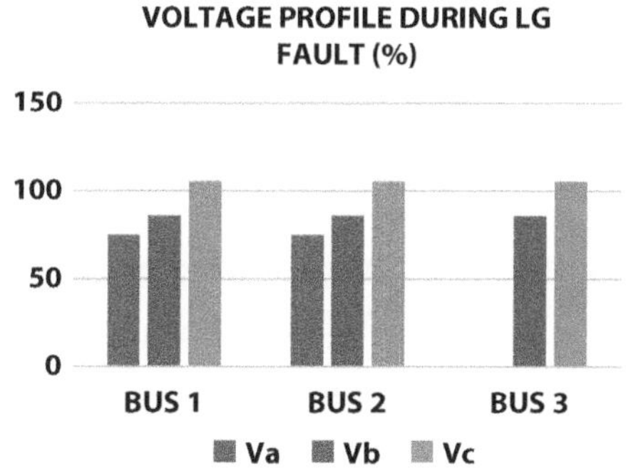

Figure 11.25 Voltage profile during line-to-ground fault at bus 3 in grid-tied mode.

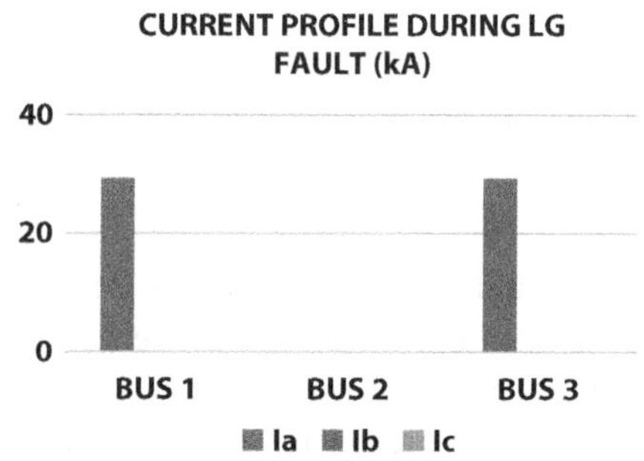

Figure 11.26 Current profile during line-to-ground fault at bus 3 in grid-tied mode.

and the current during the LG fault is tabulated in Table 11.20. The voltage profile and current profile during fault are shown in Figures 11.25 and 11.26, respectively.

11.6.4 Loss of One PV Generator in Grid-Tied Mode Operation

The operating scenario where one of the DG unit is lost in the grid-connected mode is simulated in ETAP platform. This condition illustrates the scenario where the solar PV plant may cease to generate power during night time or has to be disconnected due to malfunctioning of the unit. The microgrid simulation model in ETAP is illustrated in Figure 11.27. The load flow report is presented in Table 11.21. The load flow report aids in the identification of buses which violate the upper and lower voltage limits.

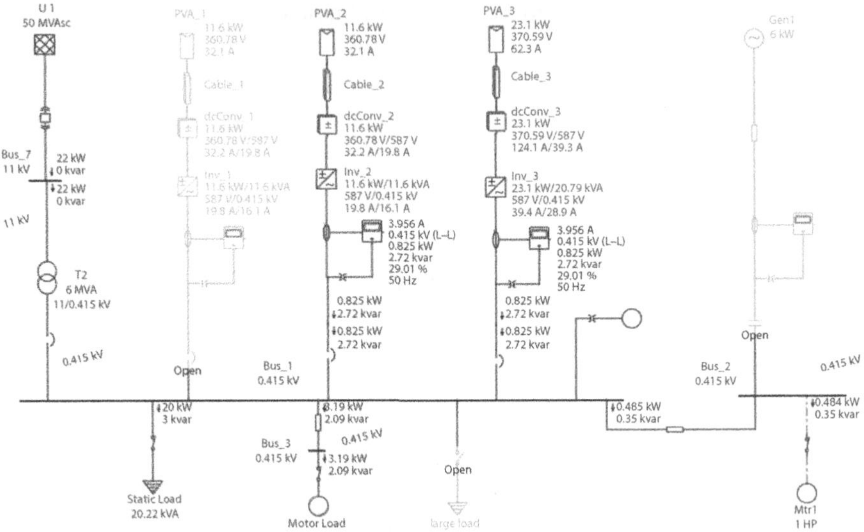

Figure 11.27 Load flow model in electrical transient analysis program during loss of PV 1 in grid tied mode.

It can be observed from Table 11.21 that the bus voltage is maintained constant during load flow, and also there is no overloading of inverters even without the diesel generator being put in service. This is one of the major advantages of operating the microgrid in grid-integrated mode. During loss of any generation unit, the designed microgrid system can meet the load demands without overloading its components. This is validated by the generated load flow report.

Table 11.21 Load flow table during loss PV 1 in grid-tied mode.

Bus_ID	Nominal KV	Type	Volt (kV)	kW loading	kvar loading	Amp loading
Bus_1	0.415	Load	0.415	23.65	5.44	33.76
Bus_2	0.415	Load	0.415	0.484	0.35	0.832
Bus_3	0.415	Load	0.415	3.19	2.09	5.301
Bus_5	0.415	Swing	0.415	0.825	2.72	3.956
Bus_6	0.415	Swing	0.415	0.825	2.72	3.956
Bus_7	11	Gen	11	22	0.0001	1.155

Table 11.22 Summarized comparison of reports generated from electrical transient analysis program.

ETAP reports	Islanded mode operation	Grid-tied mode
Load flow report		
(a) During normal loading	Voltage at all buses within limits and no weak buses identified	Voltage at all buses within limits and no weak buses identified
(b) During peak demand periods	Voltage at all buses within limits but inverters are overloaded	Voltage at all buses within limits inverters are not overloaded
(c) Loss of one PV generator	Inverters are overloaded and the backup generator is put into service	Inverters are not overloaded and the back-up generator is also not in service.
(d) Harmonic load flow during peak demand periods	THD at all the buses are maintained within limits but individual harmonic distortion (IHD) at each bus exceeds its limit	Both THD and individual harmonic distortion (IHD) at all the buses are maintained within limits
Short circuit analysis		
(a) Unsymmetrical faults (LG fault)	Fault currents and voltage at each bus computed and reported	Fault currents and voltage at each bus computed and reported
(b) Three-phase symmetrical fault	Severe under voltage at bus reported	Severe under voltage at bus reported

11.7 Comparative Analysis of a Microgrid Operating in Islanded and Grid-Tied Mode

A summary is presented in Table 11.22 after analyzing and comparing the load flow and short circuit reports generated from ETAP for the microgrid operating in both grid-tied and islanded mode.

11.8 Conclusion

In this study, a solar photovoltaic-based microgrid system of 52.3 kW capacity is modeled and simulated using ETAP software to match the load profile of a rural area school campus. The microgrid is operated in both islanded and grid-tied mode. Load flow analysis, short circuit analysis, and harmonic analysis were performed for the microgrid system. The load flow, short circuit analysis, and harmonic analysis reports generated by the ETAP software are essential for the design of microgrid components and protection equipment such as the circuit breaker. Load flow analysis is accomplished to detect the buses which are at under-voltage or over-voltage, and the voltage alert levels are established accordingly. By comparing the generated load flow report, it can be concluded that a microgrid operating in islanded mode could not successfully meet the sudden increase in load demands during peak periods. On the other hand, a microgrid in grid-tied mode could cater to the peak demands without overloading the inverters. From the short circuit analysis report, it was observed that, among all types of faults, the three-phase symmetrical faults are very severe both in case of grid-tied and islanded microgrid. In case of loss of one of the PV generation units, the microgrid operating in grid-tied mode could satisfy the load demands without overburdening the inverters. However, the inverters were overloaded in the islanded microgrid during forfeiture of one of the PV generators even after the backup generator was brought to operation.

References

1. International Energy Outlook (IEO), 2021, https://www.eia.gov/outlooks/ieo/pdf/IEO2021_Narrative.pdf.
2. Adefarati, T. and Bansal, R.C., Integration of renewable distributed generators into the distribution system: A review. *IET Renew. Power Gener.*, 10, 7, 873–884, 2016.

3. Tomin, N., Shakirov, V., Kozlov, A., Sidorov, D., Kurbatsky, V., Rehtanz, C., Lora, E.E.S., Design and optimal energy management of community microgrids with flexible renewable energy sources. *Renew. Energy*, 183, 903–921, 2022.
4. Chatterjee, A. and Mohanty, K.B., Current control strategies for single phase grid integrated inverters for photovoltaic applications-a review. *Renew. Sustain. Energy Rev.*, 92, 554–569, 2018.
5. Chatterjee, A., Mohanty, K.B., Thakre, K., Kommukuri, V.S., Grid voltage sensorless control of single phase grid tied inverter for renewable energy systems applications. *Electr. Power Compon. Syst.*, 46, 16–17, 1795–1807, 2018.
6. Saeed, M.H., Fangzong, W., Kalwar, B.A., Iqbal, S., A review on microgrids challenges & perspectives. *IEEE Access*, 9, 166502–166517, 2018.
7. Kamel, A., Alaam, M.A., Azmy, A.M., Abdelaziz, A.Y., Protection coordination for distribution systems in presence of distributed generators. *Electr. Power Compon. Syst.*, 41, 15, 1555–1566, 2013.
8. George, S.P. and Sankar, A., Optimal settings for adaptive overcurrent relay coordination in grid-connected wind farms. *Electr. Power Compon. Syst.*, 48, 12–13, 1308–1326, 2020.
9. Ballal, M.S., Kulkarni, A.M., Suryawanshi, H.M., Methodology for the improvements in synchrophasor based System Integrity Protection Schemes under stressed conditions. *Sustain. Energy Grids Netw.*, 26, 100465, 2021.
10. Sharma, A. and Panigrahi, B.K., Phase fault protection scheme for reliable operation of microgrids. *IEEE Trans. Ind. Appl.*, 54, 3, 2646–2655, 2018.
11. Sharma, A. and Panigrahi, B.K., Interphase fault relaying scheme to mitigate sympathetic tripping in meshed distribution system. *IEEE Trans. Ind. Appl.*, 55, 1, 850–857, 2019.
12. Elbana, M.S., Abbasy, N., Meghed, A., Shaker, N., μPMU-based smart adaptive protection scheme for microgrids. *J. Mod. Power Syst. Clean Energy*, 7, 4, 887–898, 2019.
13. Joseph, A., Smedley, K., Mehraeen, S., Secure power distribution against reactive power control malfunction in DER units. *IEEE Trans. Power Delivery*, 36, 3, 1552–1561, 2021.
14. Kadukar, P.R., Shete, P.S., Gawande, S.P., Transient analysis of distributed generation AC microgrid using ETAP. *IEEE International Conference on Current Trends towards Converging Technologies*, Coimbatore, India, pp. 1–6, 2018.
15. Chen, X. and Zhang, G., Harmonic analysis of AC-DC hybrid microgrid based on ETAP. *IEEE 8th International Power Electronics and Motion Control Conference*, Hefei, China, pp. 1–5, 2016.
16. Gbadamosia, S.K., Nwulu, N.I., Sun, Y., Harmonic and power loss minimization in power systems incorporating renewable energy sources and locational marginal pricing. *J. Renew. Sustain. Energy*, 10, 055501, 2018.

17. Prabhu, J.A.X., Nande, K.S., Shukla, S., Ade, C.N., Design of electrical system based on short circuit study using ETAP for IEC projects. *IEEE 6th International Conference on Power Systems (ICPS)*, New Delhi, India, pp. 1–6, 2016.
18. Ullah, M.A., Qaiser, A., Saeed, Q., Abbasi, A.R., Ahmed, I., Soomro, A.Q., Load flow, voltage stability & short circuit analyses and remedies for a 1240 MW combined cycle power plant using ETAP. *International Conference on Innovations in Electrical Engineering and Computational Technologies (ICIEECT)*, Karachi, Pakistan, pp. 1–6, 2017.
19. Makola, C., Roux, P.L., Jordaan, J., Design, analysis, and operation of photovoltaic power in a microgrid with an EESS. *IEEE PES/IAS Power Africa*, Nairobi, Kenya, pp. 1–5, 2021.
20. Liu, G., Application of ETAP in distributed power supply and micro-grid interconnection. *4th International Conference on Intelligent Green Building and Smart Grid (IGBSG)*, Hubei, China, pp. 1–5, 2019.
21. Wang, Y., Ravishankar, J., Phung, T., A study on critical clearing time (CCT) of micro-grids under fault conditions. *Renew. Energy*, 95, 381–395, 2016.

12

EV Charging Infrastructure in Microgrid

K. Narasimhaiah Achari[1]* and Pallavi Gajbhiye[2]

[1]Department of E.E.E., Ashoka Women's Engineering College, Kurnool, A.P., India
[2]Department of Electrical Engineering, UIT, Barkatullah University, Bhopal, M.P., India

Abstract

The regular working hours of a commercial building can be divided into two parts: the morning section and the afternoon phase. Since they arrived in the parking lot/charging stations, the majority of electric vehicles (EVs) can be charged during these two times. Photovoltaic (PV) units can meet the EV charging requirements both with and without the battery energy storing system (BESS), which keeps the total power constant. By effectively allocating PV power for EV charging, it is possible to improve the load profile with a lower peak demand and eliminate the other negative effects of EV charging on the microgrid (MG). When a hybrid AC/DC MG is compared to a traditional AC and DC MG, the first one reveals numerous advantages. This design involves minimal AC/DC conversion, allowing it to integrate diesel generator (DG) units and loads of both AC and DC types. Separate AC and DC buses are required, resulting in less synchronization work. It also allows for simple voltage transformation, which is accomplished *via* DC/DC converters on the DC side and transformers on the AC side. As a result, the system gains improved overall efficiency as well as reliable and cost-effective operation. Due to the various charger configurations, EV charging is a difficult task. If 1- or residential EV charging penetration exceeds the network's authorized capacity, phase voltages and currents may exceed the allowed range of imbalanced circumstances in a low voltage distribution network or MG. Furthermore, each charging configuration has its own set of standards,

**Corresponding author*: chari.narasimhaiah@gmail.com

P. Sivaraman, C. Sharmeela and P. Sanjeevikumar (eds.) Microgrids for Commercial Systems, (307–344) © 2024 Scrivener Publishing LLC

which are further subdivided into slow and fast charging. This chapter provides an overview of the different EV charger designs and control methods. Three different charging methods have been considered in this, including DC fast charging, 3-AC fast charging, and 1-AC residential/ domestic charging. Each charging configuration and control method simplifies the EV charging infrastructure with MG. The gradual use of EVs and the rapid expansion of DC and AC types of renewable resources open the way for hybrid AC/DC MG operations. The primary goal of this chapter is to provide an overview of the EV charging infrastructure in a commercial hybrid AC/DCMG, with an eye toward EVs' potential future involvement. It also includes charger configurations, individual chargers for V2G control mechanisms, and their effects on MG operations. This chapter will serve as a resource for industry, researchers, and graduate students, particularly in the emerging areas of EV charging infrastructure integration with the MG.

Keywords: Electric vehicle (EV), EV charging infrastructure (EVCI), microgrid (MG), on-board charger (OBC), energy management system (EMS)

12.1 Introduction

It is well known that vehicles with fuel ignition predominated in the past. Nearly 15% of the total global carbon dioxide emissions are caused by transportation using conventional fuels, which produce vehicle exhaust (greenhouse gas). On the other hand, advancements in power electronics and modified batteries allowed researchers to shift their focus from ignition vehicles to electric vehicles. Electric vehicles (EVs) have a great potential to drastically reduce the effects of transportation on the environment and the consumption of oil. It was first considered as an important idea to reduce air pollution and, of course, fuel consumption in the middle of the 19th century. Most importantly, a battery operation does not release any type of gas and is either entirely or partially powered by the motor. Modern battery technology now allows for small, portable units to travel farther distances for longer periods of time. It comes in a variety of styles with capacities ranging from 8 to 48 kWh on the market. The global EV market is predicted by Bloomberg to reach 11 million units by 2025 and 30 million units by 2030, making India the ideal base for EVs. The Indian government recently started a number of programs to support electric mobility [1–3].

The diffusion of charging stations, which is a field of interest for research and industry, encourages the integration of EVs into power systems. These

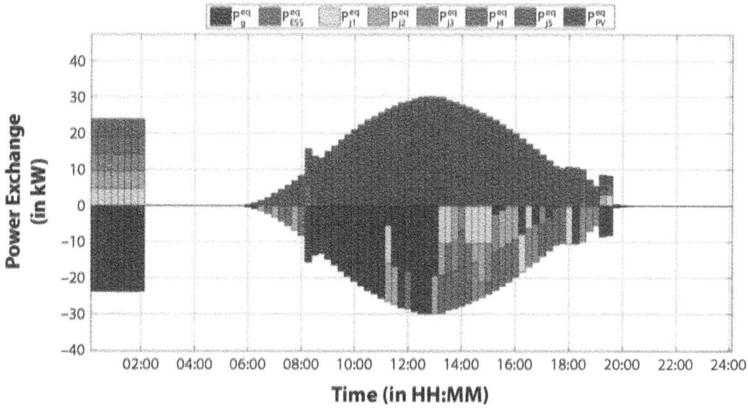

Figure 12.1 MG power balance on a sunny day.

stations can perform smart charging and benefit from the distribution of generation sources in the microgrid (MG) background. In other words, MG is a collection of distributing generation sources, loads, and energy storage units. It increases power system reliability by causing fewer problems with AC power integration. One of the tasks in developing smart MGs for EV needs is to recover the local source exploitation and synchronization of different sources [4, 5].

Figure 12.1 depicts the power balance at each interval on a sunny day. It should be noted that during the peak price period (8 AM–2 PM), the majority of PV power production is transmitted to the external grid, which is then used to charge EVs in the late afternoons. EVs generate revenue at night by supplying energy to the main grid. The energy storing technology has very little impact on outcomes, as it solely concerns the financial considerations (battery degrading cost), and the method will not select this, as it is not practical at the current price. DC loads and AC loads with converters (ACwC), such as refrigerators and air conditioners, dominate modern power systems.

12.2 An Overview of EV Charging Infrastructure

MGs are used in modern power systems because they have a variety of distributed energy resources (DERs) such as PV and wind systems, storage systems, different loads, and EV parking, and they also suggest control and management techniques with precise limitations, as shown in Figure 12.2. The MG is made up of a low-voltage distribution system that connects to

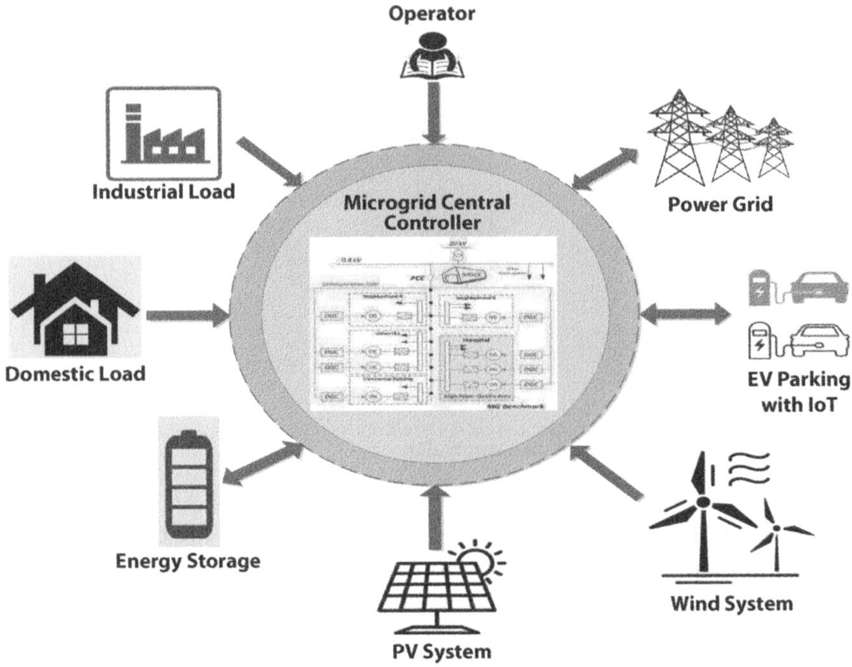

Figure 12.2 A typical microgrid network.

the grid *via* a point of common coupling (PCC). To put it in another way, the upstream power network is linked at a single point known as PCC. Electrical power is generated directly by micro-sources within the MG to meet the load demand, or it can be sent from an upstream power supply in a grid-connected manner. The MG central controller (MGCC) is the primary control point between the MG and the upstream grid, and its first duty is to maximize MG working. By introducing energy scheduling, EV parking acts as an aggregator, allowing EVs to participate in the reserve condition. The EV charging station and the renewable energy system are directly linked to the MG in order to achieve balance and coordination and thus reduce the negative impact on the grid. EV is an interesting performance for frequency management because it regulates quickly, and grid reliability is dependent on this quick response.

Major investments are required to develop EV charging infrastructure, and there are currently few business models available for private investment. Its availability is a large and complex problem, and building a charging network would be prohibitively expensive. Many researchers discovered that its availability varies greatly by location. Less infrastructure is a major issue in the evolution of the EV market. Its installation is permitted in both residential and public areas, and public charging is an essential component of the EV ecosystem [6]. Its development is dependent on the number of EVs targeted and the project's funding. Many EVs are limited to the maximum charging power they can accept due to limitations in converting AC power from the grid to DC to charge the battery. The EV charge period and intensity are sometimes used as known input for programming actions. The ultimate issue influencing charging infrastructure development decisions is the need for charging. When it comes to charging stations, technology must fully satisfy the customer's immediate need.

A weighting factor is also proposed within this model to optimize the EVs in the parking space, which is defined as follows:

$$W_{Ch}^{i,t} \propto \left\{ RBC^{i,t}, \pi_{Ch,EV}^{i}, \frac{1}{RT^{i,t}}, \frac{1}{AOB^{i}} \right\} \tag{12.1}$$

$$W_{Dch}^{i,t} \propto \left\{ SOC^{i,t}, \pi_{Dch,EV}^{i}, RT^{i,t}, \frac{1}{AOB^{i}} \right\} \tag{12.2}$$

$$RBC^{i,t} = 1 - SOC^{i,t} \tag{12.3}$$

where $RBC^{i,t}$ *and* $SOC^{i,t}$ is the remaining battery capacity and the state of charge of ith EV in time t.

$RT^{i,t}$ is the remaining time of charging or discharging period of ith EV in time t.

AOB^{i} is the age of ith EV's battery.

$W_{Ch}^{i,t}$ and $W_{Dch}^{i,t}$ are the charge and discharge weighting factors of the ith EV in time t.

$\pi_{Ch,EV}^{i}$ and $\pi_{Dch,EV}^{i}$ are the desired charging and discharging price limits of the ith EV.

The output of wind power generation can be considered as a function of wind speed. To fit the relationship among output power and wind speed, the following function can be used [7–9]. This function is described as follows:

$$P_W^t = \begin{cases} 0 \; if: V^t < V_C \; or \; V^t \geq V_F \\ \dfrac{V^t - V_C}{V_R - V_C} X \, P_R \; if: V_C \leq V^t < V_R \\ P_R \; if: V_R \leq V^t < V_F \end{cases} \tag{12.4}$$

where P_R is the rated electric power

V_C is the cut-in wind speed

V_R is the rated wind speed

V_F is the cut-off wind speed

$$P^t_{PV} = \eta \ X \ S \ X \ I\{1\text{–}0.005 \ X \ (T_a - 25)\} \tag{12.5}$$

where η is the conversation efficiency of the solar cell array

S is the array area

I is the irradiation of solar cell

T_a is the ambient temperature

Since many years ago, the government and private companies have been constructing these buildings, which have nearly 200,000 different types of stations worldwide. It is a crucial factor in the expansion of the EV market. Its legal standing varies from nation to nation and city to city. Its accessibility varies greatly, and there is no global standard for the number of charging stations and EVs required per charge point. Type of charger, charging type, access type, location and number of charging stations, budget, pricing, electrical system, energy management, local context, and resident survey information are all crucial for installing the EV charging infrastructure. A typical charging infrastructure consists of the various basic hardware elements listed below: The main and crucial component of a charging station is the power electronics configuration, which powers the on-board charger (OBC) of electric vehicles. The charge controller in a charging station will control the on-and-off time and the charging function as well as power usage metering and real-time and event data storage. Network control establishes a connection between the charging station and larger network. In order for the manager to review, monitor, and manage the use of the device, the station must be able to communicate with the network using a telecommunications device. Additionally, they control who has access to the charging stations. Installing an EV charging infrastructure generally involves the following:

Step 1: First determine the approach to install the EV charging infrastructure.
Step 2: Gather the information and perform the assessment.
Step 3: Design, documentation, and contract.
Step 4: Installation process.
Step 5: Project wrap-up and the ongoing consideration.

As shown in Table 12.1, the region and charging speed can affect the type of plug and socket used to charge the EV. The majority of EV connector standards in North America and Japan have five pins and a mechanical lock. Europe uses the seven-pin connector and benefits from three-phase AC. Teslas have the type 2 plug in Asia and Europe. The leading markets for EVs in the Netherlands and Norway have 10 times as many public charging stations per capita as the average market, while China and California have three to five times the average. Most charging takes place at home in many parts of North America, Europe, and Asia. In California, EV owners typically charge their vehicles at work, and there is typically one public charger for every 25 to 30 EVs. Private charging is uncommon in the Netherlands, where one public charger is the norm for two to seven EVs. This ratio, which ranges from three to six in China, had the highest rapid charging percentage. There appears to be no optimal global ratio for the number of EVs per charge point when different housing, population density, EV uptake, and public charging availability are taken into account. According to surveys and reviews of prior research, the EV market share increases if the ratio of EVs per charge point decreases from 25 to 5 EVs per charge point.

The different locations of the charge points illustrate the sector's various segments, and it is advised to use a variety of charging speeds in each one. Those who are accustomed to off-street parking prefer home charging, which takes place in a driveway or garage. A slow or fast charge point is

Table 12.1 Comparison of chargers.

Area	**USA**	**Europe**	**China**	**Japan**	
Standard	SAE	IEC	GB/T	CHAdeMO	**TESLA**
AC	 J1772	 62196-2	 20234.2	 J1772	
DC	 J1772	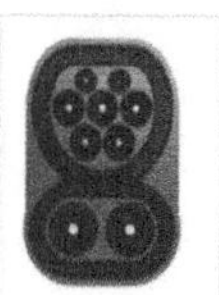 62196-3	 20234.3	 CHAdeMO	

located in this and connected to the home's electrical supply. Employees typically use workplace charging, which takes place in the parking lots at the workplaces. As an alternative to charging the vehicle at home, a slow or fast charge point is typically used in this scenario. Consumers who are travelling to locations like supermarkets, shopping malls, theaters, restaurants, tourist attractions, etc., where charging occurs only at the car parking, prefer destination charging. Charge points are utilized in a fast or rapid manner. Those who do not like off-street parking frequently use on-street charging [10–13]. This type of charging consists of charge points installed at street lights, utility poles, or parking lots that are frequently used by locals. If a driver is traveling a longer distance, en-route charging refers to charging in the highways at a highway service area or at a service station on a trunk road. Rapid charge points are present in this.

12.2.1 Charging of Electric Vehicle

The majority of electric vehicles have an OBC that transforms grid AC power to DC power, which is required to charge the battery. This solely depends on the battery's size, chemistry, and on-board electronics. In the home, at the office, or in any public location, OBCs can be charged directly using a standard home plug (slow AC) or a special charger (moderate AC). A dedicated EV charger installation is required for moderate AC charging, and this can be done without managed charging expertise [14]. The availability of EVs in parking spaces can be accessed by taking into account the required charging or discharge cost limitations with open market electrical cost. Each EV in a parking space is undisputedly in one of the states shown in Figure 12.3.

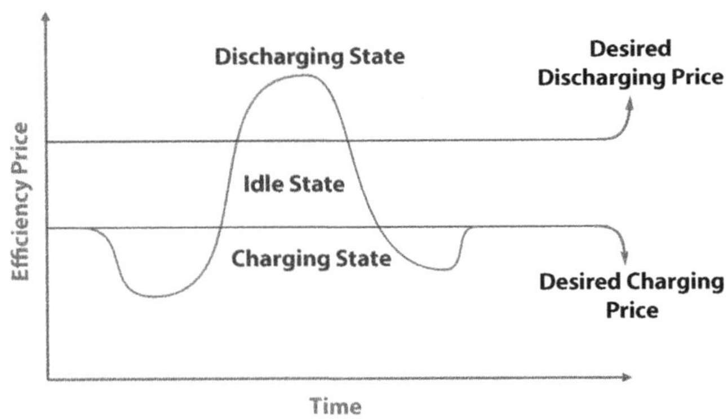

Figure 12.3 Typical electric vehicle states.

In addition, there are four different charging modes for AC and DC charging or, to put it another way, the mode of communication between a vehicle and a charging infrastructure is how charging infrastructure is defined. In mode 1, there are no safety precautions and a 120 or 240 V charge up to 16 A on a single circuit. It is also known as "dumb charging," and since the EVSE and EV cannot communicate with one another, it is not recommended. With safety precautions, mode 2 uses a shared or dedicated circuit to charge devices at 120 or 240 V up to 32 A. Temperature restrictions, over-current safety, grounding detection, and a pilot data line are additional procedures. Usually, it is utilized for home charging. Mode 3 enables 240 V charging at any ampere range on a dedicated circuit, with the same safety procedures as mode 2 and the ability to coordinate charging. Finally, mode 4 is a 400-V DC fast charging mode that requires more innovative safety and communication procedures.

12.2.2 Electrical Vehicle Charging Categorization

EVs can be charged using either AC or DC power. If an electrical distribution system provides AC, it is converted into DC for a battery with the aid of a converter. In an AC EVSE, power is sent to the OBC for conversion to DC. In the case of a DC EVSE, power is externally converted and supplied to the battery directly, skipping the OBC. Since local storage and EV batteries rely on DC, they can combine to form the DC MG. Different charge point speeds can use AC, but only rapid or ultra-rapid charge points can use DC. For the various charging speeds, EVs have a variety of connector types. A simple power electronics, usually wall- or pedestal-mounted, that regulates and securely supplies grid AC power to a vehicle's battery constitutes an EV charging station [15]. There are numerous chargers available to supply the various currents and voltage levels required by the various battery types. These chargers have power outputs between 500 W and 500 kW. It consists of the various levels listed below.

Level 1 Type of Charging

It has a standard electrical opening, a three-prong domestic plug, and 120-V AC power. It adds 8 km of range per hour and is the slowest charging type. The EV battery typically takes 8 to 30 h to fully recharge and is suitable for locations where vehicles are parked for extended periods of time.

Level 2 Type of Charging

This kind of charging requires an AC plug with 240 V. Depending on the size of the EV battery, the full recharge time ranges from 4 to 10 h. It increases range by 30 to 50 km per hour. The charging stations that fall under this level 2 category are very real and designed to be used for charging at home, at the office, and in public areas during off-peak times. There are options for network or non-network charges in this category:

i) Network charging station
A charging network component is linked to another station *via* a wireless signal or a hardwire connection. They allow for advanced control and support a variety of functions such as payment collection, internet reservation system, remote monitoring, communication display screen, charging station report, user action control, and so on. It comes in two varieties:

a) Non-controlled network charging station
They are known as "smart chargers," and they allow for load management by maximizing available power and distributing it evenly among the stations that are linked to it. If the maximum available power is 120 A and three EVs are linked, each vehicle will receive 40 A current without tripping the circuit.
b) Controlled network charging station
These are known as the "intelligent chargers." The management of a circuit's load as well as the monitoring of building demand is naturally possible so that it does not exceed the peak demand. It helps to avoid incurring incremental demand charges.

ii) Non-network charging station
These stations do not use an Internet connection, and the equipment's primary function is to obtain the charge of an EV battery without providing any additional information. However, if a meter is added, some stations will be able to monitor the energy. A key can also be used to integrate power sharing and access control features. Because these stations do not discriminate between operators, it is unlikely that the power consumed by each vehicle can be tracked. They are charging standard fees, which are not appropriate for a multi-unit residential building, and require some upgrading to meet the upcoming charging desires at an additional cost.

Open Charge Point Protocol
This is a network of charging stations, similar to a cell phone and a cell phone network. It is, in essence, an application protocol used to communicate between the charging station and the central management system.

The main task is to create an application protocol that allows charging stations and management from different vendors to communicate with one another. In other words, it allows communication between any compliant EV charging hardware and software, regardless of which retailer sold or installed it. Thus, if an EVSE provider goes out of business, EV owners can switch to another open charge point protocol (OCPP)-based network, which allows load management among many OCPP-compliant charger hardware.

DC Fast Charging

Because of their high charge rate capability, chargers that provide DC to the battery while bypassing the on-board converter are referred to as DC or DC fast chargers. It necessitates a dedicated grid infrastructure and the involvement of a distribution company (DISCOM) in the design and planning stages. Connecting and working is noticeably more exclusive. Its 480-V DC power supply qualifies it as quick charging. These stations can charge a battery of an electric vehicle by up to 80% in 25–30 minutes, with an additional 100 km/h range. The DC fast charging (DCFC) station is best suited for driving applications that require quick recharging. DCFC stations consume a significant amount of power, and during the charging process, they can cause stimulating spikes in the electrical system. Because of factors such as installation costs, site civil work, etc., DCFC options are generally not advisable for a multi-unit residential building [16].

12.2.3 Characteristics of Electric Vehicle Supply Equipment

The electric vehicle supply equipment (EVSE) is a technical term for an EV charging station or charge point. In other words, it is a fundamental component of EV charging infrastructure. It serves as a bridge between the power source and the vehicle charging port. It has various power ratings based on charging requirements in order to obtain the necessary input power for the charging infrastructure. It will undoubtedly necessitate appropriate planning and, of course, dedicated infrastructure at various stages on the distribution grid side. The EVSE safely transfers power and exchanges data between the electrical circuit and the vehicle. The EVSE processes local electrical power and employs a control system to charge EVs *via* wired connection in a secure manner [17]. This control system performs various functions such as user verification, charging authorization, data recording, network management, and data privacy. In the basic configuration of EVSEs, a minimal intelligence level is required for advanced data collection and charging, which supports the distribution company's needs and constraints. Sometimes an EVSE has a bidirectional control feature that allows you to change the amount of power that is inserted into the battery.

12.2.4 Smart Charging and Interoperability of Charging

Smart charging uses the EV charging cycle to meet the needs of both the power system and the EV owner. If EVs were charged in an unregulated manner, peak demand on the grid could increase because the charging pattern could coincide with current peak demands, contributing to overloading. Its main feature is that it is located at the crossroads of the electricity market and E-mobility. To support the EVs' smooth integration into the grid and provide flexibility, this charging requires a close coordination between the two. Smart charging may incorporate solar and wind energy into the grid by adapting the EV charging profile to resource availability, as shown in Figure 12.4. As seen, several smart charging techniques would be used, depending on power system characteristics such as the proportion of renewable energy output, load profile, and availability of interconnections. This charging creates a positive feedback loop for E-mobility by incorporating renewable sources.

The market for e-mobility, which includes EV drivers, operators, and providers, is not dependent on the electricity market. Charge point operators (CPO) and mobile service providers (MSP) play critical roles. CPO specializes in demand response services, and MSP acts as a go-between for CPO and the EV driver. EVs can provide a variety of power system services solely through smart charging. EVs can be used as both a load and

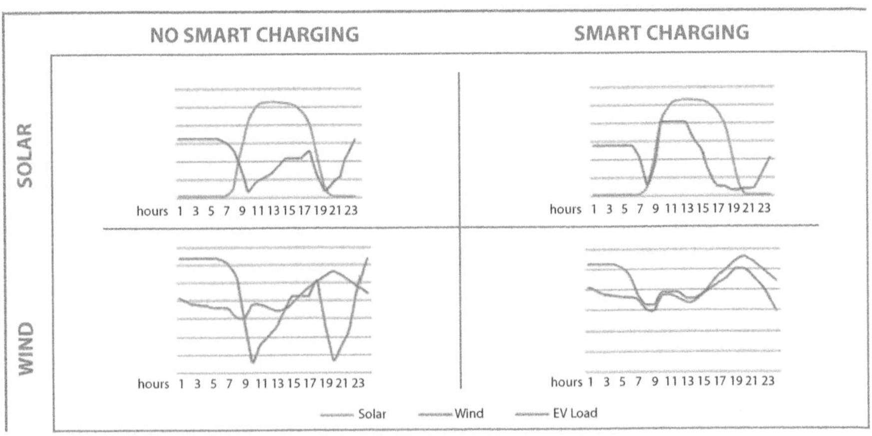

Figure 12.4 Smart charging for solar and wind generation.

a source by using grid-to-vehicle (G2V) unidirectional and V2G bidirectional, which is a very advanced technology.

EV charge points should be interoperable because it is critical for drivers to have a wide range of options as the number of EVs on the road grows. As a result, the public can easily change and use the various charge points, which also supports the sector's innovations and competitions. Interoperability and EVSE compatibility are supported by some standards. The Bureau of Indian Standard, India's national standard body, is in charge of developing the country's EV charging standards. The charging is controlled by the EV intelligence to save battery life, but the grid operator focuses on controlling the EV charge to reduce the counter-effects of large requests at predetermined intervals. Charging should be as simple as filling up at a gas station, and finding working charge points should be simple. Payment should be quick, and contactless bank payments should be possible with the addition of new payment technologies. The charging cost is very clear because the prices are displayed in a simple and standardized format. Charging is accessible and interoperable, as all drivers can use the charge points and are not limited to a single EV brand, and the inclusive design properties are tracked. AC power (0–25 kW) and DC power (0–150 kW) range is basically the charging session, which can be as short as 15 min or as long as 8 h, depending on the battery size and charging power [18–20].

12.2.5 Battery Specification in Different EV Segments

Only light electric vehicles are expected to be used for transportation in India over the next decade. It is made up of two wheels (scooter and motorcycle) and three wheels (passenger and cargo). Then, in the vehicle segments, light commercial vehicles can be electrified. Electric buses are also widely available. EV charging requirements are determined by EV battery specifications because power is supplied to the battery at a sufficient voltage and current to allow successful charging. Table 12.2 summarizes the typical charging characteristics for the various vehicle segments. It is important to understand that automobiles and buses frequently operate as a fleet, and the combined load of the fleet is greater than the charging load of each individual vehicle.

The battery's lifetime is an important parameter that influences the charging and discharging schedules of electric vehicles as well as the scheduling of ancillary services. Many EVs have removable batteries for charging at home or at work, and they do not always require a curb side charging system. Battery swapping is an alternative method of charging

Table12.2 Different segments' charging power and its requirement.

Type of vehicle	Battery size (kWh)	Slow charging (kW)	Fast charging (kW)
Two-wheeler	1–3	0.5–1	2–3
Three-wheeler	3–7	0.5–1	2–3
Car	15–80	3–6	20–100
Bus	100–400	7–50	50–500

a battery that is gaining popularity these days. A fully charged battery replaces a depleted battery in a vehicle during this process. This method is being tested for various EV segments, including electric two-wheelers and three-wheelers, electric cars, and even electric buses. Battery swapping has advantages over plug-in charging, but it has had many setbacks in its development as a common charging method. It is regarded as a viable solution for the commercial fleet of EVs and is classified into two types: manual swapping and autonomous swapping. Manually, a battery swapping station is essentially a stand-alone device in which the battery is placed and manually detached from a separate slot using only one hand. The manual swapping station is segmental and takes up little space. It is used in two-wheeler and three-wheeler battery requests because their battery size is small and can only be managed by two or three people. A robotic arm is required in autonomous mode, where the swapping process is semi-automated or fully automated [21]. They are more expensive and require more land. It is also used in four-wheeler and electric bus requests because their battery pack size is heavy and large, necessitating mechanical support.

12.3 Importance of Charging Station and Charge Point

When, how, and where a vehicle is charged is entirely dependent on available options, duty cycle requirements, the cost of each option, and sometimes personal preferences as well. EVs are charged at different power rates using AC and DC. This means that EV owners can only recharge using public charging stations. A public AC charging station (7 kW/charging) is available in restaurants, workplaces, and retail locations. They are naturally exposed to the complex daytime and evening usage. The more clients that attend public DCFC stations per day, the higher the peak demand. Today, Indian EVs cannot accept 50-kW fast charging despite the fact that it is recognized as a typical fast charging power in the global EV market. Furthermore, EV charging stations place a significant electrical load on the power grid system over time, and there are some risks involved if they are not installed properly. Installation must be focused by taking into account all relevant codes and standards for safe operation. Some researchers are also looking into how EV management in a MG uses load variation to charge and discharge.

The charging station refers to the high-energy EVSE, which means charging in mode 3 and 4 with multiple charging guns. The interchange ability of charging stations is improved as a result of connector standardization, and the strong plug-in systems help to reduce the waiting time. The EV charging network is made up of many standard-powered charging points. If a vehicle is parked and has access to an EV charging station, it is an opportunity to charge an EV battery. It is destination charging, in which the vehicles fully charge the battery before driving to the destination. As a result, instead of new locations as previously discussed, the charging infrastructure is planned in such a location where the vehicles are parked on a regular basis. Such a network has numerous advantages for both workers and users, ranging from ease of access to financial viability [22].

The charge point is linked to the electrical network in order to transfer and distribute power. High voltage transmission spans the country, while low voltage local distribution networks supply prime clients. In other words, the charge point is linked to the electrical supply system to charge the EVs, and as the number of EVs increases over time, so will the electrical demand. As a result, the system must have sufficient generation, network, and storage capacity. EVs can provide flexibility services such as load management on the local distribution system by selecting the charging style. These points are provided by commercial entities known as charge point operators, who have a diverse range. Some are active across the sector (at different speeds and in different segments), while others are only active in specific segments. Operators are working with the site owners to complete the agreement and other formalities for the installation of the charge point. These operators have various business models, and the current main models in public charging are as follows: Full operator, in which the charge point operator collects infrastructure charges and passes them on to the EV driver, where the charge point fee is offered by the service provider. It will be in charge of the system's operation, payment, service, and data collection. Concession is the same as full operator, with the exception of infrastructure funding provided by government grants. In Los Angeles, this model is used for on-street charging.

The use of a charging station shows whether the EV network can support the vehicle population or if more chargers are required. In other words, utilization represents the amount of time a station spends to supply a vehicle as a percentage of the time the station is ostensibly open. If a charger is used continuously for 24 h, then its utilization is 100%; similarly, if it is used once daily, then its utilization is 4%. In relation to the charging process, the following issues should be carefully considered:

- For EV charging the ease of access: If charging stations are placed where cars are typically parked, users can charge while still in the garage, saving time and reducing the distance that they must travel to access public charging.
- Normal power charging point: Normal charging stations, which have a compact network, reduce the need for high and ultra-high power points, which are expensive and bad for a battery's life if they are used more frequently.
- Charging infrastructure cost efficiency: Normal charging points cost less, require less space and supply, and reduce capital costs. They are able to connect to distribution networks, 1- and 3-V low voltage networks, and networks accessible in buildings and public areas.
- EV charging financial feasibility: Low setup costs for the charging infrastructure mean less need for government subsidies, which encourages private sector participation in charging operations. Effective charging infrastructure in the modern EV market should concentrate on the more accessible charge points.

EV sales are rising, and improvements are slowly making their way onto the market, but there are still problems with charging, such as inconsistent data, a lack of trustworthy standards in many markets, and obstacles to the ideal EV charging network. By enabling interoperability between charging networks, encouraging innovation and competition, and lowering driver costs, some standards can reduce the problems. Based on the Netherlands' effective labor, the government may feel that funding projects that support market expansion requires the use of open standards and data collection [23].

12.3.1 Classification of EV Charging Infrastructure

Depending on the environment and the needs, EVs were charged in a variety of ways. As a result, there are various types and designs of EV charging infrastructure available for use in various applications. The EVSEs are EV charger standards and specifications which vary from country to country based on the marketable EV models and electrical grid features. In order to include the necessary electric infrastructure, a utility must first invest in ready-made installations; however, these installations will not include the equipment needed to charge real plug-in vehicles. Second, a utility must complete the installations that involve the equipment and the components that prepare them for charging. It results in owning the equipment and infrastructure necessary to charge a vehicle as well as the single regulated building unit. Third, a utility can also provide financial incentives like price breaks for charging plug-in cars to the host site.

The infrastructure for charging consists of three components: From secondary distribution infrastructure to customer metering, the new installation is connected to the power grid. Through trenching and wiring, the meter is connected to the charging infrastructure, insulating material, and even the foundation. The third part is the actual infrastructure or EVSE purchase and installation. According to custom, the utility and the customer share responsibility for one component based on the linkage requirements for new equipment. Particularly protected through the customers are components two and three. The project to prepare utilities should include both parts one and two [24]. The rights and its use, as shown in Figure 12.5, are completely dependent upon the governance of the charging infrastructure. Generally speaking, there are three types of charging infrastructure: semi-public, public, and private:

- Private charging: This type of charging is specifically intended for individual EVs or fleets of EVs owned by one organization. It can be found at the dedicated parking space or the independent home of the EV owners. Ownership is divided between individual EV owners and EV fleet owners, and it is either self-operated or run by the CPO management system.
- Public charging: All types of EV users can use this type of charging. Common locations include public areas, on-street areas, charging stations, gas stations, freeways, and metro stations. The ownership of this type of charging is held by

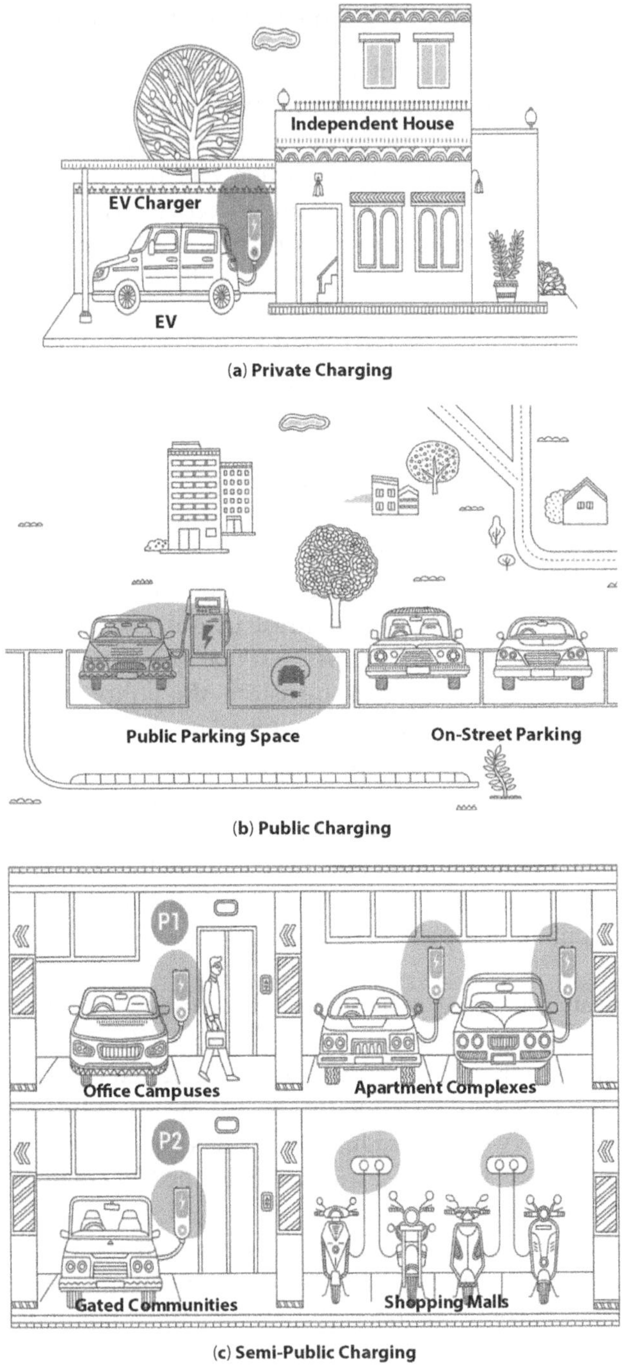

Figure 12.5 Types of electric vehicle charging infrastructure.

civic establishments, CPOs, PSUs, and host properties, and its operation is based on CPU management.
- Semi-public charging: For a limited set of EV users, this type of charging is a typical charging method. It can be found at apartment complexes, office buildings, shopping centers, gated communities, universities, and government structures, among other places. The semi-public charging is owned by host belongings, original equipment manufacturers, and CPOs, and it is operated in accordance with the CPO management system.

Because some types of charging illustrate the hybrid features, these are not the standard or fixed categories—for instance: If EV fleets are in motion, the charging infrastructure that is maintained by the fleet owners for captive use may also be made available for public charging as a paid service. Similar to this, the parking areas at bus depots and metro stations, where charging infrastructure is located, are classified as public or semi-public charging areas depending on whether they are accessible to both transit users and all EV users. Fewer shareholders are typically involved in private charging, which requires less regulatory compliance.

12.3.2 Operating Groups for EV Charging Infrastructure

The government develops funding programs that specifically target challenging market segments like curb side charging stations, multi-level residences, and direct fast charging in order to support these changing needs as the EV market expands quickly. Many governments are considering the electrification of transportation as a crucial step toward their objectives related to the environment, air quality, and energy dependence. Recognizing the need for charging stations in an established market is beneficial. It is also possible to implement the V2G concept using MG optimization without EVs, where the support for demand response is provided through operational costs and battery degradation costs. The following categories apply to the stakeholders: automotive, energy, GIS operators and supply chain, parking access supply chain, car share, fleet, EV users, community education, non-profit organizations, property developers, local transport authorities, and regional transport bodies.

Government agencies at the state and local levels are in charge of ensuring the efficient operation of the charging infrastructure. There is currently no system in place for the various agencies to work together. The desired synchronization among the various government agencies can be supported by an operating group. It is made up of all pertinent nodal and executive organizations, including the State Nodal Agency, DISCOM, municipal corporations, and urban development authorities. Additionally, the State Electrical Regulatory Commission (SERC) and the transportation agency may be mentioned. High-ranking energy development administrators may be part of an operating group under the direction of the chief secretary to ensure the desired interdepartmental synchronization. It is advised to set up a town committee to charge the EV in large towns and urban areas, run by the officers of the municipal corporation or by the head of the serving DISCOM. Within the urban local body and DISCOM, a nodal officer is designated to charge the EV during the execution process. For the local representatives to become more knowledgeable and aware of the EV charging infrastructure, capacity building is necessary.

12.3.3 Charge Point Operator and E-Mobility Service Providers

The various policies to lessen the impact of transportation on the environment are used to guide the operation of the power system. The goal of managing EVs in a smart city environment is to achieve maximum return and extreme EV necessity fulfilment. The day-to-day management of the public EV charging infrastructure and semi-public charge services is handled by CPO and e-mobility service providers (e-MSP). The electrical markets must therefore function properly because the current competitive wholesale and trade market is not always favorable. There are wholesale electrical markets in many nations, but because of competitive balancing and the absence of retail markets, the regulated services are managed centrally by a transmission system operator. Additionally, the distribution system operator looks to MSP and ICT firms for assistance in providing the smart charging services. The clients also take action to ease the technical cost and maintain grid assistance [25]. These providers are able to access the charge process thanks to the demand of e-mobility customers. This charging request is implemented on a charging infrastructure that is maintained by either e-MSP, if it also serves as a charge spot operator, or by a third party, charge spot operator.

Setting up the framework architecture, protocols, and procedures falls under the purview of CPO and e-MSP. The effective use of EV charging services by the customers is ensured by the centralized management of the charge services and their communication with the DISCOM. The network of EV charging stations is supported by the CPO arrangement for public and semi-public use. This allows the chargers to be used independently or to work on behalf of the owners of charge points. Customer-owned and self-owned charge point networks can be combined with a variety of activities and CPO can achieve this in parallel. The responsibilities of the CPO include the following:

- Planning and permission
  - To design the best EV charging installation, consider the accessibility of each location and the energy requirements to charge the EV there.
  - Coordinate with the nodal and implementing authority to obtain the essential connections, authorizations, certifications, and permissions for the EV charging facilities.
- Installation and commissioning
  - Acquire the EVSE hardware necessary to meet the specification, which is based on the demand for charging, the charging pattern, and the required charging infrastructure.
  - Set up a centralized software management system to control the backend network, which includes remote metering, EV charger classification (based on location and type of charger), user registration, and permission record supervision.
- Working and billing
  - Balance the working processes, such as scheduling charge accessibility, collecting returns, tracking charger usage in real time, load balancing, performance evaluation, etc.
  - Establish the service charging for EVs based on market ideologies or adherence to government regulations.
  - As required by law, provide DISCOM and other government agencies with the specific information.

Due to their management benefits, public energy utilities are crucial in the majority of the world's regions for the implementation and operation of the infrastructure for public charging. Additionally, the majority of CPOs are private organizations, which currently strengthens the public charging market in India. On the other hand, by connecting them with the various EV charge networks run by CPO, e-MSP provides charge services to the EV operators. The EV operators are assisted by e-MSP in placing and completing transactions at EV charge points. Occasionally, CPO can also serve as an e-MSP.

12.3.4 Availability and Management of Charging Data

Because charging infrastructure is developing quickly, access to accurate and useful data is somewhat limited in different markets. The drivers are finding it difficult to find a place to charge because of the time crunch. Because they will see a decrease in station usage, it is crucial for the charge operators. The government then believes it cannot manage direct investment effectively. Finally, the availability of charging for customers must be guaranteed for the auto drivers as well. The infrastructure for charging can store a wide range of information, including location, type, working status, and usage. While some markets offer a variety of services to drivers that promote station information, some data may be incomplete or out-of-date at any given time. Government (Alternate Fuel Data Centre in the United States), non-profit (Europe, Open Charge Map), for-profit (Canada Charge Hub, United States Zap-Map), automakers, and charge networks (Charge Point in the United States and China, State Grid Corporation) are just a few examples of the diverse stakeholder services. These services typically offer drivers maps or mobile apps, and some even provide open access to data.

As previously stated, EVs and EV chargers continuously generate battery state of charge (SoC) data, kilowatt hour in charging process, utility price signal, demand response signal from a demand management system, and other details that may be useful to the network operator. The rate at which the EV will be charged at any given time will be determined by the SoC and the charger rating, as shown in Figure 12.6. The charge rate is shown in per unit, with the absolute value normalized by the appropriate limiting factors in both curve sections. However, a lack of information about charging station maintenance and operation may be an issue, frustrating the user. Many new charging stations are now connected to the Internet and provide

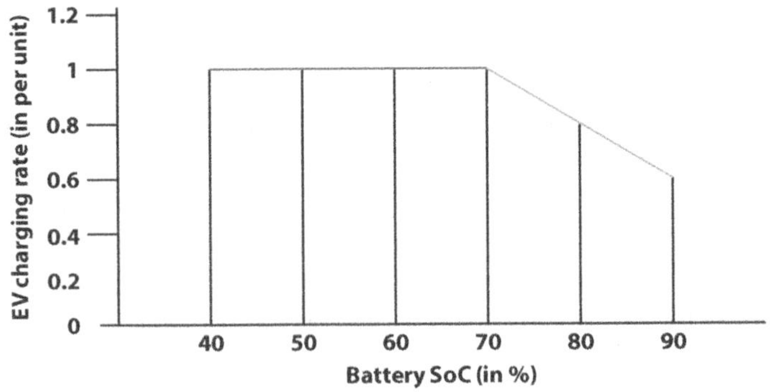

Figure 12.6 Electric vehicle charging chracteristics.

real-time information about the station's status and problems, which are then integrated into the online station and finally located for customers. A specially designed software can manage the charge and communicate with the charging station *via* the network [26]. The software network validates the deployment and EV charging station design and simplifies data flow between the charging station, its control, and its network control center. The function allows the operator to design, complete, and upgrade the software as well as set the driver access to charge, pricing, bill management, and user reports.

12.4 EV Integration to the Microgrid

An acceptable charging infrastructure facility requires an available, reliable, and reasonable electrical supply. Although a low-tension electrical distribution infrastructure is used for the rapid EV charging network, it is possible to provide an electrical connection to EV charging. A charging infrastructure distribution method consists of standard charging points. This confirms that the majority of charging stations are connected to low-tension electrical networks. Similarly, interfacing between EVs and MG necessitates the use of power electronics equipment, which provides power flow. According to some researchers, an energy management system (EMS) plays a critical role in integrating EVs into a workplace MG through optimization techniques. In general, it is observed that the indicator presents artificial, instantaneous, and operational data on EV performance. As a result, they can practically exploit for a variety of subjects such as MG operator, EV holder, and EV aggregator. It will advance the electric vehicle's contribution to energy management activities, exploitation levels, and the associated economic effects. Energy management in DC MG demonstrates unusual features and functions. To use EVs for mobility, daily operational costs must be optimized, which includes both the MG operator's and the EV owner's perspectives. If the EV is in the parking mode, the effective V2G application, which enables bidirectional energy exchange analogous to the storage device, is highlighted [27]. As previously stated, the smart charging concept improves the communication between EVs and the power system. EV aggregator is introduced for smart charging of EVs.

The connection between V2G and the electrical grid is depicted in Figure 12.7. A V2G system, a base power source, such as a DG, and renewable energy sources (RES), such as solar and wind energy, make up a MG. Buses 1, 2, and 3 connect all of these. The base model consists of a home load and five electric car profiles. In this application, it is crucial to have a

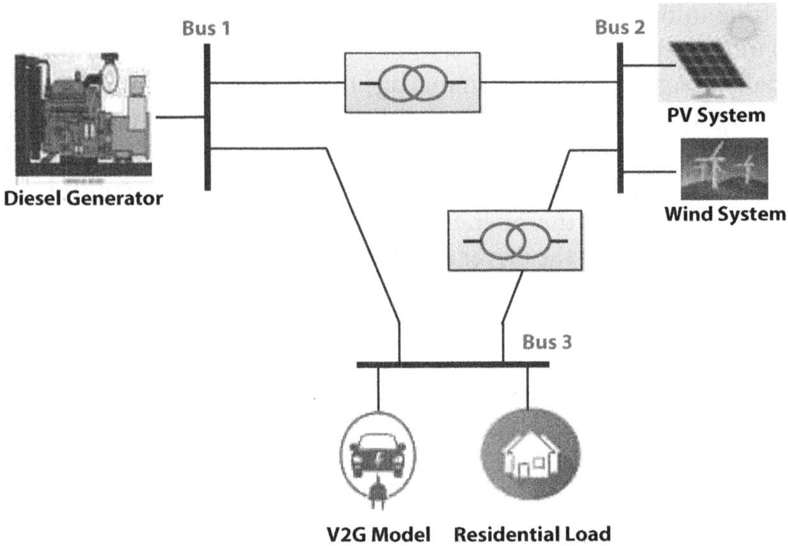

Figure 12.7 A microgrid design for domestic application.

control that allows you to enable or disable the V2G mode. Due to the fact that this charging method is bidirectional, it can both supply and consume energy between the electrical power grid and the charging station.

EV batteries can provide the quick response needed for the extra facilities, but their energy capacity is insufficient. As a result, a single EV cannot provide these services within the time frame that the power system requires. When enough EVs are gathered, they can work together to create a quick response power system and provide the services in the required amount of time. EVs are a simple source of flexibility for aggregator business models. For the EV energy facility to be practical in the wholesale market, a collection of 500 vehicles, charge points, and at least 1 to 2 MW of capacity are required. Most interconnection requests are anticipated from commercial segment fleet operators and bus transportation agencies at the beginning of the EVs' growth. The operator of the commercial charge network supports both private and commercial vehicle segments. The new loads will be high, ranging from a few hundred kilowatts to a few megawatts, depending on the number and type of vehicles. In order to support EV acceptance and ensure new and ongoing revenue for the DISCOM, it is necessary to deal with interconnection and allow requests in a timely manner.

12.4.1 EV Charge Connection's Regulatory Framework

In India, the electrical market is heavily regulated by central and state-level protocols. For EV charging, the establishment of an electrical connection is subject to a set of regulations and guiding principles, some of which are universal and some of which are tailored specifically to the charge services.

Technical Rules and Guidelines

The Central Electricity Authority (CEA) has announced some changes to the current guidelines in order to simplify the grid connectivity for charging infrastructures. The following are the primary supervisory requirements for EV charging:

Technical values to connect distribution generating sources (amendment) guidelines, 2019

- Define charging station and charging point.
- Identify EV as power generating source.
- Introduce values for the charging station, which are linked or looking for connectivity to an electrical source system.

Actions Related to Security and Power Supply System (Amendment) Guidelines, 2019

The security requirements as per the global values that should be adhered to are provided for charging infrastructures, including the universal security to charging stations, ground safety structures, fire security, charging station testing, investigation and periodic valuation, maintenance accounts, and security requirements.

State Guidelines

According to the Indian Constitution, the electrical supply and distribution system is important; thus, the guidelines at the state level find the electrical supply and connection directions. SERC's horizon is defined by the State Electricity Supply Code and the Performance Standard Guideline. It is a main controlling outline that governs the electrical supply and assembly facility *via* the DISCOM. This controlling outline varies depending on the state. The appropriate state regulations are considered when planning or installing charging facilities. Some issues in the state supply code, in particular, affect the electrical connections for EV charging.

Type of Electrical Connection

It could be one or three low-tension or high-tension levels. It will be obvious based on the required authorized load, which directly influences time and cost, to obtain the connection, tariff, and the requirement of additional challenging infrastructure such as distribution transformer (DT). The applicant's high-tension assembly appeals the sophisticated installation and monthly demand charges, which include additional time for energization and require ancillary electrical infrastructure. The sanctioned load for low- and high-tension connections varies significantly between states.

Power Supply From the Existing Network

The governed rules have some suggestions for power supply from the existing network, which commissions the EV charging connection in terms of cost and time. Getting a connection from an existing network (without any expansion) is obviously easier and less expensive than getting a connection from a distributing system extension. In addition, network expansion takes time, and some claimants must share the cost.

System Expansion Requirement for New Connection

Upgrading the system to stretch the allowed threshold of 70% for the new charging infrastructure assembly is recommended for capacity utilization of the nearest feeder. Distributing network extensions is a one-of-a-kind and time-consuming task. To obtain a power connection for EV charging, a distributing charging network in public places can augment the cost and time by authorized load on the lower side and a few charge points in each region. Those locations are where a large number of charge points are required or mandated by building codes and other government directives. These facilities raise the high operative and capital cost. SERC and DISCOM should implement EV charging as a new client desire, distinct from the current client classes. The supply code is also modified to allow for a reasonable and dependable electrical supply to the charging infrastructure [28].

12.4.2 Electricity Tariff

It is an important economic regulatory tool available to state governments, and it differs depending on the state. As previously stated, charging an EV is an additional accumulation in a DISCOM's user list. It was implemented as a separate client group in the overall tariff plan, simplifying the supply connection to charge an EV. To implement EV-related electrical tariffs, a variety of profits are required. Tariff is intended to send pure price signals to the evaporator, which can then manage the load profile to charge an EV. For CPO, this tariff is a major working cost that provides charging facilities, and charging in public places can have an impact on the corporate side.

In EV charging, a separate tariff group allows the state government to propose "EV only" inducements to encourage acceptance with structures such as:

- Excused demand charge: For electrical connections, a demand or fixed charge is collected on a sanctioned load in order to connect or maximize the demand for power, which is listed at billing time and must be paid regardless of actual power use. If, at the outset of EV adoption, there is a low demand for charging, demand charge exemptions can help the business set up the charge points.
- Reduced energy charge: Energy charge is a variable parameter in electricity tariffs that is applied to the total energy volume or electrical power consumed at billing time. For CPOs, this reduced energy charge reduces operational expenditure, and EV users benefit from lower charging costs. To charge an EV, 21 states and union territories have proposed specific tariffs that include lower power charges and waived demand charges until March 2021.

12.4.3 Technical Challenges for DISCOMs

DISCOM is in charge of providing the electrical connection for the EV charging infrastructure. It provides users with relevant information as well as a price signal that indicates how, when, and where they can charge. Its management can include demand side management, renewable generation, energy efficiency, and other technologies that clients face. It must develop new internal programs to support client interconnection desires for all charge station types. It carries out the electrical source instructions and guidelines in order to interact with the various client electrical groups. The important job for DISCOM is to forecast load growth, which includes EV load as well as expected EV charging demand in the future. The user's behavior or charging demand pattern is also unpredictable for DISCOM. As a result, many challenges await DISCOMs as EV penetration rises:

- Charging demand patterns/peak power increase: In any case, if EV penetration increases as a result of uncontrolled charging, MG operation will undoubtedly be disrupted. If multiple EVs are randomly connected and charging at the same time to draw rated power, it may result in unbalanced phase voltage, an increase in peak demand, poor power quality, and harmonic problems in the distribution network.

- Power quality issues: As we already know, in EV charging, AC power is converted into DC power using power electronic converters. The high frequency pulse-based switching in this conversion method may cause voltage and current distortions in the system. These harmonics have an impact on DISCOM's distribution assets, reducing their lifespan and causing false tripping in the case of a protection system. The most vulnerable distribution assets are relays, switch gear, transformers, power cables, capacitors, and meter equipment.
- EV charging infrastructure (EVCI) integration issues: To accelerate EV adoption, an effective deployment of EVCI based on a sustainable operation model is required. Currently, the low volume of EVs on the road results in low asset utilization of EVSE at charging stations. There are many rapid technical transitions in EV and battery-related EVSE standards these days. As a result, from the investor's perspective, it appears to be less risky to deploy EVCI due to less utilization of their asset at the moment.

12.4.4 Electrical Supply Arrangement for Charging

Stakeholders describe electrical supply-related topics such as time of use (ToU) and energy services, market arrangement, monitoring, forecasting and planning, charge point optimization, and regulation. As we all know, electricity is ubiquitous and available for electrical charging almost anywhere and everywhere. Its cost per actual power unit may be less than that of gasoline, and it is intentionally considered lower than gasoline due to the high efficiency of electric motors in comparison to internal combustion engines. Renewable and low-carbon runny oils, on the other hand, are comparatively indefinable, and electricity generated from them accounts for a growing share of global electricity generation.

EV charging has the potential to use a large amount of energy while posing no significant risk to the power grid; however, this issue will be addressed by establishments as the market grows. In March 2021, the Delhi government directs the marketable and formal constructions of more than 100 vehicles parking capacity, with 5% set aside for charging purposes only. It houses educational institutions, offices, shopping malls, and movie theaters, among other things. Another major concern is the impact of DC fast charging on local distributing infrastructure. These stations also use a large amount of energy for a short period of time. It implies that additional exclusive advancements are required at a low-use rate. This issue may become more serious as expertise grows. To maintain the upcoming venture and support, it is also critical to ensure the efficient EV charging network and meet the driver's expectations. The charging station should be placed in ideal locations on a local level to maximize the return on investment. It will increase EV usage, avoid traffic and parking issues, and reduce the load on the electrical grid [29].

There are numerous ways for EV owners or CPOs to organize the electrical assembly for an EV charge point or charging services with multiple points. Depending on the requirements, the EV owner or CPO can select the best option. The very first step in arranging an electrical source for this EV charging is to calculate the required energy demand in kilowatts. It corresponds to the sum of the rated input requirements of all charge points, which is essentially the share of the designed installation at a specific site. The energy required for battery charging is equivalent to concurrently charging the total number of batteries stored in the charging system. If the required energy demand is already known, the EV holder or CPO can choose between three options to supply the charging infrastructure: first, electrical power from the existing assembly; second, a new electrical connection arrangement; and third, captive renewable energy generation.

12.5 Industrial Microgrid and Subsystem

As we all know, commercial and residential buildings account for nearly 32% of the total power production in today's world. Furthermore, buildings are responsible for 30% of CO_2 emissions. Renewable energy combined with deployed EVs is being used as a viable method to reduce CO_2 emissions. This method can ensure energy security, but for reliable load balancing, some additional ventures are required to increase the capacity in every segment such as generation, transmission, and distribution, which helps to satisfy the EV charge demand. The rapid control and excellent performance of EVs have attracted interest as a frequency management method. Figure 12.8 depicts an industrial microgrid (IMG) concept made up of a number of manufacturing units linked to a RES and storage system. When compared to the conventional grid, the capacity of DER associated with an IMG is low. As a result, they are referred to as micro-sources.

PV units can generate power anywhere there is sunlight, including marketable or even housing areas, for EV charging. As a result, combining PV

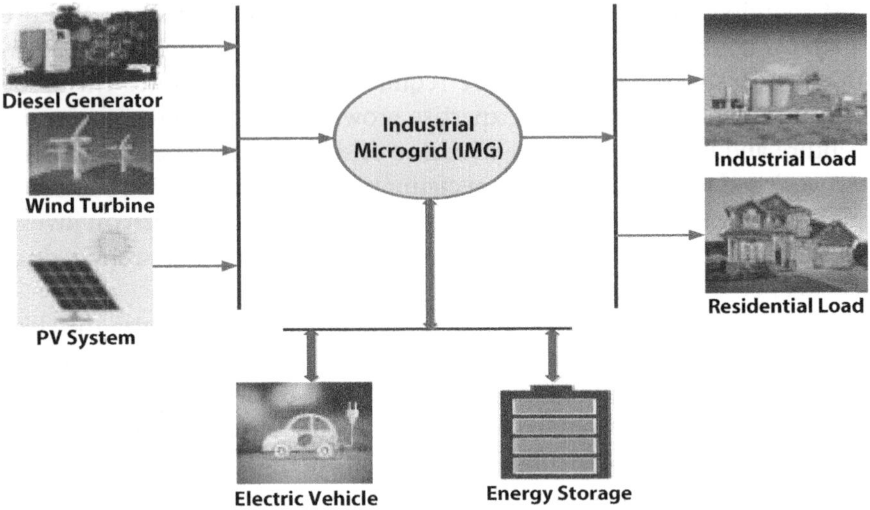

Figure 12.8 Industrial microgrid concept.

units with EV charging stations is a collective method to increase charge demand throughout the day while decreasing addiction or reliance on the current electrical grid system. The working hours for commercial buildings are generally 9 AM–1 PM in the morning and 2 PM to afterwards in the afternoon. During these hours, the majority of EVs are parked or charged at a charging station. The PV system without BESS meets the EV charging demand. In general, it is extremely difficult to keep the grid stable while optimizing large-scale charging uncertainty and heavy load characteristics. These characteristics can be improved by lowering the peak demand, and other negative effects on the distributing grid can be reduced by accurately dividing PV energy for this EV charging [30].

12.5.1 V2G Frequency Control Method

Customers and IMGs both benefit from V2G technology. EV owners can sell power to IMG during peak hours while also charging their own vehicles more cheaply and resolving IMG stability issues. The basic idea behind the V2G framework is to use EV batteries as a source of intermediary energy, as shown in Figure 12.9.

If power transfer >0, EV operates in charging that is G2V mode.
If power transfer <0, EV operates in discharging that is V2G mode.

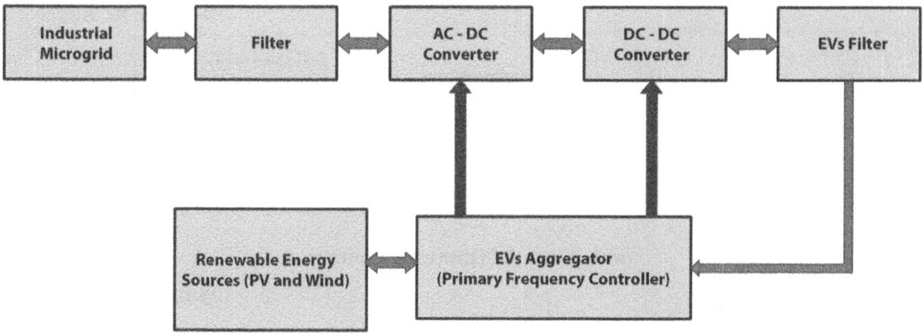

Figure 12.9 V2G system block diagram for an industrial microgrid.

If PT is 0, EV discharges very quickly, causing a frequency change. As a result, the methods for charging or discharging EVs served as the foundation for both approaches. If an EV is in a discharging state, IMG sends the necessary information to the EV aggregator.

Many articles propose and discover a hybrid AC or DC MG to integrate the benefits of both AC and DC electrical infrastructure. A hybrid AC or DC MG has several advantages over a conventional AC or DC MG. Such a structure requires less AC/DC conversion to integrate the DG unit and both AC and DC loads. EV charging is a difficult task due to the numerous charger configurations, which include a single AC charger, three AC chargers (slow and rapid), and DC fast chargers in a hybrid commercial MG system—for example, if the penetration level of 1- or home EV charging exceeds the network's rated capacity, current and voltage in the low-voltage distributing network or MG may exceed the allowed range of unstable conditions. To overcome this limitation, various configurations of EV fast chargers, as shown in Figure 12.10, can be used from either an AC or DC bus system. It requires little synchronization work by using a separate AC or DC bus. It also provides simple voltage transformation capability *via* transformers on both the AC and DC sides *via* DC-to-DC converters. As a result, overall effectiveness is enhanced through consistency and cost-effective working. In commercial MG, in addition to the hybrid configuration, there are various configurations of chargers like DCFC and 3- AC charger (slow and fast) as well as 1- AC charger, making the task of charging an EV challenging. The frequency must change synchronously with the bus voltage for hybrid AC or DC MG. The impact of EV charging on AC or DC bus voltage and frequency is identified and calculated under typical operating conditions. With the aid of the obtained results, which lessen the adverse effects of massive EV charging on the developing MG atmosphere and the likely grid investments, the significance of coordinated charging is highlighted.

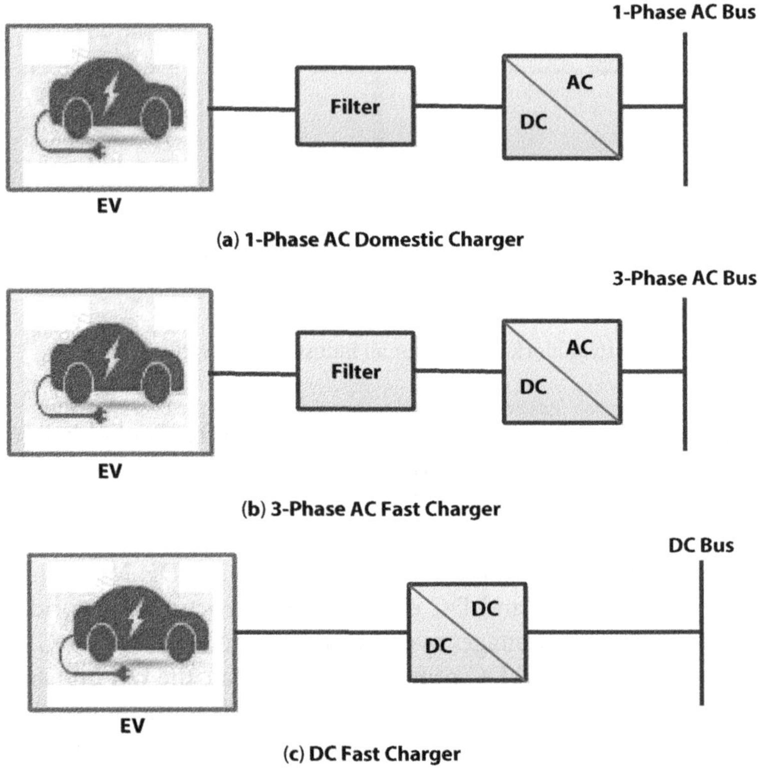

Figure 12.10 Typical electric vehicle charger configuration.

12.5.2 The DC MG Structure

A DC MG is containing a PV unit, EV charge device, BESS, grid connected network, and the MG EMS as shown in Figure 12.11. The function of the system components are as follows:

- PV system: PV modules and functional DC to DC converters make up its construction. This converter can be used in this structure with the aid of the MPPT technique. The actual output power of a PV system depends on both the ambient temperature and solar irradiance.
- EV charge device: It consists of a DC charger and an electric vehicle battery, allowing for better control of a battery's charge energy and current. These days, batteries can also be powered continuously. In the constant power charging mode, the charge time is determined by the charge capacity and charge power.

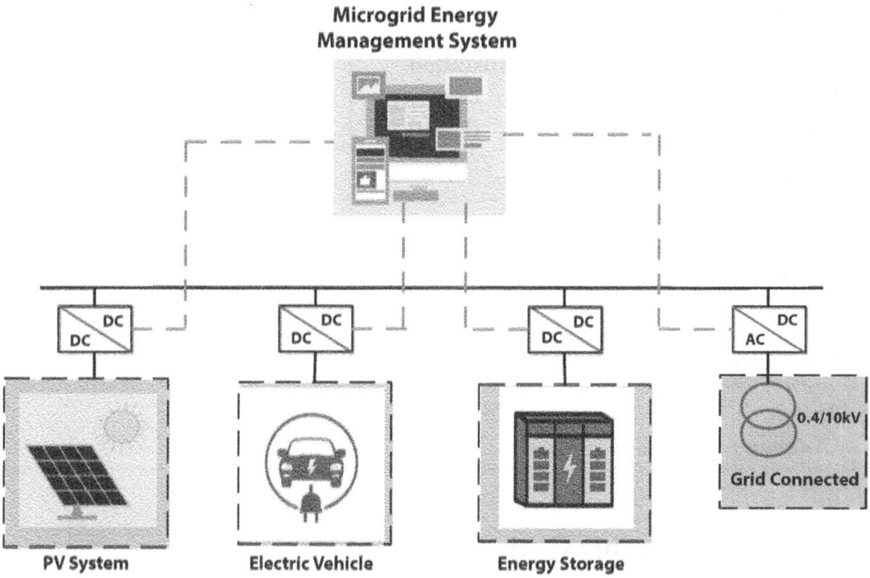

Figure 12.11 A DC microgrid system through an energy management system.

- Energy storing system (ESS): Li-ion battery and a DC-to-DC converter make up its components. The load shifting, PV intensity, and power fluctuation leveling functions are all completed by this storing system.
- Grid-associated system: It consists of a DT and a bi-directional AC-to-DC inverter. It transfers power between the DC MG and the distributing grid while it is in operation and keeps the DC bus voltage under control and stable.
- MG energy management system: This configuration manages and controls each module's working process in DC MG. This arrangement controls how much power is used to charge EVs as well as how much is supplied by PV units, energy storage systems, and distribution systems based on monitoring data. This plan helps to manage PV power dissipation, EV charging efficiency, and MG balance.

Table 12.3 shows the connected sizes for DC MG equipment accordance with the best and additional increase in PV and ESS size to support system consistency in estimation methodology.

Table 12.3 The DC microgrid device installed sizes.

Device name	Installed capacity (kWh)	Installed capacity (kWh)	Max/min SoC (in per unit)	Converter size (kW)
PV unit	40	–	–	40
ESS	30	60	0.95/0.25	30
Charge station	–	–	–	10
Utility grid	–	–	–	50
Electric vehicle	–	24	1.0/0.2	–

12.5.3 EV Charging Optimal Control Strategy

EV charge control scheme is applied with the help of MG energy management system.

- PV generation prediction: It aids in the creation of the strategies plan, and depending on the forecast's duration, it might be for the ultra-short term, short term, or long term. If the environmental conditions are the same, there is a strict correlation between the output in the future and the history.
- EV charging demand analysis: The demand for EV charging is closely related to daily mileage, power consumption in kilometers, and battery capacity. According to a survey conducted by the US Department of Transportation, domestic vehicles nationwide have an average mileage that follows a log-normal distribution, and its probability density function is as follows:

$$f_D(d) = \frac{1}{d\sigma_D\sqrt{2\pi}} exp\left[-\frac{(\ln d - \mu_D)^2}{2\sigma_D^2}\right] \tag{12.6}$$

where μ_D is trip distance expectation and σ_D is the standard deviation.

- EV charging economic analysis: The cost of electricity and the efficiency with which the equipment functions affect the economy of EV charging. The charge compensation is linked with the peak valley electrical price if an electric vehicle is being charged from the grid. The EV's energy competence is increased if energy from the PV unit is used directly for charging. Therefore, the best way to charge an EV is with energy directly from the PV unit.

- ESS control strategy: The energy storage system works during the EV charging process to improve the amount of PV dissolved in MG to mitigate variations caused by PV production and to increase EV charge efficiency. The ESS should store the energy from the valley and release it during the peak period to make up for the lack of power in order to complete its three tasks.
- EV charge control plan: If there is no optimization control over EV charging, then the PV energy portion will not be utilized for EV charging, while EV charging uses the distributing network to obtain energy during peak hours, increasing the peak load demand and the charging costs. Therefore, it is essential to improve the EV charging control in order to achieve economic optimization and decrease the peak load with the aid of MG energy management. MG EMS is used to implement the EV charge optimal scheme, and the optimization procedure is shown in Figure 12.12.

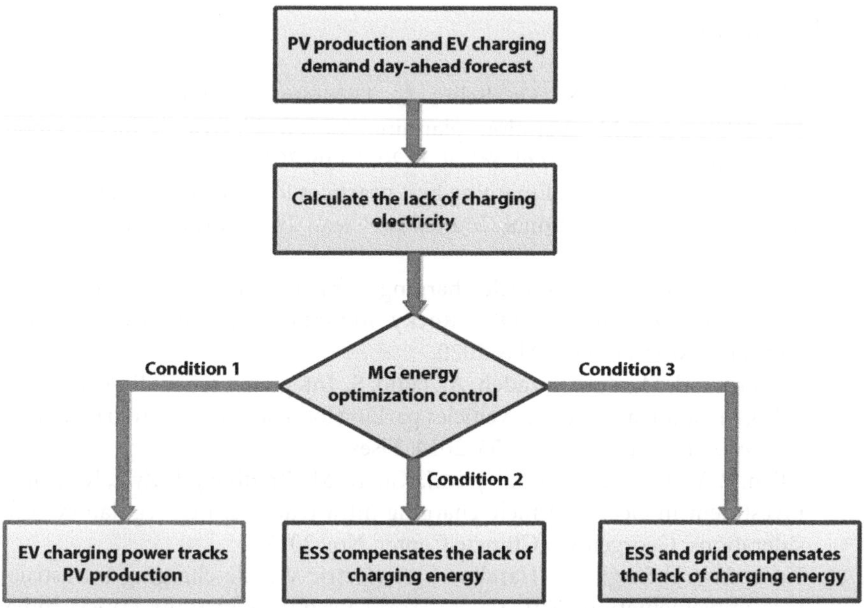

Figure 12.12 Electric vehicle charging optimization control strategy.

12.6 Summary

The RES utility will decrease during peak demand by properly designing the EV infrastructure with a hybrid AC/DC MG. This chapter provided information on various AC and DC supply-based EV charging infrastructures for EV users. Based on the scopes and new trends, it also demonstrates the effects of various EV charging techniques in a commercial hybrid AC/DC MG. The different EV charger designs and control mechanisms were covered in this chapter. Three different charging techniques were examined in this study: 3-AC fast charging, 1-AC residential/domestic charging, and DC fast charging. Every charging configuration and the related control method enhance the MG EV charging infrastructure's convenience. There are many benefits to a hybrid AC/DC MG over a conventional AC and DC MG. Because this design does not require as much AC/DC conversion, it can accommodate both AC and DC loads as well as DG units. Less synchronization work is needed because separate AC and DC buses are needed. Additionally, it supports straightforward voltage transformation using transformers for the AC side and DC/DC converters for the DC side. As a result, the system's dependability and overall efficiency, as well as its cost-effectiveness, all increase.

References

1. Aluisio, B., Bruno, S., De Bellis, L., Dicorato, M., Forte, G., Trovato, M., DC-microgrid operation planning for an Electric Vehicle Supply Infrastructure. MDPI, *Appl. Sci.*, 9, 1–20, 1 July 2019.
2. Hall, D. and Lutsey, N., Emerging best practices for electric vehicle charging infrastructure, International Council on Clean Transportation (ICCT), Oct 2017.
3. Fitzgerald, G., Electric vehicle charging infrastructure, Lighthouse Discom Programme A Niti AAYOG, Rocky Mountain Institute Collaborative, Jagabanta Ningthoujam, May 2020.
4. Honarmand, M., Zakariazadeh, A., Jadid, S., Integrated scheduling of renewable generation and electric vehicles parking lot in a smart microgrid. *Energy Convers. Manage.*, 86, 745–755, 2014, Elsevier.
5. Allen, P., Van Horn, G., Bradley, M.J., Goetz, M., Bradbury, J., Zyla, K., Utility investment in electric vehicle charging infrastructure: Key regulatory considerations, Georgetown Climate Center, Nov 2017.
6. Kant, A. and Singh, R., Handbook of electric vehicle charging infrastructure implementation, Niti Aayog, World Resources Institute (WRI) India, Version 1.

7. Pradhan, G.B., Mathur, A., Joseph, V., Datta, A., Rodrigues, N., Krishan, R., Sharma, J., Kumar, M., Rawat, A., Sharma, A.K., Vyas, S., Electric vehicles: Perspective of discoms and stakeholders prepared for distribution utilities forum, 2020.
8. CMA Competition & Markets Authority, Electric vehicle charging market study, 23 July 2021.
9. Electric-vehicle smart charging, International Renewable Energy Agency (IRENA), 2019.
10. Zhou, S., Zhuang, Y., Wu, Z., Gu, W., Yu, P., Du, J., Luo, X., Planning and real-time pricing of EV charging stations considering social welfare and profitability balance. *CSEE J. Power Energy Syst.*, 7, 6, 1289–1301, Nov 2021.
11. Guide to electric vehicle charging in multi-unit residential buildings, Delphi Group, Pollution Probe, 2020.
12. Shamiur Rahman, Md., Hossain, M.J., Rafi, F.H.M., Lu, J., EV charging in a commercial hybrid AC/DC Microgrid: Configuration, control and impact analysis. *Proceedings of Australasian Universities Power Engineering Conference (AUPEC)*, 25-28 Sept 2016, pp. 1–6.
13. Shengjun, W.U., Qingshan, X.U., Qun, L.I., Xiaodong, Y., Bing, C., Optimal EV charging control strategy based on DC microgrid. *3rd International Conference on Power and Energy Systems Engineering (CPESE)*, 8-12 Sep 2016, vol. 100, Energy Procedia, ScienceDirect Elsevier, pp. 243–247.
14. EV towards an excellent user experience, 2019, https://es.catapult.org.uk/consultation/electric-vehicles-towards-excellent-user-experience/.
15. Sabhahit, J.N., Solanke, S.S., Jadoun, V.K., Malik, H., Márquez, F.P.G., Pinar-Pérez, J.M., Contingency analysis of a grid of connected EVs for primary frequency control of an industrial microgrid using efficient control scheme. *Energies* Article, MDPI, 15, 9, 1–24, 2022.
16. Abousleiman, R. and Scholer, R., Smart, "charging: System design and implementation for interaction between plug-in electric vehicles and the power grid. *IEEE Trans. Transp. Electrification*, 1, 18–25, 2015.
17. Kamankesh, H., Agelidis, V.G., Kavousi-Fard, A., Optimal scheduling of renewable micro-grids considering plug-in hybrid electric vehicle charging demand. *Energy*, 100, 285–297, 2016.
18. Liu, Z.P., Wen, F.S., Ledwich, G., Optimal planning of electric vehicle charging stations in distribution systems. *IEEE Trans. Power Deliv.*, 28, 1, 102–110, 2013.
19. Roy, J.V., Leemput, N., Geth, F., Buscher, J., Salenbien, R., Driesen, J., Electric vehicle charging in an office building microgrid with distributed energy resources. *IEEE Trans. Sustain. Energy*, 5, 4, 1389–1396, 2014.
20. Liu, X., Wang, P., Loh, P.C., A hybrid AC/DC microgrid and its coordination control. *IEEE Trans. Smart Grid*, 2, 2, 278–286, 2011.
21. Nejabatkhah, F. and Li, Y.W., Overview of power management strategies of hybrid AC/DC microgrid. *IEEE Trans. Power Electron.*, 30, 12, 7072–7089, 2015.

22. Rahman, M.S., Hossain, M.J., Lu, J., Utilization of parked EV-ESS for power management in a grid-tied hybrid AC/DC microgrid, in: *The 25th Australasian Universities Power Engineering Conference*, 2015.
23. Leskarac, D., Moghimi, M., Liu, J., Water, W., Lu, J., Stegen, S., Hybrid AC/DC microgrid testing facility for energy management in commercial buildings. *Energy Build.*, 174, 563–578, 2018.
24. Loh, P.C., Li, D., Chai, Y.K., Blaabjerg, F., Autonomous operation of hybrid microgrid with AC and DC subgrids. *IEEE Trans. Power Electron.*, 28, 5, 2214–2223, 2013.
25. Akbari, M., Golkar, M.A., Tafreshi, S.M.M., Voltage control of a hybrid AC/DC microgrid in grid-connected operation mode, in: *Proc. IEEE PES ISGT India'11*, pp. 358–362, 2011.
26. Lasseter, R.H. and Paigi, P., Microgrid: A conceptual solution, in: *Power Electronics Specialists Conference, 2004. PESC 04. 2004 IEEE 35th Annual*, vol. 6, pp. 4285–4290, 2004.
27. Kurohane, K., Senjyu, T., Yona, T., Urasaki, N., Funabashi, T., A hybrid smart AC/DC power system. *IEEE Trans. Smart Grid*, 1, 2, 199–204, 2010.
28. Loh, P.C., Li, D., Chai, Y.K., Blaabjerg, F., Autonomous control of interlinking converter with energy storage in hybrid AC-DC microgrid. *IEEE Trans.*, 3, 49, 1374–1382, 2013.
29. Khan, F.A., Pal, N., Saeed, S.H., Review of solar photovoltaic and wind hybrid energy systems for sizing strategies optimization techniques and cost analysis methodologies. *Renew. Sustain. Energy Rev.*, 92, 937–947, 2018.
30. Sahoo, S.K., Sinha, A.K., Kishore, N.K., Control techniques in AC, DC, and hybrid AC-DC microgrid: A review. *IEEE . Emerg. Sel. Top. Power Electron.*, 6, 2, 2017.

13

Operation and Control of EV Infrastructure for Microgrid

Tamilarasu Viswanathan[1], B. Gunapriya[2], M. Mathankumar[1*], S. Suryaprakash[1] and M. Mohamed Iqbal[3]

[1]Department of Electrical and Electronics Engineering, Kumaraguru College of Technology, Coimbatore, India
[2]Department of Electrical and Electronics Engineering, New Horizon College of Engineering, Bengaluru, India
[3]Department of Electrical and Electronics Engineering, PSG Institute of Technology and Applied Research, Coimbatore, India

Abstract

New electric vehicle charging infrastructure is proposed in this chapter to implement the enhanced operation and control of microgrids for commercial practices. The development of intended solutions was initiated with the study of different sources of expansion in the microgrid between 2018 and 2021. The prediction value was computed for 2022 to 2025. The four charging levels (level 1, level 2, level 3, and level 4) are classified from the literature to utilize the existing resources in an optimal way. Changes identified in the charging infrastructure are reflected in the existing process through charging levels based on the number of electric vehicles and environmental conditions. The voltage profile in a day is computed for each charging level to normalize the number of electric vehicles utilizing the same infrastructure to improve the commercial systems. The mentioned parameters are loaded in the new forecasting model with a minimum convolution level. The proposed forecasting model's level of improvement demonstrates that, with the existing charging infrastructure, improved microgrid operation is possible. Location-wise classification is performed for the state of charging for all the days in a year with possible profit improvement. Furthermore, the charging delay in each level of charging was simulated with an incremental cost saving for different electric vehicle counts. Finally, the total electric vehicle counts revealed enhancements in a day after the computation of actual and expected movement. The energy

**Corresponding author*: mathankumarbit@gmail.com

P. Sivaraman, C. Sharmeela and P. Sanjeevikumar (eds.) Microgrids for Commercial Systems, (345–380) © 2024 Scrivener Publishing LLC

saving improved by approximately 5% to 15% for every 5-kWh charging, which shows the future adaptability of the proposed method.

Keywords: Electric vehicles, charging station, PV panel, cost optimization, microgrid, commercial system, energy savings, infrastructure facility

13.1 Introduction

The microgrid is fueled by a variety of capacity sources, including energy storage systems, tiny gas turbines, solar panels, and wind energy [1]. A facility for charging electric vehicles is also provided by the microgrid. Electric cars plugged into a charging station operate as vehicle-to-grid devices that can provide power to the microgrid or control how quickly they charge. The charging station may therefore be perceived as a variable load or producing unit. The microgrid's capacity expansion planning is done to increase the capacity of the battery energy storage system, solar panels, wind turbines, and micro-turbines. Through a long-term plan, this capacity increase is carried out for a 6-year planning horizon. The hourly functioning of the micro-turbine, energy storage system, and electric car charging station, respectively, are all optimized as part of the short-term strategy. The efficient use of dispatchable resources in the short term lowers the cost of planning and effectively contributes to the long-term strategy for cost minimization. The wind system expansion, which pays for the expansion costs, undergoes the biggest expansion. When dealing with uncertainty, the model also requires additional resources, which raises the cost. Urban areas are facing a dangerous scenario because of rising atmospheric adulteration brought on by the exponential rise of urbanization, industrialization, and the growing number of on-road automobiles. A distributed generating and transportation infrastructure powered by alternative fuels is urgently needed to lessen the consequences of this growing problem. Using a switching mechanism from one trading market to another, an ideal architecture for an energy management system for public electric car charging stations connected with the community microgrid is suggested in this study. The proposed framework is implemented by creating a mixed-integer, nonlinear problem with multiple trading locations. The objective is to maximize the selling price of the excess energy while minimizing the cost of energy to charge the vehicle and fulfil community load demand [2]. The use of electric vehicles has significantly expanded during the last 10 years. Utilizing vehicle-to-grid and grid-to-vehicle strategies can utilize temporary energy storage devices for numerous purposes. The quantity of energy drawn from the electrical grid may be cut back with the help of renewable energy sources, but the electricity produced from these resources varies and is unreliable. Although it can be a crucial part in reducing these harmful impacts, the technology is still highly costly. As a result, this study investigates the viability of employing a vehicle temporarily to meet system requirements. To determine the number of vehicles that reduce the overall consumption of electrical energy pulled from the grid, a charging/discharging algorithm is provided [3]. By comparing the amount of grid electricity consumed before and after the construction of charging stations, the results demonstrate the energy savings of the recommended methodology. Installation of charging stations minimizes the energy consumed from the grid during working hours in this case study. This demonstrates an increase in the percentage of locally generated energy that is used, thus lowering the amount of energy pumped into the grid for later use and lowering grid storage costs. The microgrid has become a framework *via* which energy sources might be annexed for effective network functioning because of the increasing integration of renewable generation and advancements in the power system toward smartness. However, the sources must be carefully chosen to work together in a synergistic manner to eliminate sporadic obstacles to the fruitful production integration of renewable generation and advancements in the power system toward smartness. The load-following hybrid solar photo voltaic (PV) and small hydro-microgrid with a grid-isolated electric car charging system is presented in this study. Based on the nodes' local measurements, a decentralized multi-agent smart voltage network with reactive power compensation automatically manages and monitors the network limitations. When load demand is at its highest, the solar system supports hydropower, and when irradiation is at its lowest level, hydropower charges solar storage. Individual electric car charging is set within the energy balance during surplus production [4]. The public grid is put under additional strain due to the rapid growth of electric cars, which also generates load swings and slows down high penetration. This study proposes a real-time rule-based system for direct current microgrid-powered electric car charging stations. An electrochemical storage system, a link to the public grid, and solar sources make up this microgrid concept. The management model of the charging station incorporates discrete events, whereas the charging station model is based on data-driven modeling. Most of the drivers' preferences are considered since this study focuses on the power management technique of a charging station under power limitations. The architecture of the charging system is given, and typical issues that arise while charging are covered [5]. Numerous voltage concerns, such as voltage swings in the power system, have been brought on by high penetrations of renewable energy sources in microgrids. Through connections between energy storage systems and power electronics, these variations may be managed. As a result of recent advancements in electric automobiles, technologies can now replace traditional storage devices. This study suggests an integrated islanded microgrid that supports voltage management while facilitating energy storage. A voltage controller that adjusts voltage by injecting or pulling active power from the charging station has been modeled based on the active droop feature. To concurrently meet the demands of microgrids, regulated charging and discharging are used to accomplish the needed active power management. A power distribution control strategy has also been developed [6]. The market for electric vehicles is expanding, which is a significant step toward achieving urban mobility that is carbon-free. The integration of renewable energy production into an urban microgrid in accordance with an optimized model of energy management addresses upcoming network and smart city concerns. This project intends to define a charging station powered by a microgrid powered by photovoltaics as an intelligent infrastructure designed for recharge in an urban area. Interactions, the public power distribution system, and the nearby building can all be made easier by this method. The study offers crucial components to inspire the stakeholders to develop in accordance with social expectations, urban planning, and the objectives of sustainable development. The study also emphasizes cross-disciplinary research positions that show the necessity of a holistic approach to stay central [7]. In the contemporary power system, a hybrid microgrid-powered charging station lowers transmission losses with better power flow regulation. The hybrid microgrid, however, cannot effectively use the associated renewable energy sources when battery electric vehicle charging is not coordinated with it. Additionally, a multiport charging facility is included in the development plans for the next charging stations, overloading the power grid. [8] examines the following technical issues: energy management strategy, converter control, and effective utilization of multiport charging from a photovoltaic source as well as maintenance of bus voltage in the face of utility grid overloading, which can be caused by either local load or insufficient power through its energy storage unit. The use of fossil fuels and the presence of car fleets in isolated microgrids, like those found on islands, present serious environmental problems. Electric vehicles and distributed energy resources based on renewable energy have been introduced in several locations to address this. They do, however, raise issues with planning and operations. This paper's goal is to present a two-level microgrid problem. The first issue takes into account a charging method that maximizes the utilization of renewable energy sources while minimizing the charging costs. The outcomes show the advantages of investing in more PV generation and in the charging method from an economic and environmental standpoint [9]. Benefits to the economy, environment, and society encourage grid-connected microgrids to spread quickly. The main objective of commercial microgrids continues to be their economic viability. A stationary electrical energy storage device is frequently at the heart of microgrids, enabling islanding and efficient main grid use control. Because of this, an earlier study concentrated on the size of stationery. The economically ideal fixed capacity has been significantly altered with the introduction of large-scale electric car charging hub microgrids. Research on how the charging technique affects the economically optimal capacity of stationery is still lacking, even though it is well thought out to aggregate and then compensate for stationery. [10] suggests a technique for making scheduling decisions that incorporate a mixed-integer linear programming model to close this gap. Globally, the building of microgrids is encouraged as a solution to the issues of energy disparity in island regions and the consumption of fossil fuels. To ensure economic viability in the deployment of a microgrid system, it is crucial to assess the capacity of renewable energy sources and storage systems. There are island regions in certain microgrids where the policy is to consider both the installation of the microgrid system and the augmentation of electric cars. The necessary capacity may rise, though, if the load pattern differs from the solar radiation pattern. To avoid the requirements of vehicle-to-grid technology and real-time pricing, it devised and tested a method for lowering the cost of a microgrid system by utilizing a regulated charging load [11]. In the future, the number of charging stations will explode due to the rapid growth of car electrification. In the meantime, the power supply is experiencing large power variations due to the increased demand for charging power and the stochastic charging behavior patterns. More auxiliary power reserves, adequate demand-side energy management, and other technological measures are required to guarantee system stability and the supply–demand balance of the electrical market. Furthermore, with the right system management techniques, the increased penetration level of renewable energy sources in the conventional power system may be used to support the power peak brought on by the charging demand. [12] outlines several significant problems and challenges while providing an overview of recent research on charging stations. Micro-grid has evolved into a micro-grid technology research field, with benefits for systems that rely heavily on new energy generation. [13] presents a hybrid energy storage technique that combines flywheel and battery energy storage, using a micro-grid system of solar power-generating electric car charging stations as the research object.

High-frequency power fluctuations and some low-frequency powers are stabilized using flywheel energy storage. To keep the bus voltage balanced, the reference power is balanced using a battery. The microgrid's composition and different operating modes are examined, and a hierarchical coordinated control based on voltage monitoring stages is suggested. The microgrid's composition and different operating modes are examined, and a hierarchical coordinated control based on voltage monitoring stages is suggested to work under various operating conditions, such as fluctuating input power, shifting charging conditions, and battery over discharge. [14] suggests a productive planning algorithm for placing intelligent electric car charging stations in rural areas. A collection of distributed generators and charging stations are jointly allocated and sized in the planning issue to balance the supply with the overall demand of normal loads and charging. The deployment and operating costs and related greenhouse gas emissions are two competing goals that must be minimized while meeting the technical requirements of the microgrid. The planning algorithm determines the ideal locations and sizes of the charging stations and units to achieve this. To accomplish this, an iterative solution of a multi-objective mixed-integer non-linear program is used. Using a non-dominated sorting genetic algorithm, an outer sub-problem determines the positions and dimensions of the units and charging stations. [15] aims to enhance the energy management performance of various microgrid systems under the uncertainty associated with electric car charging considering the growing expansion of decentralized power systems and electric automobiles. To reduce the transmission losses, operational expenses, and carbon emissions of various microgrid systems, a multi-objective optimization model is developed. First, a novel approach to anticipating electric car charging demands is suggested, based on a back-propagation neural network enhanced by deep learning for short-term memory. Based on the forecast data, a two-layer solution algorithm is suggested, consisting of a modified consistency algorithm for quick economic scheduling at the single microgrid layer and an adaptive multi-objective evolutionary algorithm based on decomposition and differential evolution at the multiple microgrid layer. [16] presents a novel decentralized control method for charging stations based on a medium-voltage direct current bus. This is integrated into a microgrid with a photovoltaic system, a battery energy storage system, a local grid connection, and two units of fast charging. The main contribution resides in the cited model based on fuzzy logic that includes the state of charge of the battery as a control variable. This control contains two independent fuzzy logic systems whose function is to maintain the voltage and the battery within proper thresholds and to keep the power balance stable among the units of fast charge and the rest of the components. The new control method was tested in many operating situations under various conditions of sun irradiance and initial battery charge, with the goal of demonstrating its correct performance in all scenarios considered. The microgrid changed its operating mode to either discharge more power at a higher price or charge more power at a lower price to assure economically sound operation [17]. The cost of system stability depreciation was compared with commercial pricing. Bus gesturing is utilized as a control signal to achieve an independent, decentralized functioning of the microgrid at the transitional bus voltage regulation level. [18] is used to optimize the size and architecture of an autonomous microgrid. The aim is to show that the algorithm may be used to construct more dependable and affordable microgrids. An autonomous grid is created with this objective in mind. Meteorological information as well as the building's annual energy use have been monitored and documented. A microgrid topology made up of photovoltaics, wind turbines, battery energy storage devices, and diesel generators is suggested based on the data. To attain the low cost and reliability goals with real autonomous features, it is essential to determine the capacity of these components that work best together. On the one hand, oversizing that can result from arbitrary evaluations without any size optimization will raise the expenses of the initial investment, ongoing operations, and maintenance. To simulate a standalone polygonation power plant for a strip mall, an in-house simulation program has been created. A solar panel array has been connected to batteries, unitized regenerative polymer electrolyte membrane fuel cells, and diesel generators as main storage and backup systems, respectively. Then, it was determined whether the same plant structure and control method could be used in eight different temperature zones across multiple states [19]. The findings demonstrate that, excluding a reduction in the initial investment costs using incentives, the deployment of these systems is still unfavorable, even under the most advantageous conditions in terms of daily radiation and average temperature. It is not anticipated that the microgrid will be out of commission for a protracted period if severe weather, such as a hurricane, tsunami, or other similar disasters, causes the microgrid to be isolated from the main power system. Making the electrical system completely dependable to provide service for the consumers is also not economically feasible [20]. The strength of the study resides in its capacity to identify the ideal compromise between operational expense and grid resilience. [21] creates power references in real time for the developed vehicle storage controllers and offers a centralized solution to the economic dispatch problem of parking lots. In contrast to the conventional vehicle storage controller, the proposed optimization-incorporated distributed vehicle storage controller makes use of a sparse vehicle-to-vehicle communication network and may transition between decentralized and distributed control modes depending on the scenario. By using interlinking converters, the power flow between the AC and DC sub-grids is controlled. Combining droop control strategies in the power control loop enhances the control structure. This improvement makes possible the simultaneous adjustment of AC and DC bus voltages. To verify the effectiveness of the created architecture with a genuine business network and loads, case studies are conducted. Plug-in electric car use in microgrids has the potential to promote environmental sustainability and reduce on-peak demand, in addition to improving economic efficiency. As a result, a two-stage energy management method is discussed in this research for the contribution of demand response programs to commercial building microgrids. To optimize energy management [22], this study incorporates the volatility of power costs into a model-based, predictive control-based strategy. To best manage energy in the commercial building, the optimization issue first examines the use of wind power. Second, the control section receives the total charged power reference that was computed in this stage so that it might be assigned. Distribution networks are getting more complex than ever due to the diversification of energy generation, which is always expanding, and the decreasing cost of energy storage. The distribution networks and end users can both profit from this complexity when it is used to create managed microgrids. Microgrids are entire designs of hardware and software for functions and users rather than just distributed generating and energy storage systems like solar panels and battery banks [23]. To offer reliable, effective, and sustainable electricity to the end user, all the various components must be built to cooperate. This research introduces an experimental hybrid microgrid testing facility for high-efficiency distribution systems integrating networks. This testing facility offers a research testbed for examining various microgrid system elements. Clean renewable energy may be used in microgrid systems, which can run independently and provide electricity to isolated locations that are inaccessible to the main grid. Even though many power providers are now interested in microgrids, others contend that renewable energy is not as ecologically beneficial as it is portrayed to be. Through a life cycle evaluation, this research examines the environmental effects throughout the course of a microgrid's life cycle and its energy payback period [24]. Regular blackouts happen on a weekly or even daily basis, interrupting the provision of electricity. A grid's unreliability can be caused by a variety of factors, including poor network or infrastructure, inclement weather, and problems with electricity generation. However, the issues that sites confront because of an unstable grid are typically the same [25]: output loss, decreased staff productivity, and potential equipment damage. Power outages on crucial production lines frequently result in damage to the work-in-process as well as raised operating expenses. A business case is presented to show how microgrids are a viable substitute for a glass manufacturing facility that frequently experiences outages. The design and implementation of a reliable photovoltaic-based microgrid data gathering system that can concurrently monitor several systems is the main focus of this article. Three solar systems make up the microgrid: off-grid, hybrid, and grid-assisted systems. To detect and record electrical and climatic factors, a system for connecting and constructing a collection of sensors was devised. For calibration, voltage signal conditioning was used, which helps to eliminate errors caused by analog signals. The design primarily aids in determining the microgrid power plant's solar energy potential considering the remote community's energy requirements [26]. Additionally, while comparing the logged data, the simultaneous monitoring of the three systems guarantees that the external working circumstances are the same. The standards introduce the global grounding system, which guarantees that none of the interconnected grounding systems will experience dangerous touch voltages. In practice, this results in the creation of an intrinsically safe grounding system because it allows for touch voltages up to the conventional limit value to be applied continuously. For the utilities to be responsible for confirming its presence, the linkage of the grounding systems through the cable shields of the utilities' distribution helps to create a global grounding system. A grounding system must meet safety requirements in relation to the utility's disclosed point-of-delivery specifications [27]. The grounding systems in urban and commercial regions frequently interfere with one another and are connected by common exterior conductive elements. Data center-integrated commercial buildings that are converted to microgrids benefit from improvements in cost, dependability, and resilience as microgrids have developed from early prototypes to more mature technology. To make extensive microgrids commercially viable, microgrid design and economically scaling distributed energy resources are becoming more challenging. To maximize all the benefits that a hybrid microgrid architecture may provide, an optimal size formulation has been given in this research. Energy storage, photovoltaics, and power electronics devices are co-sized for cost-effective, grid-connected, and reliable island operations [28]. To achieve the highest level of cost reductions, time-of-use energy charges and peak demand costs are specifically modeled. The advantages of the suggested technique are illustrated by numerical results from a genuine commercial building load. Although the number of electric vehicles in fleets is rising, widespread adoption would not be achievable without a suitable charging infrastructure. Such infrastructure should be built in a way that considers both the environment in which it is built and the driving habits of electric car drivers. If these factors are ignored, there is a chance that the needs of the users may not be met, leading to increased expenses. First, let us quickly go over the challenges generated by the installation of charging infrastructure, specifically the technical, financial, and user acceptance issues. The objectives of the literature's infrastructure location models are next examined. maximizing the service offered at a given cost or minimizing the infrastructure charges for a certain level of service [29]. The future of road transportation will likely be dominated by electric vehicles, which also have a great chance to save the earth from the potentially catastrophic effects of global warming. Compared with internal combustion engines, electric vehicles have several advantages, particularly in terms of greenhouse gas emissions, energy efficiency, and dependence on fossil resources. Despite this, the quick adoption of electric vehicles is hindered by range anxiety and the lack of adequate charging infrastructure. In-depth reviews of the literature are presented on the most recent infrastructure deployments, difficult problems with their implementation, charging power levels, various charging techniques, and the effects on the electric grid [30]. The move to electric vehicles has been aided by developments in battery technology and their production methods as well as the depletion of fossil fuel resources, energy dependence, rise in fuel prices, and environmental issues brought on by fossil fuel vehicles. Various internal and external variables have affected how quickly some nations have absorbed the new technology compared with others. The country's industry, legislation, R&D initiatives, charging infrastructure, battery market, and infrastructure were all assessed. To highlight the flaws in the current charging infrastructure, a map of a charging station density was created [31]. The infrastructure should be improved much more, and related incentives should be addressed in more detail. Electric vehicles are becoming more and more popular as a new form of transportation, offering a huge opportunity for the electrification of transportation. This is due to the increase in vehicle driving range and the drop in vehicle costs. However, most of the current research in charger planning are focused on finding the best site from a global optimization standpoint [32]. A predict-then-optimize diagram is provided to anticipate the charging demand using a multi-relation graph convolutional network architecture and to optimize the competitive resource allocation strategy to achieve a competitive strategy in the future. To find the Cournot competition equilibrium with the convergence analysis, a new parallel computing approach is proposed. As a new mode of transportation, they are becoming more and more popular as a significant chance for the electrification of transportation because of their longer driving range and lower vehicle costs. However, most of the current research in charger planning is focused on finding the best site from a global optimization standpoint. To find the equilibrium of the competitive situation using convergence analysis, a new parallel computing approach is proposed. To determine each service provider's charger number in each zone, the optimal charger sizing problem is also considered [33]. The outcomes demonstrate the potency of this work's suggested strategy, which may bridge the knowledge gap between data and the best charging infrastructure planning. [34] proposes a bi-level optimization algorithm for modeling the design and utilization management problem of electric vehicle charging infrastructure with user-equilibrium choices. The lower level seeks to reduce the users' journey times and charging costs, while the top level seeks to reduce overall facility deployment costs and increase income from charge collections. The bi-level mixed-integer non-linear program is solved using an iterative method that provides theoretical lower and upper bounds for the bi-level model and solves it to global optimality. To demonstrate the convergence of the algorithm in a limited number of iterations, a set of criteria is assessed. The findings show that the suggested bi-level model is capable of accurately identifying the ideal site, physical capacity, and demand-responsive pricing strategy for charging facilities. [35] is based on precise, annual data on the distribution of public charging infrastructure and the ownership rate of battery electric vehicles. It concentrates on regions where the initial infrastructure for public charging was set up during this time. The installation of the first public charging station sparked uptake in these mostly rural regions. Observe an average rise in the local ownership rate of electric vehicles. This work's findings are resistant to the impact of anticipation. Additionally, they are unaffected by various treatment levels. Identify public charging infrastructure as acting as a stimulant to the spread of battery electric cars, even if it cannot entirely rule out adverse impacts. One of the key factors in reducing transportation-related emissions has been identified as vehicle electrification. The secret to securing the environmental advantages from the broad adoption of electric cars is the proper location of charging facilities and control of charging operations. The distribution of geographical and temporal chances for charging electric vehicles as well as the establishment of charging infrastructure across areas are examined in this article using empirical trip trajectory data. [36] offer a mixed-integer linear programming integrated optimization platform for predicting the placement of electric car charging infrastructure in both home and non-home sites, including infrastructure costs and dynamic energy pricing. Electric vehicle charging infrastructure is widely regarded as an essential public service and strategic asset for facilitating the transition as electrification emerges as one of the most viable paths for a low-carbon transition in the transportation industry. It is important to optimize the environmental advantages with limited resources, given the considerable investments needed for deployment and the pressing need to prevent climate change. However, most recent studies pay little attention to the environmental perspective and instead concentrate on economic efficiency. As a result, [37] uses the Malmquist method and a slacks-based data envelopment analysis model. This work undertakes a determinant analysis to explain the discrepancies and give theoretical support for future advancement. It reveals that resource usage and energy structure have a substantial influence on mitigation efficiency. Since the industrial revolution, electrification has played a significant role in both technical advancement and societal growth. [38] has boosted innovation, enhanced productivity, and improved living standards in all spheres of society and economy. The energy transition to zero-carbon-emissions-based economy to combat climate change is predicted to drive electrification in the transportation, construction, and industrial sectors further and deeper during the next several decades. A data-driven, realistic model of charging demand that accounts for the wide range of charging habits of potential users with a thorough economic dispatch model, [39] analyses how charging control and infrastructure development affect charging load while assessing the grid's influence on the fast adoption of electric vehicles. It has been discovered that full electrification and predicted adoption lead to an increase in peak net power consumption. High home charging rates and locally customized settings might put pressure on the grid. Switching to unregulated daylight charging can instead cut down on the amount of non-fossil fuel production, ramping, and emissions. This work's findings compel decision-makers to implement charging infrastructure that encourages a switch from at-home charge to daytime charging and to adjust utility tariffs to account for generation-level consequences. To reduce the transportation sector's carbon impact is turning to electric mobility as a crucial priority, but the uptake of electric vehicles is sluggish, mostly because there are not enough reliable charging stations [40]. The fascinating question in building charging infrastructure is which organization should make the investment. Here the automotive supply chain is examined, and this work presents distinct strategies for building charging infrastructures for manufacturers that invest in setting up the infrastructure with funding. The development of the infrastructure for charging requires the most work. Additionally, social welfare is at its peak in these two situations, which defies logic given that the Model has greater backing. The most effort is required in developing charging infrastructure. Furthermore, social welfare is at its peak in these two conditions, which defies logic given the Model's larger support. A feasible alternative to gasoline-powered vehicles for reducing pollution and reliance on fossil fuels. Range anxiety, a small operating range, a lengthy charging process, and a lack of adequate charging infrastructure have all hampered adoption. High-end batteries and chargers can increase market share and minimize charging demand and time, respectively. The primary contribution of this study is the development of a realistic modeling framework that takes intercity needs into account and uses reasonable assumptions and parameters that have been calibrated through numerous stakeholder meetings. It also considers the effects of various technological advancements on infrastructure requirements [41], so future research seeking accurate data can make use of the study's parameters and conclusions. Infrastructure for charging electric vehicles will be essential to the decarbonization of the transportation industry over the coming decades. Power electronics converters' function as an energy transition technology in the broad acceptance and effective utilization of sustainable energy sources will therefore be further enhanced. However, this profound transformation is not without its difficulties. To name a few, these difficulties include the reliability of power converters and the stability of the power system due to the loss of system inertia [42]. Some of these difficulties are already present, such as the slow deployment of charging infrastructure and competing charging standards. However, there is a natural shift toward power systems that use more electronics and batteries. There is an urgent need for new business models since low profit margins have become a substantial barrier to infrastructure investment and operation for electric car charging. However, non-binding regulations and ambiguous roles provide a social conundrum that makes promoting charging stations in urban residential areas challenging. Through an analysis of potential incentive systems involving the government, operators of the charging infrastructure, real estate firms, and owners of electric vehicles, the viability of solving this conundrum was investigated. [43] develops a theoretical model based on the tactical interactions between various actors in promoting charging stations in urban residential areas by utilizing evolutionary game theory. According to this work's findings, the ideal situation is one in which all parties collaborate closely to spread awareness about charging stations. Electric vehicles encourage an emission-free mode of transportation by reducing pollution. An effective and efficient nationwide network of charging infrastructure may be necessary for the widespread adoption of electric automobiles. An effort has been made to explore and assess the risk variables in this research work. The risk variables are categorized into primary groups, including financial, market, political/legal, and operational risks, with the aid of a literature analysis and feedback from industry experts. This method is used to identify the risk variables that are connected to them and rank them appropriately [44]. The usage of solar power generation might be improved by the widespread integration of battery energy storage systems in distribution networks, which would also help to reduce the negative impacts of electric car fast charging. An innovative deep reinforcement learning-based power scheduling method is presented in [45]. The network consists of electricity arbitrage from the main grid, power generation, and quick charging demand. To increase operator profit while preserving voltage limitations, forecasted power generation and smart charging demand are necessary. The two main obstacles to the adoption of electric car charging infrastructure are accessibility and pricing. While significant attempts are being made to find such socio-demographic and transportation characteristics that link to the present distribution of charging stations, it analyzes the existing charging infrastructure. This work's findings show that population density and the density obstructing adoption and decarbonization targets are not connected. On the contrary, there is a strong bias against low-income people in the allocation of charging stations [46]. However, the existence of roadways was positively associated. The findings highlight the requirement for equitable and reasonable policy frameworks for implementing charging infrastructure. The transportation sector's increased use of conventional gasoline led to the fuel's quick depletion. As a result, the key study focus during the past 10 years has been the electrification of the transportation sector. In recent years, it has been observed. The increase in traffic density had a significant negative impact on the current electrical system and affected grid dependability. Some remedies must be adopted to mitigate the negative consequences of an increase in charging stations to meet demand [47]. The difficulties with adoption are also explored. The difficulties are also explored with adoption of future developments in the industry, such as the use of renewable energy sources and the advantages of car-to-grid technology. In order to promote the widespread use of personal electric vehicles, a sufficient charging infrastructure is essential. Urban floods, whose frequency and severity are rising because of climate change, might be a barrier. Here it is investigated how flooding-related disruptions affect public charging networks regionally [48]. While there is no discernible difference in the battery's capacity to support regular urban driving patterns, it does see an excessive strain on the chargers, both close to and unexpectedly far away from the flooded areas. [49] gives information on the circumstances in both developed and emerging countries. The perception of the experts may have constraints on a focused group study. A crucial step in fostering electric vehicle adoption is the creation of a public charging infrastructure network. This will help to lower greenhouse gas emissions associated with the usage of conventional vehicles and enhance the local environment by lowering air pollution. The feasibility, effectiveness, and city strategic goals for building a network of public charging infrastructure are discussed in [50]. It is used to visualize how these initiatives might benefit the public in the long run and in a sustainable manner. According to the findings, the analysis of the charging infrastructure network implementation strategy shows that a variety of public and commercial stakeholders must be involved. In addition, pertinent and essential patterns and ideas are offered, which could assist other public managers in scaling up the use of this service and business model. The objective of this work is to study the existing expansion in energy resources. Four different levels of charging communication are introduced between the source and electric vehicle facilities, hourly changing level locations are studied based on the locality for incremental cost saving, energy improvement is based on voltage profile combined with the number of electric vehicles, and the performance of commercial EV charging stations on microgrids is simulated. Additionally, the commercial systems on a weekly and monthly EV handling delay for effective resource management is performed. Various incremental cost savings of electric vehicles are computed for arrival and departure history.

13.2 Proposed Electric Vehicle Charging Infrastructure for Enhancing Microgrid Operation

The study of the year-by-year expansion level of microgrid resources with the use of electric vehicles increased the projected value for the year 2018 in all resources, as shown in Figure 13.1 (a) to (h). Figure 13.1 (a) to (d) represents the available capacity of renewable energy resources for the years 2018 to 2021; for every year, projected value and the installed capacity of PV, wind turbines, conventional generator, and battery storage are shown—for example, the predicted value of PV is around 125% of existing, but the actual capacity is only 25% in 2018, but the wind turbine installed capacity is 40% compared with the project value of 35% in the same year because of market requirements. From the available data in the energy department, the projected and installed capacity, respectively, is not

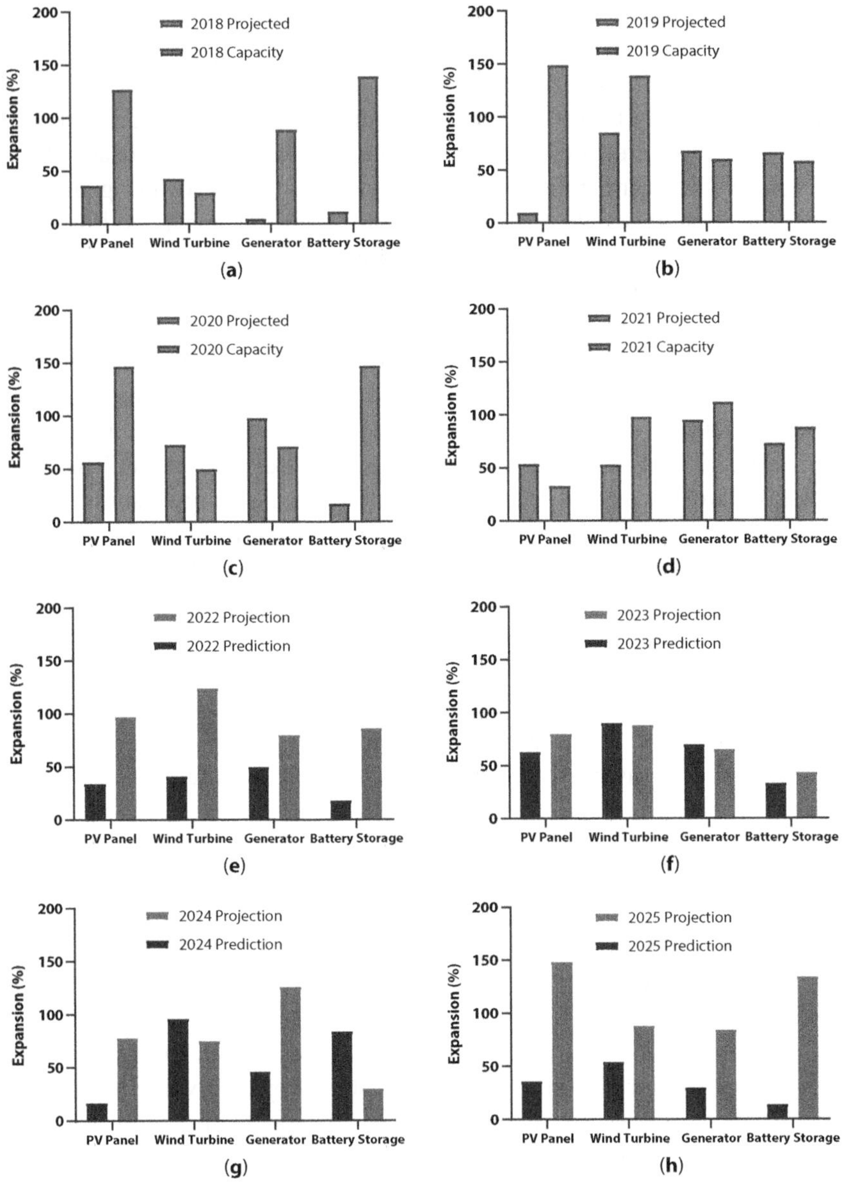

Figure 13.1 The use of electric vehicles increases the level of microgrid resources year after year: (a) 2018, (b) 2019, (c) 2020, (d) 2021, (e) 2022, (f) 2023, (g) 2024, and (h) 2025.

to be linear. Similar to 2018 to 2021, the future predicted and projected installation capacity, respectively, presented in Figure 13.1 (e) to (h) follows non-linearity. In the forthcoming year, renewable energy resources will maintain the project level in comparison with generators and battery storage. After 2021, the projection values match the predicted values for most of the resources, indicating that the electric vehicle requirements vary the microgrid resources equally with real-time energy requirements. There needs to be a new level of electric vehicle charging infrastructure to improve the existing system.

The arrangement of different sources connected to the DC bus with EV facility through charging requirements and updated microgrid status is shown in Figure 13.2. The level 1 charging facility is equipped with a charging facility with energy source control units through glassy communication, balancing the EV status such as DC bus and number of arrivals and departures from the EV facility. The number of arrivals and departures facilitates the utilization of sources based on the availability of the

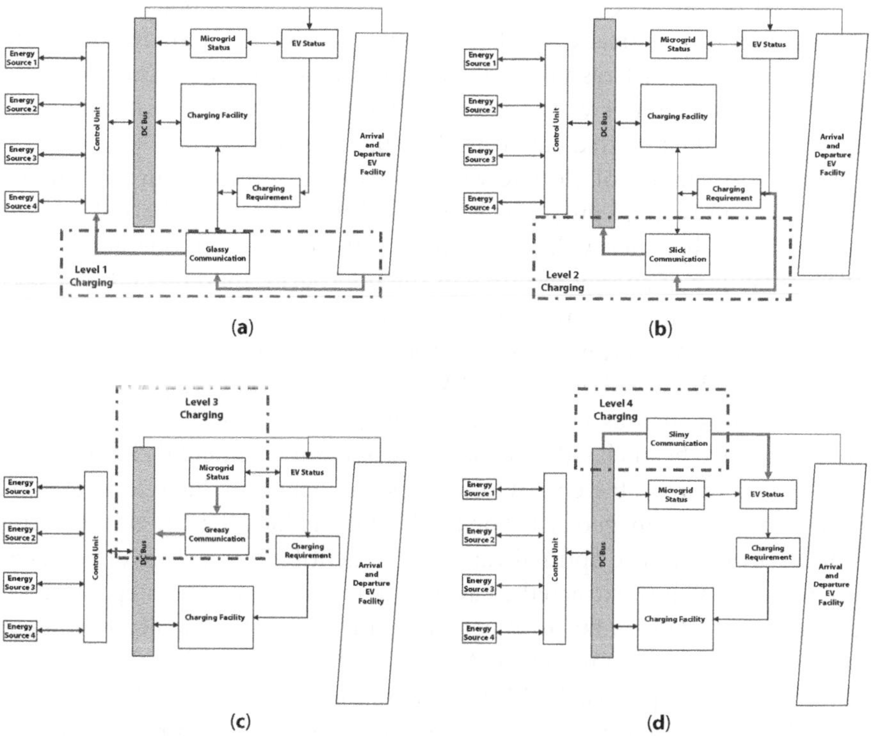

Figure 13.2 System-level electric charging for battery and control: (a) level 1, (b) level 2, (c) level 3, and (d) level 4.

different sources. In general, synchronizing between the energy source and the location of the EV station is not considered for level 1 charging. Hence, the changing nature of these commercial systems is not found, but a generalized improvement is achieved. Level 2 charging is the same as the glassy communication, but the number of computed variables is reduced with the generalized DC bus status and charging replacement unit in the EV station. The full process is done with minimal variation, not disturbing the energy sources and microgrid. The number of improvements is also limited with the charging unit and fails to meet the wide range of requirements. The gray communication feature was introduced as a level 3 charge between the existing system with microgrid status and the DC bus. In practice, the number of computations is higher than all other levels. Additionally, variable-level fine-tuning is possible without considering more than one block of control. The power fluctuations and control units are reduced to a small number of variables. Slick communication bypasses the existing microgrid and EV status with the DC bus and contacts the EV facility directly with the vehicle control count. The number of variables is limited to four, and cost minimization is the primary objective of level 4 charging. In practical cases, the system accepts the rapid change in the requirements of fast responses in the commercial system. The proposed system is designed to switch between the different levels of control. It optimizes the EV facility in a day from the existing study.

The optimization of electric vehicle charging infrastructure in microgrids for commercial purposes is initiated with the flow chart shown in Figure 13.3. This energy flow follows an immense loop of operation from the availability of energy resource in the microgrid. Incremental cost saving and energy improvement is balanced for every EC count. This adapts for every level of charging assessing the vehicle count at that charging station. The study considers various EV charging station load profiles in various locations such as villages, rural villages, towns, and cities. Different locations have been classified as seasonal and environmental changes vary cost-wise and available energy resources are identified. Each level of control is uniquely managed based on the EV facility location and the charging station priorities. The optimal variation is identified in each level to match the proper DC bus extracted to avoid the loss drop step by step. Some of the energy resources fail to meet the entire year's EV facility fulfilment from equations (13.1) to (13.3). All the blocks mentioned in the standard microgrid are improvised with the techniques mentioned in [5], [9], and [14]. Power transfer among the resources is combinedly monitored with level and location points. The selection of the level between each combination is computed with the formula as follows:

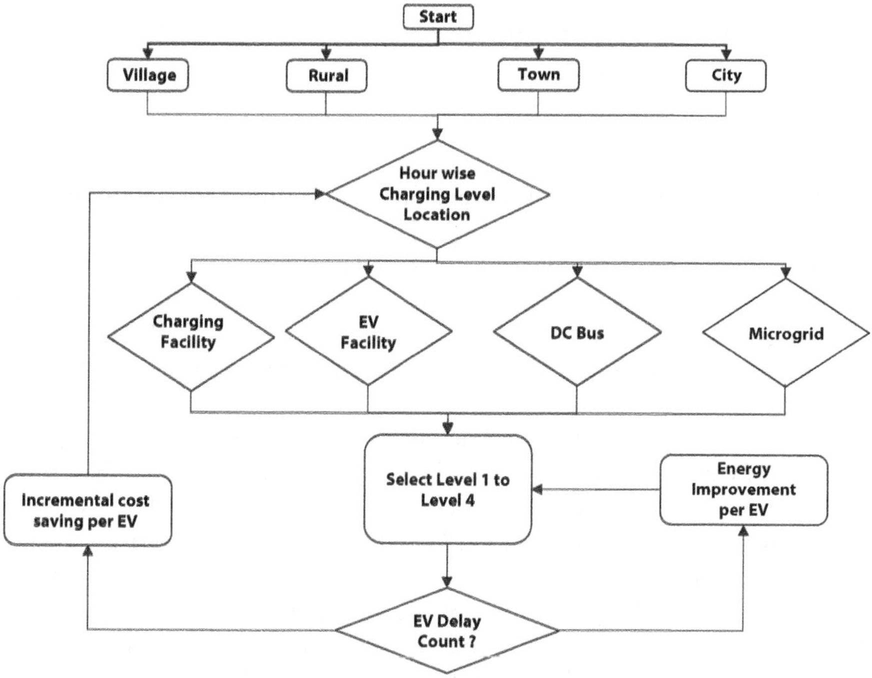

Figure 13.3 Flow chart of the proposed electric vehicle charging scheme with microgrid for commercial systems.

$$P_{MG} \underset{P_{EV}}{\overset{P_{DC}}{\rightleftarrows}} \left.\boxed{P_d + P_a}\right|_{t_d} + \left.\boxed{P_d + P_a}\right|_{t_a} \tag{13.1}$$

The power variation at each level was shifted and introduced as in commercial systems.

$$P_{MG} \underset{P_{EV}}{\overset{P_{DC}}{\rightleftarrows}} \left\langle \begin{matrix} P_{L1} \\ P_{L2} \\ P_{L3} \\ P_{L4} \end{matrix} \right\rangle \tag{13.2}$$

The total number of electric vehicles counts towards handling the maximum number of resources. It is used at any point in time, as expressed with each level of hour variation.

$$E_{C(EV)} = \left\langle \left| \frac{\max\left(E_{D(EV)}\right) + \min\left(E_{A(EV)}\right)}{\min\left(E_{delay(EV)}\right)} \right| + \left| \frac{\max\left(E_{D(EV)}\right) + \min\left(E_{A(EV)}\right)}{\min\left(E_{delay(EV)}\right)} \right\rangle \right. \tag{13.3}$$

The systematic sequence of power sharing in commercial systems is represented by Equations (13.1) to (13.3). Furthermore, each level has its own communication protocol that is represented as the commitment function. Initially, the level 1 power condition was the DC bus combined with the charging unit from the EV facility. The number of functions decided by the instant electric vehicle availability in the entire system varies with the time and location of the stations.

In this work, the type of electric vehicle is assumed to have the same capacity of voltage profile for the study. The complete variation in handling the electric vehicle oscillates between the positive and negative voltage profile as presented in Figure 13.4. When the number of electric vehicles reaches a maximum count of 100, the voltage profile is considered as a unity profile to identify the probability of the variation as simulated in Figure 13.4 (a). The voltage profile sticks out on the hour-wise vehicle count in the blue dots mapped from the red spherical lines. It indicates that the availability of voltage in the charging station makes an impact on the hour-wise handling for each level which is not unique. Hence, the solution should be different and capable of considering the restraints in the mentioned power equations. Figure 13.5 (a–b) depicts the level of charging available at stations throughout the year for all days and locations. The modeled value represents charging that reaches a maximum of 90% from 35% during the night after charging from 35% to 55% in the evening. The drastic discharge occurs in the morning and can reach up to 15% throughout the day, with a slight improvement of 5% in the afternoon. The extension of charging level leads to simulated values in terms of vehicle delay with incremental cost saving as presented in Figure 13.6 (a–d). This boundary identifies the microgrid operating minimum and maximum limits of energy vehicle count for selecting the incremental cost savings. The balance between the electric vehicle counts and the hour represents a discrete time for studying the operation. Continuous variations are difficult to study and inform the gap between the voltage profile and the minimum electric vehicle count for level 1. The voltage profile for level 2 and level 3 charging schemes is mostly positive, indicating that the method frequently corrects the availability of the DC bus via microgrid or charging facility. The number of electric vehicles counted at an average value of 50 indicates the minimum

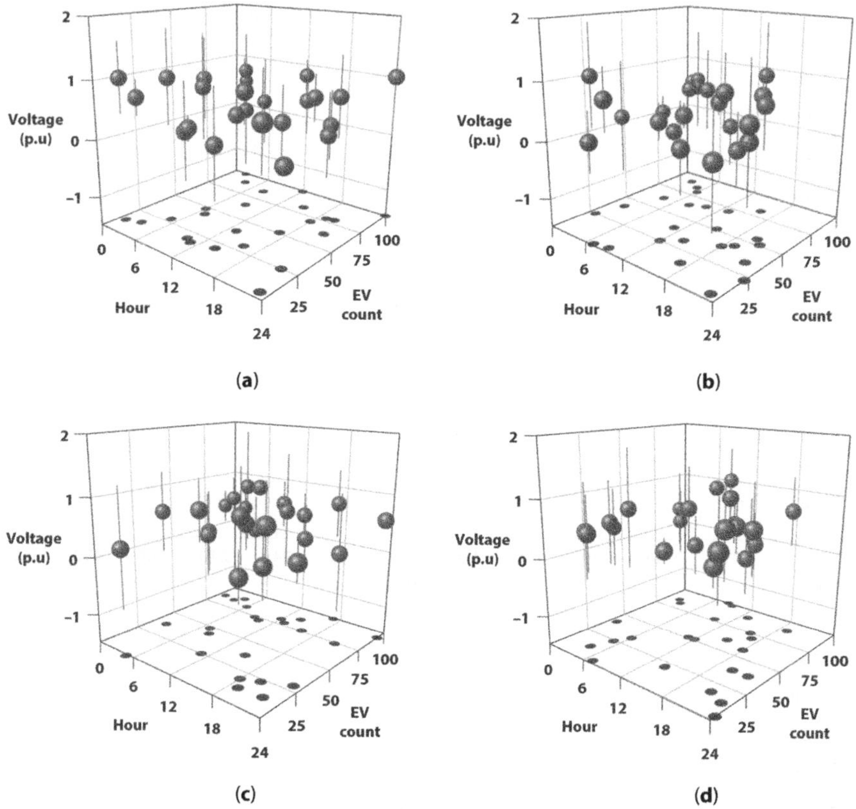

Figure 13.4 Voltage profile for the number of charging electric vehicles in a day: (a) level 1, (b) level 2, (c) level 3, and (d) level 4.

charge duration executed as the test case for further study. The optimization of multiple criteria is decided based on the charging units' availability. It is controlled on an hourly basis of arrival and departure times.

13.3 Implementation of Proposed Commercial EV Charging Stations on Microgrids

The performance comparison of commercial EV charging stations on microgrids as different levels of proposed charging with their combination of maximum of two sets is tabulated in Table 13.1. The voltage profile of the entire DC bus system is simulated and summarized, as is the error improvement from existing methods [7, 9, 12, 14–16], the maximum 3-h

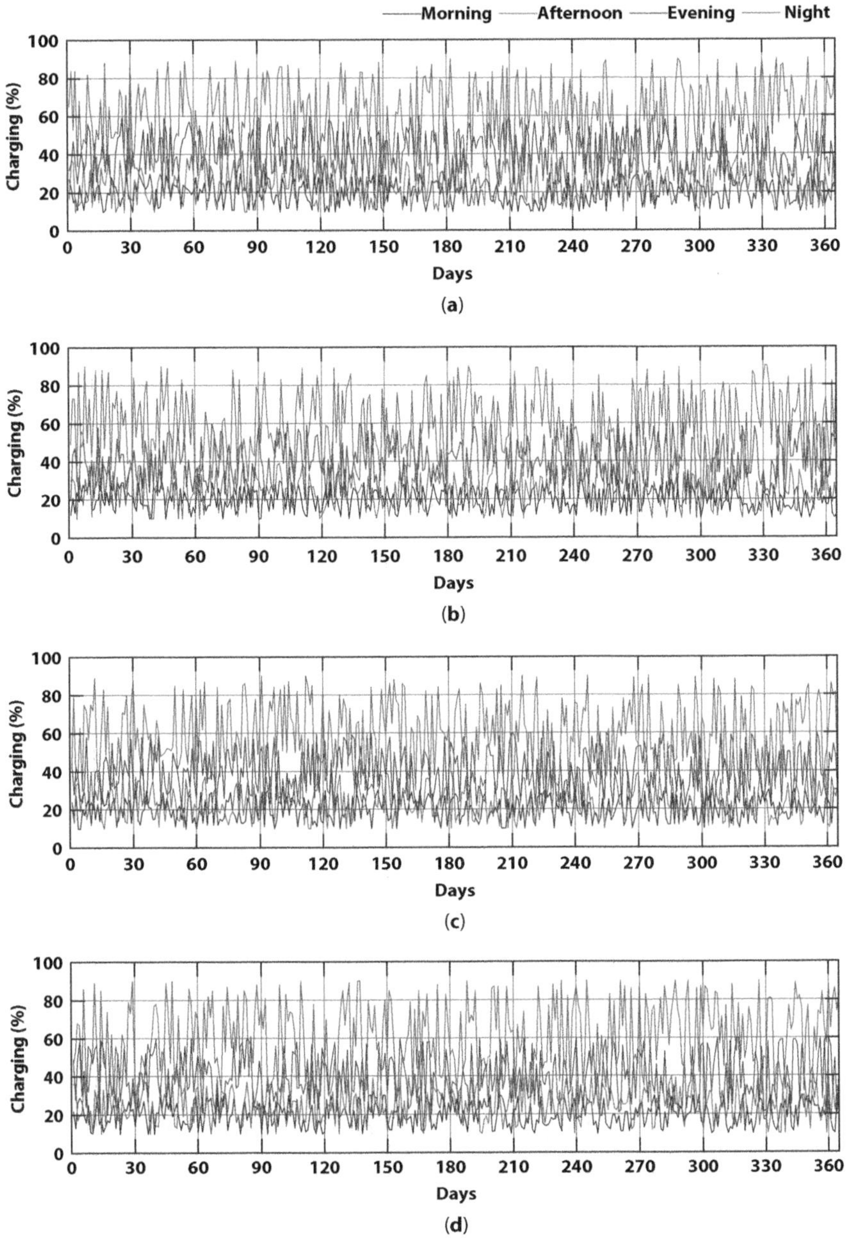

Figure 13.5 State of charge at various EV charging stations: (a) village, (b) rural, (c) town, and (d) city.

Table 13.1 Performance comparison of commercial EV charging stations on microgrids.

Proposed method	Voltage profile (p.u.)	Error improvement	Peak hours	Minimum EV count	Maximum EV count	Minimum kWh	Maximum kWh	Incremental cost saving (Rs/kW)
Level 1	0.744	49.11%	2, 7, 22	6	41	1	44	19
Level 2	0.976	39.77%	2, 8, 12	3	33	14	35	39
Level 3	0.857	46.19%	5, 7, 20	11	28	10	40	15
Level 4	0.266	13.89%	2, 9, 11	14	48	9	50	10
Level 1 + level 2	0.588	38.32%	3, 9, 16	10	35	2	44	26
Level 1 + level 3	0.145	14.65%	4, 9, 11	6	31	5	31	19
Level 1 + level 4	0.345	12.53%	4, 6, 12	12	23	12	31	8
Level 2 + level 3	0.156	13.65%	1, 8, 15	7	47	6	39	30
Level 2 + level 4	0.899	18.97%	2, 9, 11	12	30	11	42	37
Level 3 + level 4	0.499	3.12%	4, 8, 11	2	31	24	39	19

Table 13.2 Commercial systems on a weekly and monthly EV handling delay for effective resource management.

Week	Day	Morning		Afternoon		Evening		Night	
		Minimum delay (minutes)	Maximum delay (minutes)	Minimum delay (minutes)	Maximum delay (minutes)	Minimum delay (minutes)	Maximum delay (minutes)	Minimum delay (minutes)	Maximum delay (minutes)
1	1	2.15	3.32	0.93	5.87	0.93	5.87	0.93	5.87
	2	2.33	5.31	0.1	5.72	0.1	5.72	0.1	5.72
	3	2.27	3.33	1.74	5.7	1.74	5.7	1.74	5.7
	4	2.87	4.1	0.8	5.06	0.8	5.06	0.8	5.06
	5	2.74	5.86	1.96	5.58	1.96	5.58	1.96	5.58
	6	2.28	5.72	2.06	4.23	2.06	4.23	2.06	4.23
	7	2.94	3.51	1.52	3.86	1.52	3.86	1.52	3.86
2	8	2.59	5.3	0.43	3.61	0.43	3.61	0.43	3.61
	9	1.66	5.34	1.41	4.96	1.41	4.96	1.41	4.96
	10	2.34	4.31	0.75	3.71	0.75	3.71	0.75	3.71
	11	0.34	4.83	0.99	3.15	0.99	3.15	0.99	3.15

(Continued)

Table 13.2 Commercial systems on a weekly and monthly EV handling delay for effective resource management. (*Continued*)

Week	Day	Morning		Afternoon		Evening		Night	
		Minimum delay (minutes)	Maximum delay (minutes)	Minimum delay (minutes)	Maximum delay (minutes)	Minimum delay (minutes)	Maximum delay (minutes)	Minimum delay (minutes)	Maximum delay (minutes)
	12	2.2	3.17	0.66	5.9	0.66	5.9	0.66	5.9
	13	1.21	5.5	0.19	4.58	0.19	4.58	0.19	4.58
	14	2.64	4.58	0.16	3.92	0.16	3.92	0.16	3.92
3	15	1.99	4.1	2.73	3.41	2.73	3.41	2.73	3.41
	16	2.01	3.56	1.72	4.48	1.72	4.48	1.72	4.48
	17	2.05	3.92	0.49	5.77	0.49	5.77	0.49	5.77
	18	2.52	5.44	1.84	4.49	1.84	4.49	1.84	4.49
	19	0.88	3.64	2.84	4.3	2.84	4.3	2.84	4.3
	20	1.75	5.65	1.18	4.84	1.18	4.84	1.18	4.84
	21	0.39	4.74	0.7	5.03	0.7	5.03	0.7	5.03

(*Continued*)

Table 13.2 Commercial systems on a weekly and monthly EV handling delay for effective resource management. (*Continued*)

		Morning		Afternoon		Evening		Night	
Week	**Day**	**Minimum delay (minutes)**	**Maximum delay (minutes)**	**Minimum delay (minutes)**	**Maximum delay (minutes)**	**Minimum delay (minutes)**	**Maximum delay (minutes)**	**Minimum delay (minutes)**	**Maximum delay (minutes)**
4	22	1.07	4.93	2.76	4.32	2.76	4.32	2.76	4.32
	23	2.06	4.35	2.71	4.93	2.71	4.93	2.71	4.93
	24	1.42	3.79	0.05	5.23	0.05	5.23	0.05	5.23
	25	0.98	5.46	0.6	4.99	0.6	4.99	0.6	4.99
	26	0.95	4.65	2.31	3.82	2.31	3.82	2.31	3.82
	27	0.31	4.3	0.29	5.91	0.29	5.91	0.29	5.91
	28	0.61	4.78	2.08	3.1	2.08	3.1	2.08	3.1
	29	2.87	5.29	0.86	5.61	0.86	5.61	0.86	5.61
	30	1.77	4.44	1.7	5.8	1.7	5.8	1.7	5.8

peak capability, at any instant of time, maximum vehicle count, and minimum and maximum number of energy units utilized with incremental cost savings.

The level 1 charging method provides the lowest kilowatt-hour usage of any method due to the availability of resources clearly reflected in the charging facility. Hence, the maximum number of vehicles is much higher than with any other method. The peak hours are during the early morning and in the evening. The error improvement is 49.11% better than the existing literature [14], and for the same configuration, the voltage profile is reduced by 0.744 p.u. for handling the large number of electric vehicles at level 2. Similarly, the microgrid-based charging control schemes, such as level 3 and level 4, dominate in achieving the minimum EV count. They have a proportionately lower cost than the other methods. The maximum of 50 kwh per hand within the 14- to 48-car peak period is a drastic improvement in the literature reported so far. The minimum number of EV counts between three and 28 oscillates between the incremental cost savings with other configurations as small as 8 to 10 Rs/kW during the peak hours, mostly in the morning and evening. To enhance the existing operation, the features are studied by hour irrespective of the charging unit, which is almost common for all locations on all days of the year. The peak variation in the operating hours often frames a pattern for the utilization of the commercial facilities. The amount of charge available in a day remains the same. The state of charge in the electric vehicle facility is programmed from the data resource available in [14].

The various locations exhibit the same level of charging in the different periods of the day. For the day's availability, the control ranges from 20% to 30%, 25% to 45%, 30% to 55%, and 35% to 85%. These characteristics look the same because of the four different locations that share the same distance of about 50 km as mentioned in the literature. The voltage profile of the microgrid and energy resource capacity also remain the same for the study, except that the vehicle count in each facility is independent of the location. Analysis of 365 days requires in-depth study for hourly-based state of charge requires more data but often increases the complexity of the overall commercial system. According to the day-by-day study on the state of charging, the performance impacts are less than 4% to 22% in the overall system, as reflected in the existing works. The computation of state charge is not reflected in one of the energy storage resources considered in Section 13.2. The expansion in the forthcoming years does not impact the overall microgrid with the constant DC bus utilization. Similarly, the electric vehicle facility reduces the number of electric vehicles after the arrival of the maximum number of vehicles handled in a day.

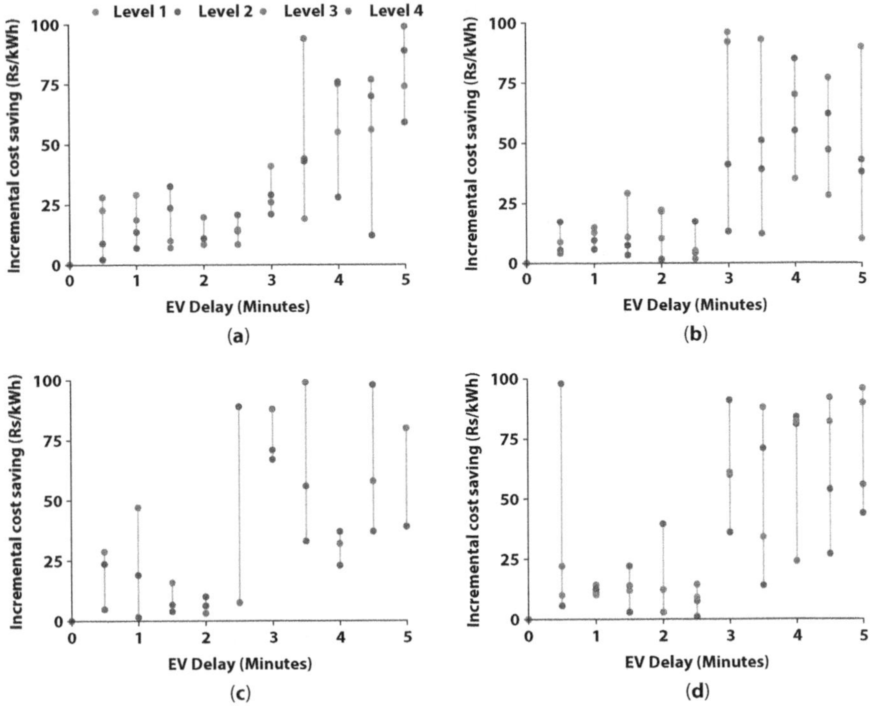

Figure 13.6 Incremental cost saving (Rs/kWh) in EV charging station handling: (a) one to 10 vehicles, (b) 11 to 25 vehicles, (c) 26 to 50 vehicles, and (d) greater than 50 vehicles.

13.4 Validation of the Proposed Commercial EV Charging Stations on Microgrids

The commercial systems on a weekly and monthly EV handling delay for effective resource management are classified with the four classifications in minimum and maximum delay as tabulated in Table 13.2. The start of the day maximizes the 0- to 2-min delay and facilitates the maximum delay of up to 6 min. The beginning and end of the week had the shortest possible delay of seconds, 3 min, and 4.5 min. There is a greater value of 2 and 4 min during the first 3 days of effective resource management, with almost no delay during the afternoon. Every charge comes with different levels of incremental cost, handling the first 10 vehicles at less than 25 Rs/kW for four vehicles. For the next 3 min, the constant delay among the vehicles will be 25 to 50 Rs/kW. The larger variation in the delay of charging pattern improvises the cost saving with the pool of solutions. The vehicles are up to 25 for the low incremental cost saving of 25 Rs/kW, with a high number of vehicles after reaching level 4. Level 3 combines microgrid and DC bus status. For larger electric vehicles, the level 2 and level 4 bypasses offer higher incremental cost savings of up to 75 Rs/kW than the other methods. The study of simulated EV handling improvement for different levels of charging achieved between 12 AM and 11 PM is presented in Figure 13.7 (a–d). The number of vehicles expected to depart and arrive varied across all levels. Actual arrival and expected arrival synchronize with departure values for the ideal case, but the commercial system neglects to come across the charging facility. The mean average shifts the peak time from any of the periods in the day reflected in the case study. The energy cost savings impact on the energy improvement for the model and simulated values are

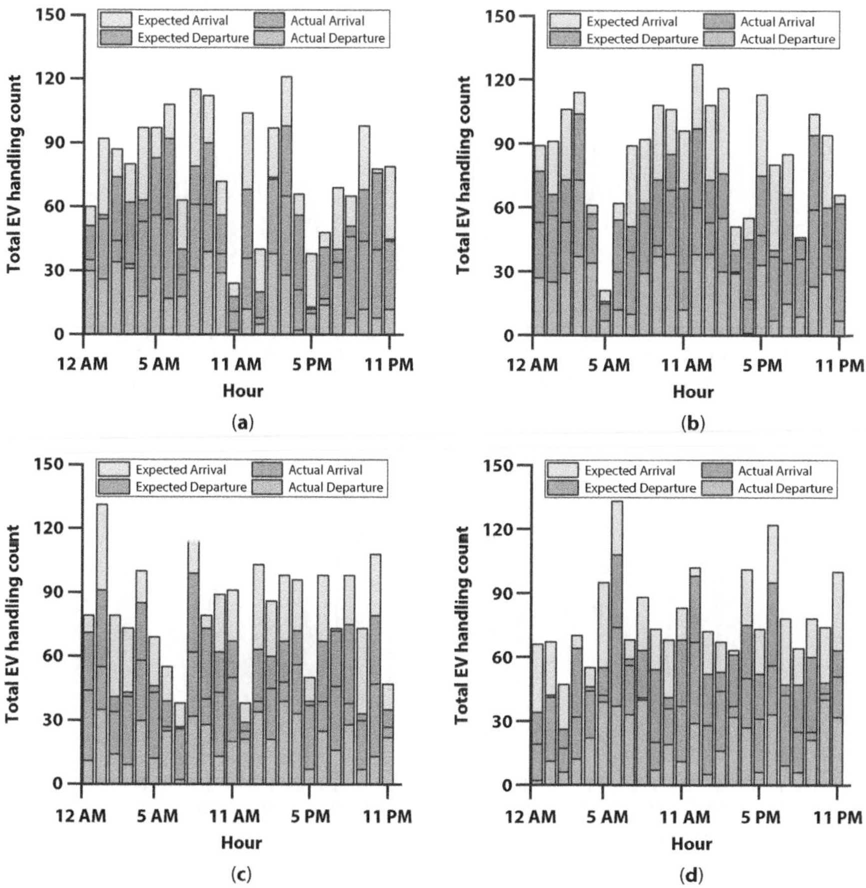

Figure 13.7 Simulated EV handling improvement for different levels of charging: (a) level 1, (b) level 2, (c) level 3, and (d) level 4.

presented in Figure 13.8 (a–d). At level 1 charging, the EV facility provides enough information to perform high-cost energy saving as modeling with 5% to 10% improvement in energy saving. The low-range energy saving with minimal improvement presented in level 2 switching states faces the low handling of electric vehicles. Low and high energy savings of 10% to 15% are achieved in direct microgrid level 4 charging improved across the board. The level 4 charging method dominates in the midrange of 20 to 40 kWh charging with 5% to 15% energy improvement. It ensures that the proposed method can easily switch over among all the levels. It improves the charging infrastructure in microgrid systems.

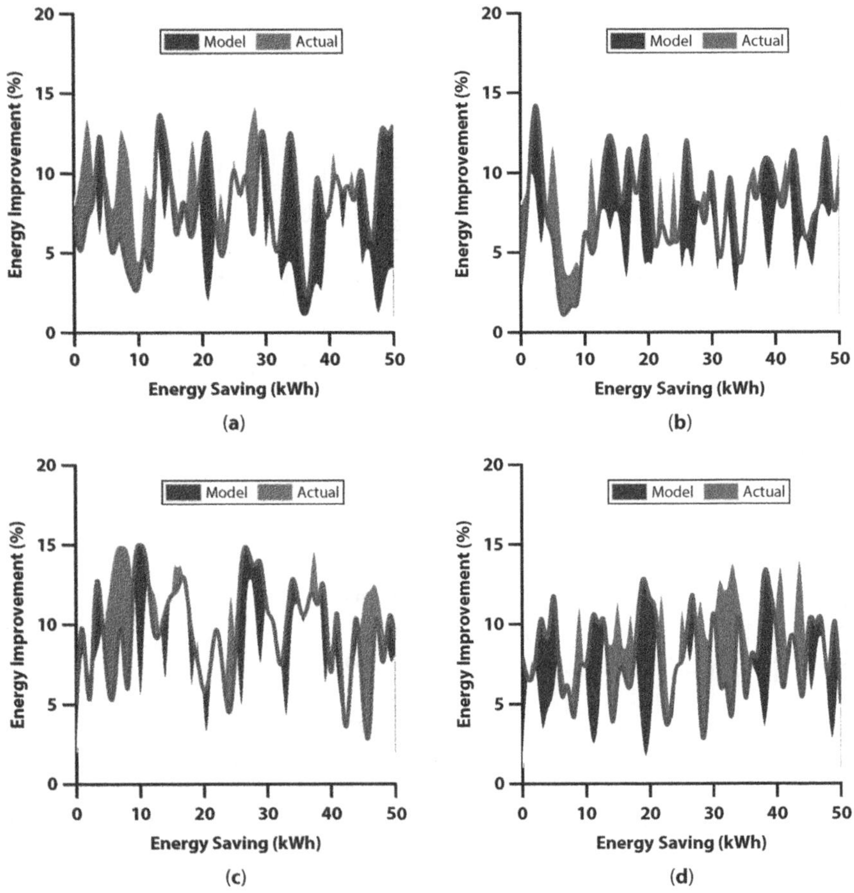

Figure 13.8 Energy saving after implementing the proposed charging: (a) level 1, (b) level 2, (c) level 3, and (d) level 4.

13.5 Conclusion

In this chapter, a new electric vehicle charging infrastructure enhancement technology with real-time energy resource management is proposed. The availability of different energy resources in the microgrid studies for the past 5 years and projected for the future 5 years, with 20% to 120% expansion as reference, is presented. The four schemes at the charging level are near to a charging facility with energy source management (level 1), a charging facility with DC bus (level 2), a microgrid with DC bus (level 3), and direct EV status (level 4) based on energy profile. The locality depends on the charging level at four different places and four different times considering the four different facilities. In total, 64 profiles select the combination of levels 1 to 4 from the EV delay count, adjusting the increment cost per cost-wise energy improvement. The combination of the DC bus and the electric vehicle charges the microgrid, including all switching levels. Maximum and minimum delays in arrival and departure energy cost savings were determined by utilizing the resources matched with the voltage profile, including the positive and negative profiles. The total enhancement of error increased from the range of 3.12% to 49.11% without any combination of peak hours, mostly in the morning and evening. a minimum of 1 kWh and a maximum of 50 kWh for the handling of 2 to 48 maximum electric vehicle counts. There was a week- and day-wise minimum delay of 0 min and a maximum delay of less than 5 min, which has not been reported earlier in any of the literature. The increased vehicle count reduced the incremental cost savings from 75 to 90 Rs/kW. For a higher number of vehicles, the delay is reduced by a low level of switching the incremental cost. The 5% to 10% improvement with 15 kWh energy saving, 0 to 5% improvement in 10 kWh energy enhancement, 0 to 10 kWh and 5% to 15% modeled value with 40 to 50 kWh with 5% to 10% energy improvement, and 25% to 35% energy saving enhances the 5% to 15% improvement of the proposed schemes. The study can be extended to the real-time scenario with various features for proposed charging adapted to the new commercial EV charging stations on microgrids.

References

1. Mehrjerdi, H., Dynamic and multi-stage capacity expansion planning in microgrid integrated with electric vehicle charging station. *J. Energy Storage*, 29, 101351, 2020.

2. Ahmad, F., Alam, M.S., Shariff, S.M., Krishnamurthy, M., A cost-efficient approach to EV charging station integrated community microgrid: A case study of Indian power market. *IEEE Trans. Transp. Electrif.*, 5, 1, 200–214, 2019.
3. AbuElrub, A., Hamed, F., Saadeh, O., Microgrid integrated electric vehicle charging algorithm with photovoltaic generation. *J. Energy Storage*, 32, 101858, 2020.
4. Olatunde, O., Hassan, M.Y., Abdullah, M.P., Rahman, H.A., Hybrid photovoltaic/small-hydropower microgrid in smart distribution network with grid isolated electric vehicle charging system. *J. Energy Storage*, 31, 101673, 2020.
5. Wang, D., Locment, F., Sechilariu, M., Modelling, simulation, and management strategy of an electric vehicle charging station based on a DC microgrid. *Appl. Sci.*, 10, 6, 2053, 2020.
6. Singh, S., Jagota, S., Singh, M., Energy management and voltage stabilization in an islanded microgrid through an electric vehicle charging station. *Sustain. Cities Soc.*, 41, 679–694, 2018.
7. Sechilariu, M., Molines, N., Richard, G., Martell-Flores, H., Locment, F., Baert, J., Electromobility framework study: Infrastructure and urban planning for EV charging station empowered by PV-based microgrid. *IET Electr. Syst. Transp.*, 9, 4, 176–185, 2019.
8. Savio, D.A., Juliet, V.A., Chokkalingam, B., Padmanaban, S., Holm-Nielsen, J.B., Blaabjerg, F., Photovoltaic integrated hybrid microgrid structured electric vehicle charging station and its energy management approach. *Energies*, 12, 1, 168, 2019.
9. Clairand, J.M., Álvarez-Bel, C., Rodríguez-García, J., Escrivá-Escrivá, G., Impact of electric vehicle charging strategy on the long-term planning of an isolated microgrid. *Energies*, 13, 13, 3455, 2020.
10. Haupt, L., Schöpf, M., Wederhake, L., Weibelzahl, M., The influence of electric vehicle charging strategies on the sizing of electrical energy storage systems in charging hub microgrids. *Appl. Energy*, 273, 115231, 2020.
11. Chae, S.H., Kim, G.H., Choi, Y.J., Kim, E.H., Design of isolated microgrid system considering controllable EV charging demand. *Sustainability*, 12, 22, 9746, 2020.
12. Wu, Y., Wang, Z., Huangfu, Y., Ravey, A., Chrenko, D., Gao, F., Hierarchical operation of electric vehicle charging station in smart grid integration applications—An overview. *Int. J. Electr. Power Energy Syst.*, 139, 108005, 2022.
13. Shen, L., Cheng, Q., Cheng, Y., Wei, L., Wang, Y., Hierarchical control of DC micro-grid for photovoltaic EV charging station based on flywheel and battery energy storage system. *Electric. Power Syst. Res.*, 179, 106079, 2020.
14. Shaaban, M.F., Mohamed, S., Ismail, M., Qaraqe, K.A., Serpedin, E., Joint planning of smart EV charging stations and DGs in eco-friendly remote hybrid microgrids. *IEEE Trans. Smart Grid*, 10, 5, 5819–5830, 2019.
15. Tan, B. and Chen, H., Multi-objective energy management of multiple microgrids under random electric vehicle charging. *Energy*, 208, 118360, 2020.

16. Garcia-Trivino, P., Torreglosa, J.P., Fernandez-Ramirez, L.M., Jurado, F., Decentralized fuzzy logic control of microgrid for electric vehicle charging station. *IEEE J. Emerg. Sel. Top. Power Electron.*, 6, 2, 726–737, 2018.
17. Yuvaraja, T. and Ramya, K., Discretionary controller for hybrid energy storage system based on orderly control considering commercial value in decentralised microgrid operation. *COMPEL-The Int. J. Comput. Math. Electr. Electron.Eng.*, 37, 6, 1969–1980, 2018.
18. Çetinbaş, İ., Tamyürek, B., Demirtaş, M., Sizing optimization and design of an autonomous AC microgrid for commercial loads using harris hawks optimization algorithm. *Energy Convers. Manag.*, 245, 114562, 2021.
19. Tribioli, L. and Cozzolino, R., Techno-economic analysis of a stand-alone microgrid for a commercial building in eight different climate zones. *Energy Convers. Manag.*, 179, 58–71, 2019.
20. Tavakoli, M., Shokridehaki, F., Akorede, M.F., Marzband, M., Vechiu, I., Pouresmaeil, E., CVaR-based energy management scheme for optimal resilience and operational cost in commercial building microgrids. *Int. J. Electr. Power Energy Syst.*, 100, 1–9, 2018.
21. Rahman, M.S., Hossain, M.J., Lu, J., Rafi, F.H.M., Mishra, S., A vehicle-to-microgrid framework with optimization-incorporated distributed EV coordination for a commercial neighborhood. *IEEE Trans. Ind. Inform.*, 16, 3, 1788–1798, 2019.
22. Tavakoli, M., Shokridehaki, F., Marzband, M., Godina, R., Pouresmaeil, E., A two stage hierarchical control approach for the optimal energy management in commercial building microgrids based on local wind power and PEVs. *Sustain. Cities Soc.*, 41, 332–340, 2018.
23. Leskarac, D., Moghimi, M., Liu, J., Water, W., Lu, J., Stegen, S., Hybrid AC/DC Microgrid testing facility for energy management in commercial buildings. *Energy Build.*, 174, 563–578, 2018.
24. Wang, R., Lam, C.M., Hsu, S.C., Chen, J.H., Life cycle assessment and energy payback time of a standalone hybrid renewable energy commercial microgrid: A case study of town island in Hong Kong. *Appl. Energy*, 250, 760–775, 2019.
25. Mehta, R., A microgrid case study for ensuring reliable power for commercial and industrial sites, in: *2019 IEEE PES GTD Grand International Conference and Exposition Asia (GTD Asia)*, IEEE, pp. 594–598, 2019, March.
26. Meyer, E.L., Apeh, O.O., Overen, O.K., Electrical and meteorological data acquisition system of a commercial and domestic microgrid for monitoring pv parameters. *Appl. Sci.*, 10, 24, 9092, 2020.
27. Parise, G., Parise, L., Vagnati, G., Grounding microgrid systems in metropolitan and commercial areas. *IEEE Trans. Ind. Appl.*, 56, 2, 1156–1161, 2019.
28. Wang, Y., Yi, Z., Shi, D., Yu, Z., Huang, B., Wang, Z., Optimal distributed energy resources sizing for commercial building hybrid microgrids, in: *2018 IEEE Power & Energy Society General Meeting (PESGM)*, IEEE, pp. 1–5, 2018, August.

29. Metais, M.O., Jouini, O., Perez, Y., Berrada, J., Suomalainen, E., Too much or not enough? Planning electric vehicle charging infrastructure: A review of modeling options. *Renew. Sustain. Energy Rev.*, 153, 111719, 2022.
30. Ashfaq, M., Butt, O., Selvaraj, J., Rahim, N., Assessment of electric vehicle charging infrastructure and its impact on the electric grid: A review. *Int. J. Green Energy*, 18, 7, 657–686, 2021.
31. Gönül, Ö., Duman, A.C., Güler, Ö., Electric vehicles and charging infrastructure in Turkey: An overview. *Renew. Sustain. Energy Rev.*, 143, 110913, 2021.
32. Li, C., Dong, Z., Chen, G., Zhou, B., Zhang, J., Yu, X., Data-driven planning of electric vehicle charging infrastructure: A case study of Sydney, Australia. *IEEE Trans. Smart Grid*, 12, 4, 3289–3304, 2021.
33. Babic, J., Carvalho, A., Ketter, W., Podobnik, V., A data-driven approach to managing electric vehicle charging infrastructure in parking lots. *Transp. Res. D: Transp. Environ.*, 105, 103198, 2022.
34. Mirheli, A. and Hajibabai, L., Hierarchical optimization of charging infrastructure design and facility utilization. *IEEE Trans. Intell. Transp. Syst.*, 23, 9, 1–14, 2022.
35. Schulz, F. and Rode, J., Public charging infrastructure and electric vehicles in Norway. *Energy Policy*, 160, 112660, 2022.
36. Li, X. and Jenn, A., An integrated optimization platform for spatial-temporal modeling of electric vehicle charging infrastructure. *Transp. Res. D: Transp. Environ.*, 104, 103177, 2022.
37. Li, G., Luo, T., Song, Y., Climate change mitigation efficiency of electric vehicle charging infrastructure in China: From the perspective of energy transition and circular economy. *Resour. Conserv. Recycl.*, 179, 106048, 2022.
38. Rivera, S., Kouro, S., Vazquez, S., Goetz, S.M., Lizana, R., Romero-Cadaval, E., Electric vehicle charging infrastructure: From grid to battery. *IEEE Ind. Electron. Mag.*, 15, 2, 37–51, 2021.
39. Powell, S., Cezar, G.V., Min, L., Azevedo, I.M., Rajagopal, R., Charging infrastructure access and operation to reduce the grid impacts of deep electric vehicle adoption. *Nat. Energy*, 7, 10, 932–945, 2022.
40. Kumar, R.R., Chakraborty, A., Mandal, P., Promoting electric vehicle adoption: Who should invest in charging infrastructure? *Transp. Res. E: Logist. Transp. Rev.*, 149, 102295, 2021.
41. Kavianipour, M., Fakhrmoosavi, F., Shojaei, M., Zockaie, A., Ghamami, M., Wang, J., Jackson, R., Impacts of technology advancements on electric vehicle charging infrastructure configuration: A Michigan case study. *Int. J. Sustain. Transp.*, 16, 7, 1–14, 2021.
42. Rivera, S., Goetz, S.M., Kouro, S., Lehn, P.W., Pathmanathan, M., Bauer, P., Mastromauro, R.A., Charging infrastructure and grid integration for electromobility. *Proc. IEEE*, 111, 4, 371–396, 2022.
43. Huang, X., Lin, Y., Lim, M.K., Zhou, F., Ding, R., Zhang, Z., Evolutionary dynamics of promoting electric vehicle-charging infrastructure based on public–private partnership cooperation. *Energy*, 239, 122281, 2022.

44. Gupta, S., Khanna, R., Kohli, P., Agnihotri, S., Soni, U., Asjad, M., Risk evaluation of electric vehicle charging infrastructure using fuzzy AHP–a case study in India. *Oper. Manag. Res.*, 16, 1, 1–14, 2022.
45. Lai, C.S., Chen, D., Zhang, J., Zhang, X., Xu, X., Taylor, G.A., Lai, L.L., Profit maximization for large-scale energy storage systems to enable fast EV charging infrastructure in distribution networks. *Energy*, 259, 124852, 2022.
46. Khan, H.A.U., Price, S., Avraam, C., Dvorkin, Y., Inequitable access to EV charging infrastructure. *Electr. J.*, 35, 3, 107096, 2022.
47. Gupta, R.S., Tyagi, A., Anand, S., Optimal allocation of electric vehicles charging infrastructure, policies and future trends. *J. Energy Storage*, 43, 103291, 2021.
48. Raman, G., Raman, G., Peng, J.C.H., Resilience of urban public electric vehicle charging infrastructure to flooding. *Nat. Commun.*, 13, 1, 1–9, 2022.
49. Kore, H.H. and Koul, S., Electric vehicle charging infrastructure: Positioning in India. *Manag. Environ. Qual.: An Int. J.*, 33, 3, 776–799, 2022.
50. Pardo-Bosch, F., Pujadas, P., Morton, C., Cervera, C., Sustainable deployment of an electric vehicle public charging infrastructure network from a city business model perspective. *Sustain. Cities Soc.*, 71, 102957, 2021.

14

Renewable-Energy-Powered EV Charging Station for Microgrid PSO-Based Controller for PV-Powered EV Charging Station

K. Punitha[1*], R. Aruna[1] and S. Kanthammal[2]

[1]Department of Electronics Engineering, P S R Engineering College, Sivakasi, Tamil Nadu, India
[2]Department of Zoology, Government Art and Science College, Sivakasi, Tamil Nadu, India

Abstract

Government policies, global warming, climate change, and a growing awareness of fossil fuel depletion have all contributed to the growth of electric vehicles (EV). With constantly changing load demands, EVs play an important role in supporting the power grid or microgrid. When EV charging stations use the current grid for charging, there is an increase in load-side demand and added strain on the grid. The reliability, efficiency of power conversion, and ease of interface with renewable energy sources (RES) of DC microgrid-based EV charging make it more efficient than AC power distribution for charging electric vehicles on the grid. A photovoltaic (PV) system is one of the most eco-friendly forms of energy for electric vehicles. This decade has seen a tremendous push in EV manufacturing as well as commercial charging stations throughout the world. An efficient charging station with a power electronics converter provides an EV with reliable, uninterruptible power for recharging its battery. Electric vehicle charging stations usually have at least one renewable energy source, a controller, a DC–DC converter, and a DC–AC inverter to connect to the grid. The use of power controller techniques in solar PV with electric vehicles can improve grid stability, reduce grid congestion, and increase solar self-consumption. This chapter discusses the solar PV-powered charging station that uses an effective controller. For PV to be operated efficiently,

**Corresponding author*: kgpunitha@gmail.com

P. Sivaraman, C. Sharmeela and P. Sanjeevikumar (eds.) Microgrids for Commercial Systems, (381–408) © 2024 Scrivener Publishing LLC

an MPPT controller based on particle swarm optimization (PSO)-optimized incremental conduction (IC) is developed. Due to the lack of knowledge about the start of the perturbation value of voltage, conventional MPPT algorithms do not offer maximum power under rapidly varying and partial shading conditions. This work uses PSO to provide IC MPPT with an optimal perturbation value of voltage to track maximum accurate power in a fast convergent manner. Using MATLAB/Simulink, the proposed approach is evaluated with and without PSO's optimum voltage for IC MPPT. It outperforms conventional IC MPPT with less oscillation and faster convergence of 3 s.

***Keywords*:** Electric vehicle, charging station, solar photovoltaic, maximum power point tracking, incremental conduction algorithm, particle swarm optimization, MATLAB simulink

14.1 Introduction

With fossil fuels becoming increasingly carbonized and depleted, users and researchers have turned to renewable energy sources (RES) like solar and wind. It has been found that combinations of photovoltaic (PV) and electric vehicles (EV) systems can reduce CO_2 emissions by 86% and energy costs by 51% [1]. EV demand on the global market is increasing day by day, which means that efficient EV charging stations are also needed. EV are charged primarily by electronic components, either on board or off board, to deliver energy from existing power supply infrastructures to the storage system [2]. Due to their weight, space, and cost constraints, on-board chargers' power rating is less, and they usually take a long time to charge the batteries. On the other hand, off-board battery chargers' rating is high, and they are less constrained by size and weight. With off-board battery chargers, EVs can benefit from fast and frequent high power charging process to extend the driving range and reduce recharging times [3]. Therefore, the selection of efficient controllers and their application in EV charging station are still a concern for researchers. These controllers need to be optimized. An optimized algorithm provides optimized parameters. This chapter presents an optimal algorithm to determine the optimal parameters for the converter for integration of a PV power generation system and an electric vehicle charging station through a common grid. The optimal parameters for a converter controller can be determined using particle swarm optimization (PSO). Swarm intelligence and movement provide the basis for the PSO stochastic optimization technique [4]. This technique relies on social interaction to solve problems. A number of particles make up a swarm that moves around in the search space searching for the best solution [5]. As our system incorporates PV, the converter should utilize maximum power point technique (MPPT). For PV panels to produce maximum power, it is very important to determine the maximum power output as it allows optimal use of a PV system irrespective of irradiation and temperature [6]. The majority of articles dealing with MPPT algorithms are based on the perturb and observe (P&O) or on incremental conductance method (IC) [7, 8]. The conventional P&O and incremental conduction (IC) MPPT methods are not suitable for varying weather conditions. Hence, MPPT methods in the last decade have been based on soft computing approaches [9–11]. For solar PV MPP monitoring, many researchers used PSO [12–14] as well as hybrid PSO [15–19]. Even though PSO exhibited better performance and a higher convergence rate than soft computing algorithms, both are generally expensive for implementation due to high memory requirement and being complex and time consuming [20–28]. The conventional P&O and IC MPPTs do not have any prior knowledge of MPP voltage when they begin their perturbation, so it takes more time to reach stabilized MPP power with more oscillation. The shortfalls are removed by using PSO, which provides IC MPPT with an optimized voltage as a starting voltage, allowing it to track MPP precisely without oscillation. Due to the use of PSO in this study offline to deliver output to the IC MPPT, the shortcomings of cost, complexity, and time-consuming issues are eliminated. MATLAB/Simulink is used to develop a PV-powered EV charging system and to demonstrate the proposed algorithm. PSO coding and overall MATLAB/Simulink block are given in the appendix for reference.

14.2 Renewable-Energy-Powered EV Charging Station for Microgrid

Figure 14.1 shows the layout of an EV charging station for microgrid (MG) system which includes a PV system, electric vehicle charging stations (EVCSs), energy storage, a DC link capacitor bank, and a load. As a complement to the power grid, MG provides reliability, local resilience, stability, and power quality [29]. A MG's goal is to reduce grid energy imports and charge EVs using excess PV energy. Electric vehicles are energy storage devices that can be discharged or charged during peak reduction, load shifting, and grid reactive power support, etc. [30]. EVs and energy storage systems will also be charged by the microgrid. In extreme weather conditions or when power is limited from the grid at any one time, MGs may be able to assist with electric vehicle (EV) charging [31]. A MG connection can be much more flexible than a central grid connection for charging electric vehicles. The cost of the system will increase due to large current

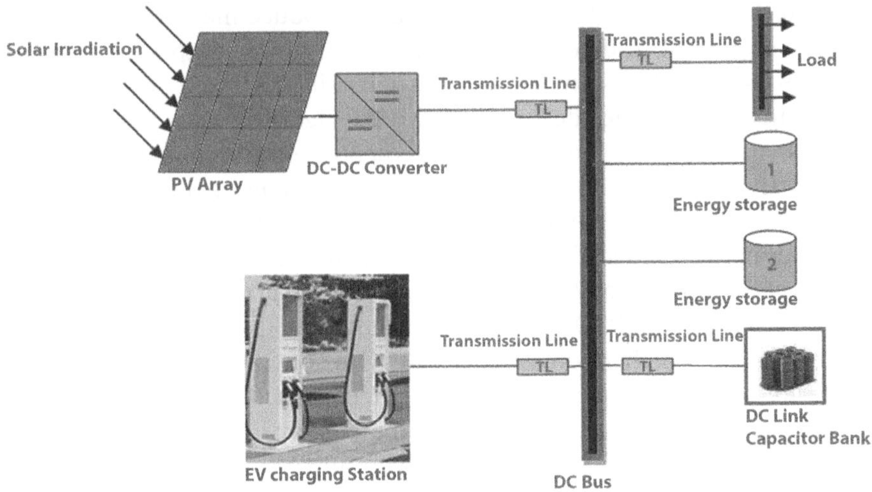

Figure 14.1 Layout of EV charging station for microgrid.

draws from the central grid, especially during peak hours. MG simplified the task of providing adequate charging facilities. EV charging became more affordable due to fast charging, which, in turn, reduced the cost.

14.3 EV Charging Station

It is expected that the number of commercial charging stations will increase as the use of electric vehicles increases. A variety of charging methods are available for electric vehicles (EVs), depending on their locations and requirements. Different countries have different standards for electric vehicle charging stations, depending on the availability of EV models and the characteristics of their electrical grids. As a result, EV charging infrastructure comes in a variety of types and is suited for a variety of uses. In order to build an electric charging station for electric vehicles, safety, efficiency, feasibility, and cost must be considered.

14.3.1 Types of EV Charging Station

EV charging stations fall into three categories—slow or level 1, fast or level 2, and rapid or level 3. It is important to note that these represent the power outputs and therefore charging speeds that can be achieved by charging an EV. Increasing the charging level decreases the charging speed. Faster charging of an EV is facilitated by higher charging power.

Depending on whether a charger is AC or DC, it comes with an associated set of connectors.

SLOW/Level 1 Charger

Chargers of this type are the most common. Using an AC plug and a standard wall outlet, the charger can deliver 3 to 6 kW of power, which can be charged between 8 and 12 h. It is possible to use these chargers at home without any additional equipment or installation. Additionally, these charging stations are the most affordable, yet they also take the longest to charge. In this category, chargers can be used overnight to fully charge vehicles in the morning. These chargers offer maximum battery life and are cheaper than fast chargers. It is possible to halve the charging time by using a charger with twice the ampere rating.

FAST/Level 2 Chargers

These are type-2 AC chargers. Depending on the model, fast chargers range from 7 to 22 kW (single-phase or three-phase 32 A). An EV with a 40-kWh battery can be charged by a 7-kW charger in 4–6 h and by a 22-kW charger in 1 to 2 h based on its unit speed and the vehicle. These chargers can be used in both commercial and residential charging stations. Fast chargers are typically found in car parks, leisure centers, supermarkets, and shops, which are the places likely to be parked in for several hours. Untethered units are therefore more flexible and could be used by any EV with the correct cable. Charging rates, when using a fast charger, will depend on the car's on-board charger, with not all models able to accept 7 kW or more. These models can still be plugged into the charge point but will only draw the maximum power accepted by the on-board charger. Almost all EVs and PHEVs can charge on type 2 units, with the correct cable at least. It is, by far, the most common public charge point standard around, and most plug-in car owners will have a cable, with a type 2 connector charger-side. The vast majority of fast chargers provide AC charging, though some networks are installing 25-kW DC chargers with CCS or CHAdeMO plugs whose connectors are depicted in Figure 14.2.

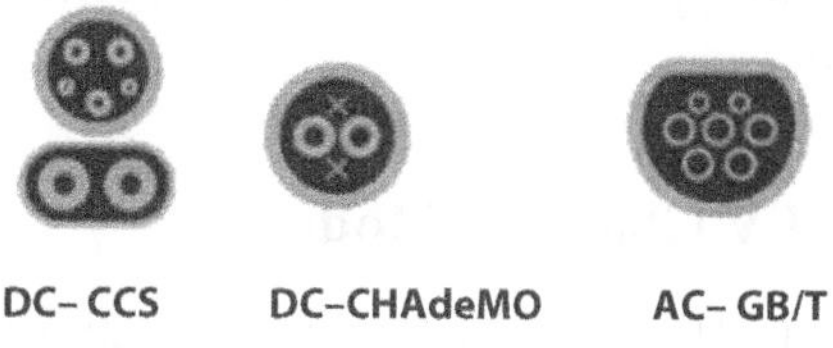

Figure 14.2 Types of plug.

RAPID/Level 3 Chargers

These are the fastest charger for any EV, usually located at locations close to main locations like public gathering places, highways, etc. The devices provide high-power direct current (DC) or alternating current (AC) to recharge a battery as rapidly as attainable. DC fast chargers, however, are typically used in industrial and commercial settings. These charges require high-powered, highly specialized equipment to install and maintain. Depending on the model and brand, EVs can be recharged to 80% within 20 min, though an average modern EV would take around an hour on a standard 50-kW rapid charge point. Initially, it will charge at a normal rate, but as the battery approaches full charge, the speed of charging decreases. As such, terms are quoted at the charge of 80%, after which the charging speed contracts significantly. It maximizes charging efficiency and helps protect the battery. These rapid chargers can be used only with vehicles with rapid charging capability. These chargers are further divided into two. They are as follows:

- **Rapid DC**—most common chargers in India. These chargers provide power at 50 kW (125 A). Connectors typically charge an EV to 80% in 20 min to an hour, depending on battery capacity and starting state of charge.
- **Ultra-Rapid DC**—provide power at 100 kW or more. These are mostly either 100, 150, or 350 kW—though other maximum speeds between these are possible. According to Zap-Map, "These are the next-generation of rapid charge points, able to keep recharging times down despite battery capacities increasing in newer EVs."

14.3.2 Types of EV Charging Cables

The International Electro technical Commission (IEC) describes three different types of EV charging cables as shown in Figure 14.3.

With type A, the cable is permanently connected to the EV, but it is detachable at the charging station. Type B specifies a cable that is detachable at both ends, and type C is a cable that is permanently connected to the charging station.

14.3.3 Types of EV Charging Modes

International standard IEC 61851 describes four different charging modes called mode 1, 2, 3, and 4.

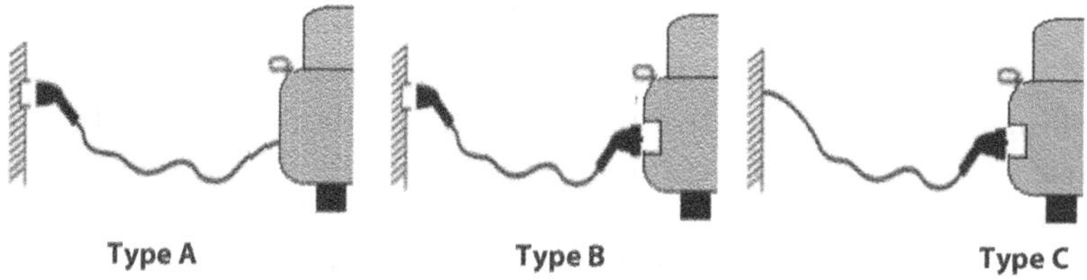

Figure 14.3 The three main types of EV charging cable.

Mode 1

Figure 14.4 EV charging mode 1.

Mode 1

With this mode, the EV is directly connected to a household socket as shown in Figure 14.4. The maximum current of this mode is 16 A, and its voltage should not exceed 250 V with a single-phase system and 480 V in the case of a three-phase network. Mode 1 is the simplest possible charging mode and does not support any communication between the EV and the charge point. This is used for charging electric bikes and scooters. This charging mode is prohibited or restricted in many countries.

Mode 2

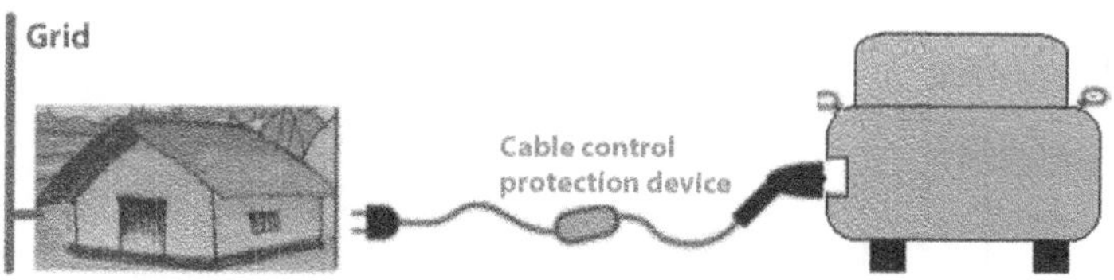

Figure 14.5 EV charging mode 2.

Mode 2 can be used with both household and industrial sockets. The maximum current of this mode is 32 A, and its maximum voltage should not exceed a 250-V single-phase or 480-V three-phase network. Household socket–outlets do not always provide electric power according to the actual standards. Besides this, socket–outlets and plugs designed for household applications might not be able to tolerate a continuous current draw at the maximum rated value. That is why connecting an EV to the socket–outlet for a long time with no control and safety functions can increase the risk of electric shock. To solve this problem, specialists developed charging mode 2 that uses a special type of charging cable equipped with an in-cable control and protection device which performs the required control and safety functions. The use of this mode 2 is also subject to restrictions in many countries which is depicted in Figure 14.5.

Mode 3

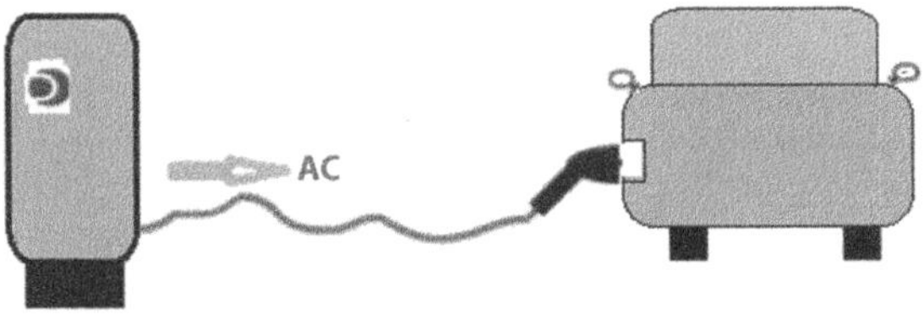

Figure 14.6 EV charging mode 3.

This charging mode shown in Figure 14.6 can deliver a maximum current of 250 A with either a 250-V one-phase or 480-V three-phase network. Moreover, it supports a mode of operation compatible with mode 2 where the maximum current is restricted to less than 32 A for both single-phase and three-phase systems. In order to charge the battery, AC current is supplied by the charging station to the on-board circuitry. Several controllers and protection functions are employed to guarantee public safety. Any of the three possible connecting cables (type A, type B, and type C) can be used in this mode.

Mode 4

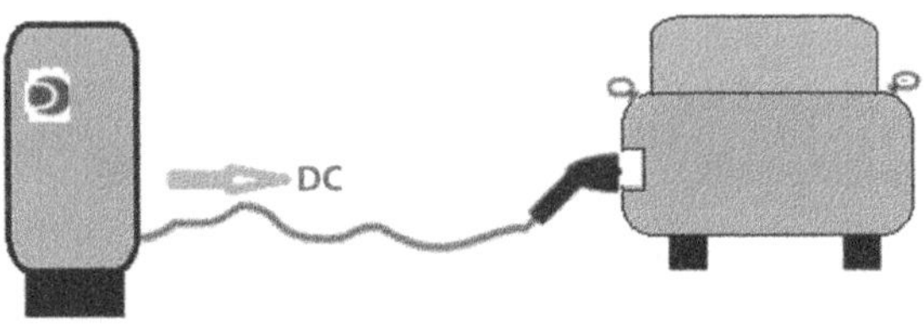

Figure 14.7 EV charging mode 4.

Off-board chargers with DC outputs are only found in this charging mode. Batteries are charged directly by DC current rather than using the on-board charger. A maximum current of 400 A can be provided by this mode at 600 V DC and is shown in Figure 14.7.

A higher level of communication and stricter safety features are required due to the high power level involved in this mode. In this charging mode, an external current converter is required to transform AC current into DC, which is connected to the charging cable outside the vehicle. In mode 4, only a case C connection can be made, where the cable is permanently attached to the charger.

14.3.4 Types of EV Charger

Generally, chargers can be categorized into two types: off-board chargers and on-board chargers. On-board chargers are used for low power applications, whereas off-board chargers are used for high power DC fast charging. On-board chargers are located inside the EV, and off-board chargers are located outside the EV. On-board chargers charge electric vehicles from AC sources; the main challenges are power limitations and charging times. Fast charging is offered by off-board chargers, and during peak hours, vehicle-to-grid charging is also an option. Almost any location with an electrical outlet will be able to charge the on-board charge. In this type of charger, a simple battery management system (BMS) is adequate. In Figure 14.8, the two types of EV chargers are illustrated in a block diagram.

14.4 System Description

The system block diagram is depicted in Figure 14.9. It consists of a PV system, a DC–DC boost converter with MPPT capability, and a DC–AC inverter for balancing power supply. In this study, uncertainty may be caused by climate change both from the load side as well as the generation side, which can be reduced by using properly optimized controllers. The growing demand for renewable energy reflects the rising cost and depletion of fossil fuels. Among these renewable sources, solar systems, known as photovoltaic (PV) systems, have interested researchers all over the world. This system is able to convert solar energy into electricity. However, the power generated by this source is varying according to the weather conditions. Therefore, a MPPT is mandatory to improve its efficiency and reduce its cost. A DC/DC converter plays a good role to transfer the PV panel maximum power to the load. The MPP operation can be achieved through adjusting its duty cycle.

However, the achievement of MPPT remains a challenging task due to the dynamic behavior of the PV array, which is significantly impacted by environmental conditions. This challenge becomes even more convoluted

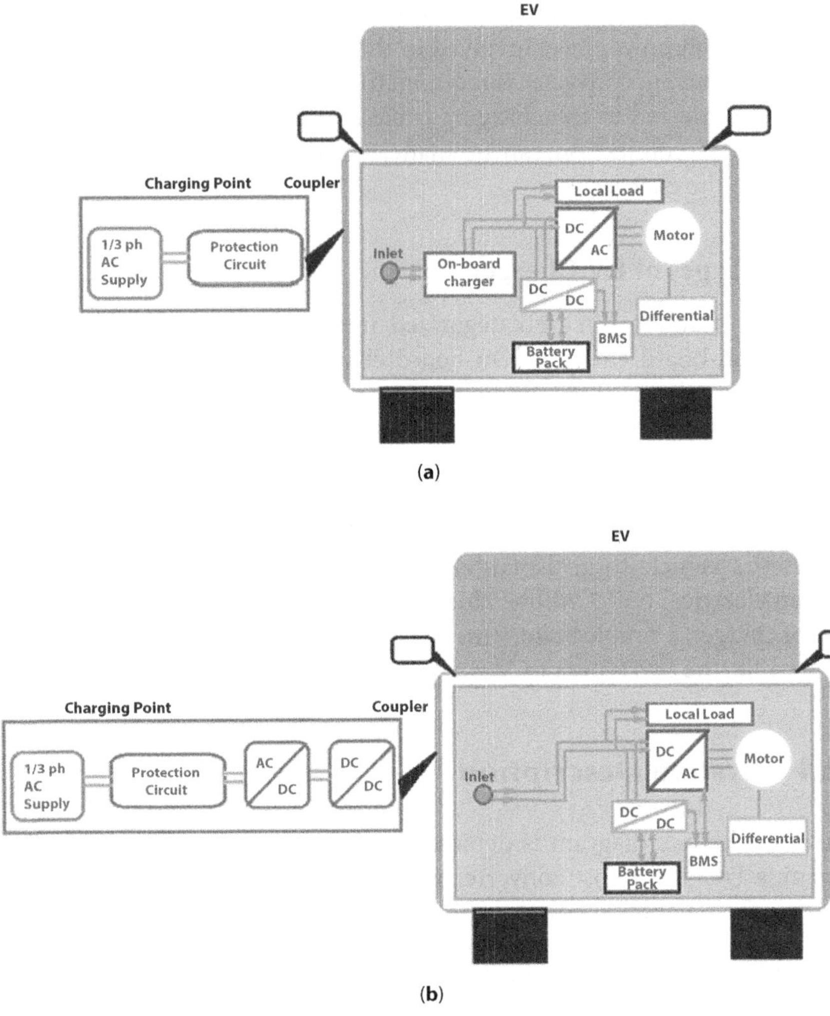

Figure 14.8 EV charger arrangement: (a) on-board and (b) off-board.

in the case of partial shading conditions (PSCs), which occur when the entire PV array is not receiving homogeneous irradiation, which can result from several factors, like clouds, trees, buildings, or even dust. Furthermore, under PSCs, the PV array's power–voltage (P–V) curve exhibits several power maxima, which result from the use of bypass diodes that protect shaded PV cells from damage caused by the hotspot phenomenon.

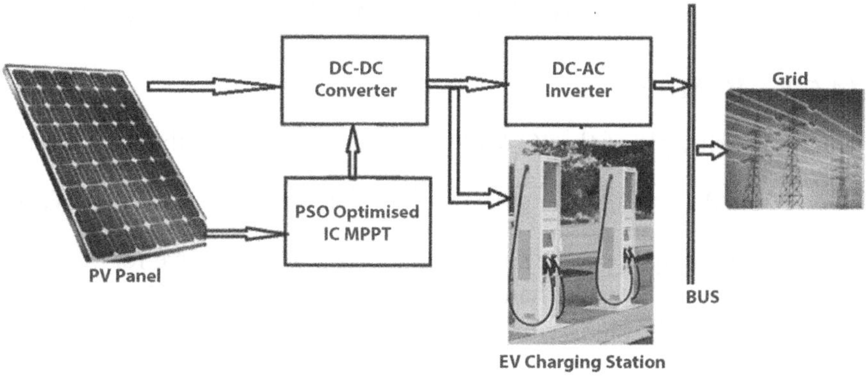

Figure 14.9 System block diagram.

Under PSCs though, conventional MPPT techniques such as P&O and incremental conductance (INC) does not differentiate between a global MPP (GMPP) and a local MPP (LMPP) since they converge to the MPP that comes in contact first, which is most likely a local MPP. Therefore, this leads to a considerable energy loss of up to 70%, far away from the desired optimal energy harvesting. To overcome this problem, the conventional algorithm needs knowledge of initial perturbation value especially under PSC. The nature-inspired optimization algorithms have become popular due to their effectiveness in dealing with complex problems, such as the non-linear behavior of PV arrays. In addition, their ease of implementation makes them attractive to address the MPPT issue in PV systems operating under PSCs. However, the nonlinear behavior of the DC/DC converter makes this technique susceptible to excessive ripple, steady-state inaccuracy when there is a broad range of environmental condition variations. On the other side, to increase the robustness and stability of the system against external disturbances such as solar irradiation, temperature, and load changes, nonlinear controllers have been developed. Even though PSO algorithm produces zero steady-state oscillation, it provides a disadvantage of confronting the slow speed of convergence, and the search space is large due to the fact that a lot of time is wasted in visiting states of poor fitness. It is therefore necessary to propose an algorithm with both the goodness of conventional MPPT and the superiority of optimization algorithms and to overcome the shortcomings of both.

14.5 Proposed PSO Optimized IC MPPT Algorithm

14.5.1 IC MPPT

The basic principle of the IC MPPT method is based on the fact that the slope of the PV array power curve $\left(\dfrac{dP_{PV}}{dV_{PV}}\right)$ is zero at the MPP, positive for values of the output power smaller than MPP and negative for values of the output power greater than MPP. Equations (14.1) and (14.2) give the condition of the IC algorithm.

$$\frac{dP_{PV}}{dV_{PV}} = \frac{d(V_{PV}.I_{PV})}{dV_{PV}} = V_{PV}.\frac{dI_{PV}}{dV_{PV}} + I_{PV} = 0 \tag{14.1}$$

$$\left.\begin{aligned}
&\frac{dP_{PV}}{dV_{PV}} > 0 \text{ if } \frac{I_{PV}}{V_{PV}} > -\frac{dI_{PV}}{dV_{PV}}, \text{left of MPPi.e.} V_{ref} = V_{ref} + \Delta V \\
&\frac{dP_{PV}}{dV_{PV}} = 0 \text{ if } \frac{I_{PV}}{V_{PV}} = -\frac{dI_{PV}}{dV_{PV}}, \text{at MPPi.e.} V_{ref} = V_{ref} \\
&\frac{dP_{PV}}{dV_{PV}} < 0 \text{ if } \frac{I_{PV}}{V_{PV}} < -\frac{dI_{PV}}{dV_{PV}}, \text{right of MPPi.e.} V_{ref} = V_{ref} - \Delta V
\end{aligned}\right\} \tag{14.2}$$

The above-mentioned principle illustrated is shown in Figure 14.10. The MPP can thus be tracked by comparing the instantaneous conductance $\left(\dfrac{I_{PV}}{V_{PV}}\right)$ with the incremental conductance $\left(\dfrac{dI_{PV}}{dV_{PV}}\right)$.

In this equation, V_{ref} is the reference voltage at which the PV array will be forced to operate. Initially, V_{ref} is set to zero. At the MPP, V_{ref} equals to V_{MPP}. Once the MPP is reached, the operation of the PV array is maintained at this point unless a change in current, I, occurs as a result of a change in atmospheric conditions leading to a variation in the MPP. The algorithm, then, tracks the MPP by applying decrements or increments to V_{ref}. The size of the increment or decrement determines how fast the MPP is tracked.

Under variations in the atmospheric conditions, this algorithm tracks the MPP by applying increments or decrements to V_{ref}. The size of the increment or decrement value is crucial for the success of this approach. If the size is large, the algorithm finds the MPP quickly but results in oscillations

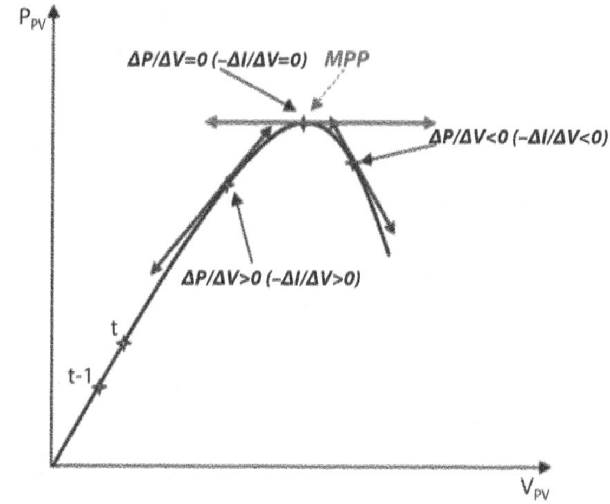

Figure 14.10 Principle of the IC algorithm.

around the MPP. On the other hand, if the size is small, the oscillations around the MPP are reduced, but the rate of convergence will decrease. The maximum power point tracking (MPPT) is employed to control the input voltage and the duty cycle of the DC/DC converter in the PV system so that the PV array delivers the maximum power. Though the IC conductance algorithm is effective in tracking the MPP without any oscillations, the algorithm may not track the MPP under rapidly varying atmospheric conditions due to its slow response.

Hence, in this work, some modifications are made in the conventional IC algorithm to speed up the MPPT process so that it is suitable for rapidly varying atmospheric conditions, the details of which are presented in the next section.

14.5.2 PSO

PSO, developed by Eberhart and Kennedy, is a metaheuristic algorithm based on the concept of swarm intelligence capable of solving complex mathematics optimization problems existing in engineering. PSO is inspired by the ability of flocks of birds, schools of fish, and herds of animals to adapt to their environment, find rich sources of food, and avoid predators by implementing "information sharing" approaches, hence, developing an evolutionary advantage. PSO is initialized with a randomly generated population of particles (initial swarm), and a random velocity is assigned to each particle that propagates the particle in search space toward optima over a number of iterations. Each particle has a memory remembering the best position attained by it in the past, which is called personal best position (P_{best}). Each particle has its P_{best}, and the particle with the best value of fitness is called global best particle (G_{best}). Suppose that the search space is D dimensional, the ith particle of the population can be represented by a D-dimensional vector $(x_i^1, x_i^2, \ldots, x_i^D)^T$. The velocity of this particle can be represented by another D-dimensional vector $(V_i^1, V_i^2, \ldots, V_i^D)^T$. The previously best visited position of ith particle is denoted by P_i, and the best particle in the swarm is denoted by P_g.

Particle: This term is used to identify an individual in the swarm (e.g., a bird in the flock or a fish in the school). *Agent* is also used in some circles. Each particle has a location in the swarm. In the optimization algorithm, each particle location represents a design point that is a potential solution to the problem.
Particle Position: This term refers to the coordinates of the particle. In the optimization algorithm, it refers to a design point (a vector of design variables).
Particle Velocity: The term refers to the rate at which the particles are moving in space. In the optimization algorithm, it refers to the design change.
Swarm Leader: This is the particle having the best position. For the optimization algorithm, the term refers to a design point having the smallest value for the cost function.

The update of the particle's position is accomplished by the following two equations. Equation (14.3) calculates a new velocity for each particle based on its previous velocity, and Equation (14.4) updates each particle's position in search space.

$$v_i^{k+1} = w \; x \, v_i^{k+1} + r_1 \; x \; c_1 \; x\left(P_{best} - x_i^k\right) + r_2 \; x \, c_2 \; x\left(G_{best} - x_i^k\right) \quad (14.3)$$

$$x_i^{k+1} = x_i^k + v_i^{k+1} \quad (14.4)$$

In the aforementioned expression, i represents the variable of the optimization vector, k is the number of iterations, and the velocity and position, respectively, of the ith variable within k iterations, and the parameter w is known as inertia that maintains a balance between the local and global search. c_1 and c_2 are acceleration constants. r_1 and r_2 are two generated random numbers which are uniformly distributed in the interval [−1, 1].

The variable P_{best} records the best position affected by the ith particle up to the exact time of measurement. The following equation indicates that this position is only recorded as P_{best} if the condition stated below is satisfied.

$$P_{best} = x_i^k \text{ if } fit\left(x_i^k\right) \geq fit(P_i) \tag{14.5}$$

The PSO's search space, where each point represents an output voltage value as a solution to the MPPT problem. The evaluation of the particles is based on the output power of the PV panel respective to the final voltage value which is indicated by fit as the fitness evaluator for the particles. The following equation shows the position matrix of the n particles which represents n solutions to the MPPT problem.

$$x_i^k = x_1^k, x_2^k, x_3^k \ldots\ldots\ldots\ldots x_i^k \ldots\ldots\ldots x_{n-1}^k, x_n^k \tag{14.6}$$

where x_i^k is the position of the ith particle at k[th] iteration.

Therefore, the algorithm must be initialized when the following equation is satisfied:

$$\left|\frac{fit(x_{i+1}) - fit(x_i)}{fit(x_i)}\right| > \Delta P \tag{14.7}$$

The PSO algorithm implementation process in this paper is as follows:

i) Initialize the size of swarm, dimension of search space, maximum number of iterations, and the PSO constants w, c_1, and c_2. Define the random numbers r_1 and r_2.
ii) Find out the current fitness of each particle in the population.
iii) Attribute the particles with random initial positions and velocities.
iv) Evaluate the fitness value of each particle.
v) Calculate the global best fitness value: current global best fitness = min (local best fitness).
vi) Update the particle velocity and position for the next iteration. Find out the current fitness of each particle: If current fitness < local best fitness, set local best fitness = current fitness.

vii) Determine the current global best fitness [current global best fitness = min (local best fitness)]: If the current global best fitness < global best fitness, then global best fitness = current global best fitness. The position corresponding to global best fitness is assigned to G_{best}.

viii) Repeat steps 6 and 7 until the maximum number of iterations is achieved or there is no improvement of the global best fitness value.

ix) Terminate the iterative algorithm when the criterion is reached.

14.5.3 PSO Optimized IC MPPT

Even though the PSO algorithm produces zero steady-state oscillation, it provides a disadvantage of confronting the slow speed of convergence, and the search space is large due to the fact that a lot of time is wasted in visiting states of poor fitness. The shortcomings of IC MPPT and PSO can be overcome by combining both in effective ways such that PSO is used offline to provide optimized voltage initial perturb value for IC MPPT to increase its conversion time.

In this study, a PSO technique is used to provide a solar array's IC-MPPT with the optimum voltage. To evaluate the proposed method, MATLAB/Simulink software is used.

The flow chart of PSO as well as IC MPPT is shown in Figure 14.11.

The result of the PSO algorithm is given to the IC MPPT for further tracking of MPP voltage. The idea behind the incremental conductance method is to increase or decrease the reference voltage (Vref) value (which is in the vicinity of the MPP voltage) based on the comparison of the instantaneous conductance with the incremental conductance. The increment size determines how fast the MPP is tracked. Fast tracking can be achieved with bigger increments, but the system might not operate exactly at the MPP and oscillate about it instead; so, there is a trade-off.

14.6 Case Study—MATLAB Simulation

To demonstrate the effectiveness of the proposed PSO-optimized IC-MPPT technique, simulations were performed appropriately. The simulation model specifications of the PV module and DC–DC converter used are shown in Table 14.1.

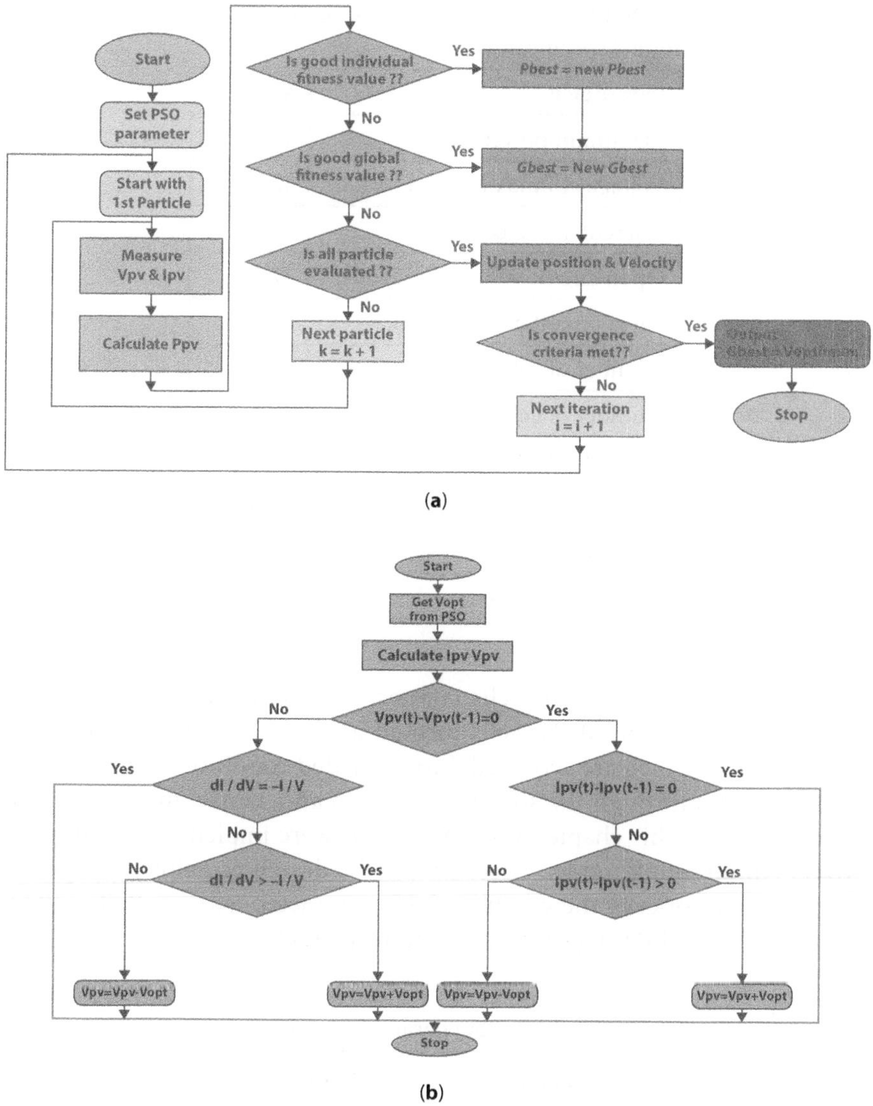

Figure 14.11 (a) Flow chart of the PSO. (b) Flow chart of the proposed IC-MPPT.

Table 14.1 PV cell and DC–DC converter model specification.

Parameter	**Value**
Maximum power (Pmax)	250 W
Voltage at Pmax (Vmax)	30.7 V
Current at Pmax (Imax)	8.15 A
Open circuit voltage (Voc)	37.3 V
Short circuit current (Isc)	8.66 A
Number of cells in series	60
Number of cells in parallel	3
Capacitor (C1)	470 μF
Capacitor (C2)	47 μF
Inductor (L)	1 mH
Switching frequency	10 kHz

A PV model had 60 PV cells connected in series which are divided into three groups of 20 PV cells formed as three PV strings with three bypass diodes which are activated during shaded conditions, allowing current to bypass through the shaded cell group and, hence, minimizing the current loss. In this chapter, the simulations were implemented using the MATLAB/Simulink model. According to the design principle, the specification frameworks of the complete PSO parameter settings, that is PSO-optimized IC-MPPT algorithm, are shown in Table 14.2.

Table 14.2 PSO parameter settings.

Parameter	**Value**
Number of particles	3
Sampling time	0.1s
Maximum iteration	3
W	0.1
C_1	1
C_2	1

Under uniform shading conditions, the current–voltage (I–V) and power–voltage (P–V) characteristic curves of the modeled 250-kWp PV of various irradiations from 50 W/m² to 1,000 K/m² are shown in Figure 14.11. Under these variable irradiation conditions, the power, voltage, and current of the PV output and the DC–DC converter output are shown in Figure 14.12 to Figure 14.14, where it can be inferred that the DC–DC converter output voltage is around 30 V, which is the optimum value clearly seen in the P–V characteristic curves in Figure 14.12. This shows that the proposed PSO-optimized IC-MPPT performed well.

Appendix 14.A1 includes the PSO MPPT coding used in this work for reference. The overall system in MATLAB/Simulink is depicted in Figure 14.A2 in the Appendix.

In order to produce partial shading effects, the irradiation values of the second and third PV strings were changed to 200, 400, 600, and 1,000 W/m², respectively. PV modules in the rest of the solar farm are irradiated at 1,000 W/m². Due to the fact that there are four patterns, there are four peaks in the P–V curve and four different steps in the I–V curve as shown

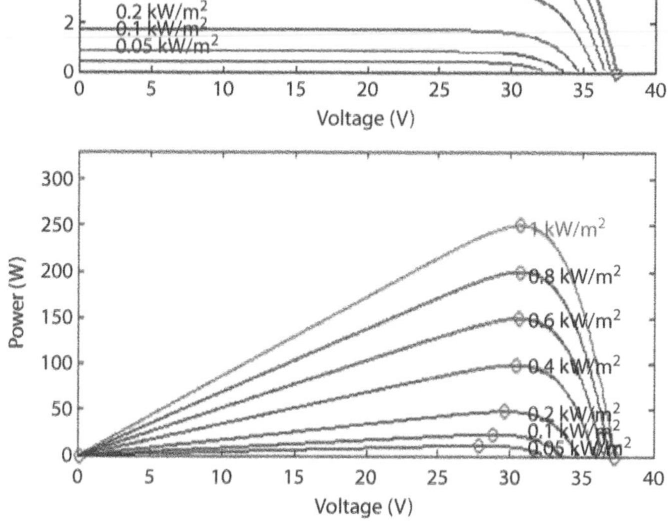

Figure 14.12 I–V and P–V curves.

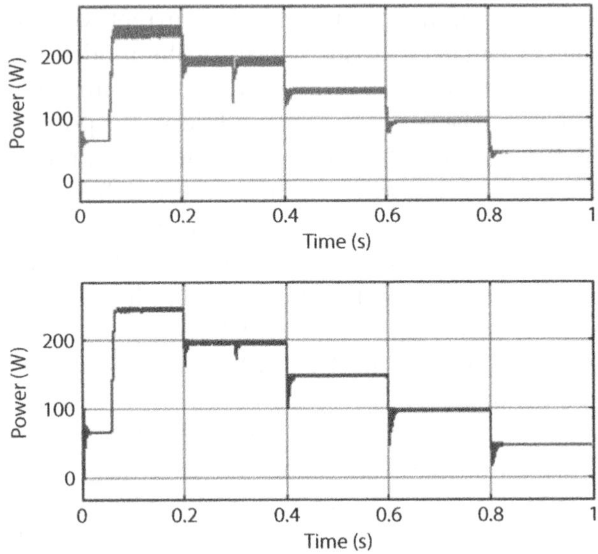

Figure 14.13 Power vs time of PV output and DC–DC converter output.

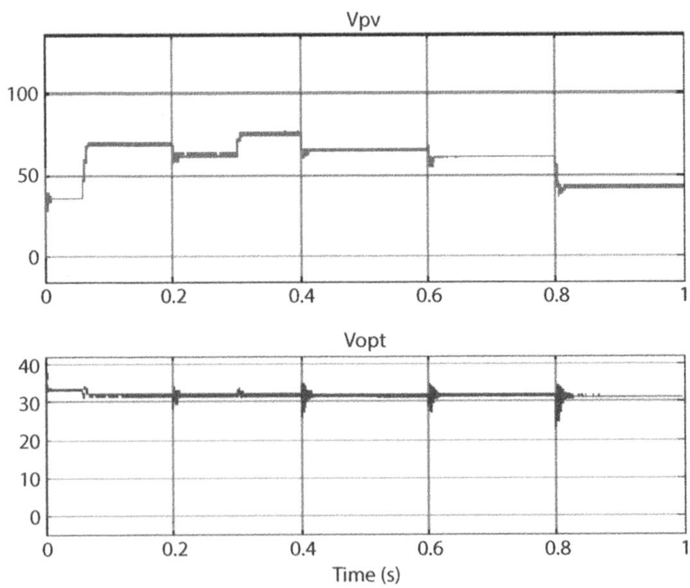

Figure 14.14 Voltage vs time of PV output and DC–DC converter output.

in Figure 14.15. In Figure 14.16, the PSO algorithm is tracking the optimal voltage with very low oscillations and fast convergence. As a start of perturbation, the PSO algorithm provides the optimal voltage to the IC MPPT. Thus, one can reduce the tracking time and increase the convergent speed. Comparing IC MPPT performance power curves with and without PSO reference voltage reveals this, which is depicted in Figure 14.17.

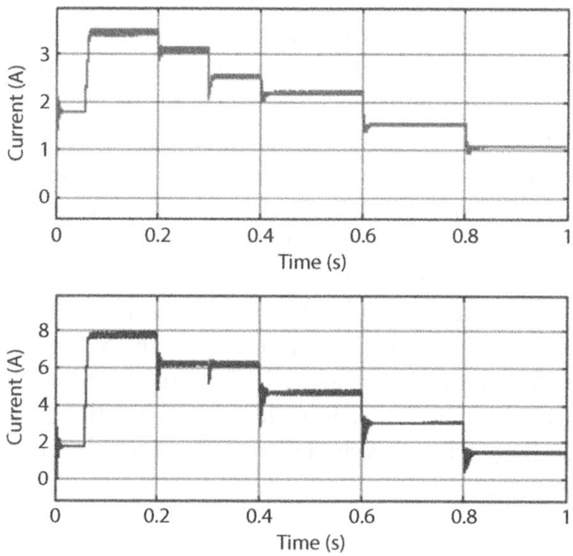

Figure 14.15 Current vs time of PV output and DC–DC converter output.

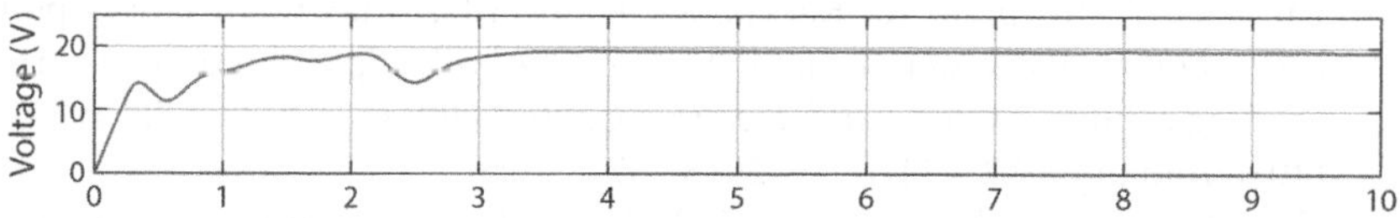

Figure 14.16 Optimum voltage turned by PSO.

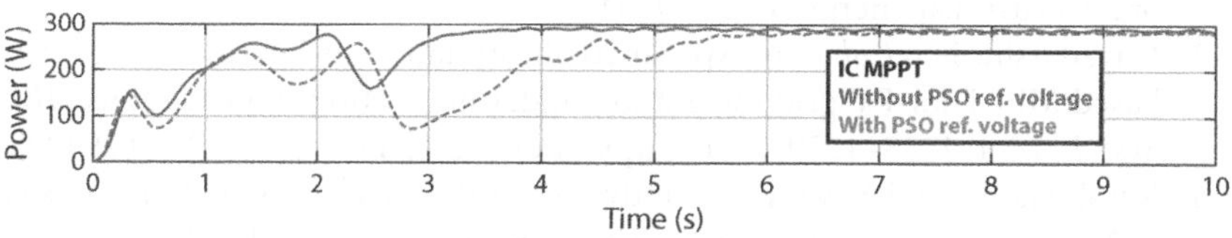

Figure 14.17 Maximum power tracking of IC MPPT with and without PSO optimum voltage.

It has been observed that IC MPPT with PSO optimum voltage tracks maximum power with less oscillation and fast convergence speed of 3 s prior.

14.7 Conclusion

It has been attempted to highlight EV charging stations, charger types, and modes as well as to develop PV-based EV charging stations in MATLAB/Simulink with improved performance. The load also consists of buck converter-controlled EVs. We have demonstrated in simulations that our optimized PSO technique based IC MPPT reaches the maximum power point voltage in a shorter time with an accurate maximum power output. By analyzing the simulation results, it can be obtained that the PSO-optimized IC MPPT has advantages of high convergence speed and high precision and can accurately track the real maximum power compared with traditional IC MPPT. In the future, another metaheuristic algorithm could be used instead of PSO, which may yield better results. The integration of solar energy and EV technology is an excellent combination, and EVs provide energy storage as well. The forecasting of PV systems will also be performed in the future in order to manage the grid efficiently, to charge and discharge electric vehicles, and to predict the consumption and generation of energy.

References

1. Farzaneh, J., Keypour, R., Karsaz, A., A novel fast maximum power point tracking for a PV system using hybrid PSO-ANFIS algorithm under partial shading conditions. *Int. J. Ind. Electron. Control Optim.*, *2*, 1, 47–58, Jan 2019, https://doi.org/10.22111/IECO.2018.25721.1056.
2. Chang, S., Cho, J., Heo, J., Kang, J., Kobashi, T., Energy infrastructure transitions with PV and EV combined systems using techno-economic analyses for decarbonization in cities. *Appl. Energy*, *319*, 119254, 2022, https://doi.org/10.1016/j.apenergy.2022.119254.
3. Priyadarshi, N. *et al.*, An experimental estimation of hybrid ANFIS–PSO-based MPPT for PV grid integration under fluctuating sun irradiance. *IEEE Syst. J.*, *14*, 1, 1218–1229, 2020, https://doi.org/10.1109/JSYST.2019.2949083.
4. Karaboga, D. and Kaya, E., Adaptive network based fuzzy inference system (ANFIS) training approaches: A comprehensive survey. *ArtifIntell Rev.*, 52, 2263–2293, 2019, https://doi.org/10.1007/s10462-017-9610-2.

5. Semero, Y.K., Zheng, D., Zhang, J., A PSO-ANFIS based hybrid approach for short term pv power prediction in microgrids. *Electri.Power Compon. Syst.*, 46, 1, 95–103, 2018, doi: 10.1080/15325008.2018.1433733.
6. Jha, M., Blaabjerg, F., Khan, M.A., Bharath Kurukuru, V.S., Haque, A., Intelligent control of converter for electric vehicles charging station. *Energies*, 12, 12, 2334, 2019, https://doi.org/10.3390/en12122334.
7. Punitha, K., Devaraj, D., Sakthivel, S., Artificial neural network based modified incremental conductance algorithm for maximum power point tracking in photovoltaic system under partial shading conditions. *Energy*, 62, 330–340, 2013b, https://doi.org/10.1016/j.energy.2013.08.022.
8. Sharma, A. *et al.*, A novel TSA-PSO based hybrid algorithm for GMPP tracking under partial shading conditions. *Energies*, 15, 9, 3164, 2022, doi: 10.3390/en15093164.
9. Lal and Singh, Single-stage utility-scale PV system with PSO based MPPT controller. *Eighteenth National Power Systems Conference (NPSC) – 2014*, 2014, doi: 10.1109/NPSC.2014.7103821.
10. Calvinho, *et al.*, Design and implementation of MPPT system based on PSO algorithm. *International Conference on Intelligent Systems (IS) – 2018*, 2018.
11. Himabindu, N., Hampannavar, S., Deepa, B., Swapna, M., Analysis of microgrid integrated Photovoltaic (PV) powered electric vehicle charging stations (evcs) under different solar irradiation conditions in India: A way towards sustainable development and growth. *Energy Rep.*, *7*, 8534–8547, 2021, https://doi.org/10.1016/j.egyr.2021.10.103.
12. Kariem, H., Touti, E., Fetouh, T., The efficiency of PSO-based MPPT technique of an electric vehicle within the city. *Meas. Control*, *53*, 3–4, 461–473, 2020, https://doi.org/10.1177/0020294019882973.
13. Mingyu, L., Xinhong, C., Bingyu, C., Random inertia weight PSO based MPPT for Solar PV under partial shaded condition. *IOP Conf. Ser.: Earth Environ. Sci.*, 585, 1, 012028, 2020, https://doi.org/10.1088/1755-1315/585/1/012028.
14. Belghith, O.B., Sbita, L., Bettaher, F., MPPT design using PSO technique for photovoltaic system control comparing to fuzzy logic and P&O controllers. *Energy Power Eng.*, *08*, 11, 349–366, 2016, https://doi.org/10.4236/epe.2016.811031.
15. Chtita, S., Motahhir, S., El Hammoumi, A., Chouder, A., Benyoucef, A.S., El Ghzizal, A., Derouich, A., Abouhawwash, M., Askar, S.S., A novel hybrid GWO–PSO-based maximum power point tracking for photovoltaic systems operating under partial shading conditions. *Sci. Rep.*, *12*, 10637, 2022, https://doi.org/10.1038/s41598-022-14733-6.
16. Gavhane, P.S., Krishnamurthy, S., Dixit, R., Ram, J.P., Rajasekar, N., EL-PSO based MPPT for solar PV under partial shaded condition. *Energy Procedia*, *117*, 1047–1053, 2017, https://doi.org/10.1016/j.egypro.2017.05.227.
17. Javed, S. and Ishaque, K., A comprehensive analyses with new findings of different PSO variants for MPPT problem under partial shading. *Ain Shams Eng. J.*, *13*, 5, 101680, 2022. https://doi.org/10.1016/j.asej.2021.101680.

18. Hassan, A., Bassa, O., Masoumb, A.S., An improved genetic algorithm based fractional open circuit voltage MPPT for solar PV systems. *Energy Rep.*, *9*, 1535–1548, 2023, https://doi.org/10.1016/j.egyr.2022.12.088.
19. Faisal, M., Hannan, M., Ker, P.J., Rahman, M., Begum, R., Mahlia, T., Particle swarm optimised fuzzy controller for charging–discharging and scheduling of battery energy storage system in MG applications. *Energy Rep.*, *6*, 215–228, 2020, https://doi.org/10.1016/j.egyr.2020.12.007.
20. Mannaa, S., Kumar Singha, D., Kumar Akellaa, A., Kotbb, H., M.AboRasb, K., M.Zawbaacde, H., Kamelf, S., Design and implementation of a new adaptive MPPT controller for solar PV systems. *Energy Rep.*, *9*, 1818–1829, 2023, https://doi.org/10.1016/j.egyr.2022.12.152.
21. Sagonda, A.F. and Folly, K.A., A comparative study between deterministic and two meta-heuristic algorithms for solar PV MPPT control under partial shading conditions. *System Soft Computing (SASC)*, *4*, 200040, 2022, https://doi.org/10.1016/j.sasc.2022.200040.
22. Hamza Zafar, M., Mujeeb Khan, N., Feroz Mirza, A., Mansoor, M., Akhtar, N., Usman Qadir, M., Ali Khan, N., Raza Moosavi, S.K., A novel metaheuristic optimization algorithm based MPPT control technique for PV systems under complex partial shading condition. *Sustain. Energy Technol. Assess.*, *47*, 101367, 2021, https://doi.org/10.1016/j.seta.2021.101367.
23. Gong, L., Hou, G., Huang, C., A two-stage MPPT controller for PV system based on the improved artificial bee colony and simultaneous heat transfer search algorithm. *ISA Trans.*, 132, 428–443, 2022, https://doi.org/10.1016/j.isatra.2022.06.005.
24. Kishore, D.K., Mohamed, M., Sudhakar, K., Peddakapu, K., Swarm intelligence-based MPPT design for PV systems under diverse partial shading conditions. *Energy*, *265*, 126366, 2023, https://doi.org/10.1016/j.energy.2022.126366.
25. Ram, J.P., Babu, T.S., Rajasekar, N., A comprehensive review on solar PV maximum power point tracking techniques. *Renew. Sustain. Energy Rev.*, *67*, 826–847, 2017, https://doi.org/10.1016/j.rser.2016.09.076.
26. Rezk, H., Aly, M., Al-Dhaifallah, M., Shoyama, M., Design and hardware implementation of new adaptive fuzzy logic-based MPPT control method for photovoltaic applications. *IEEE Access*, *7*, 106427–106438, 2019, https://doi.org/10.1109/access.2019.2932694.
27. Shi, J., Zhang, W., Zhang, Y., Xue, F., Yang, T., MPPT for PV systems based on a dormant PSO algorithm. *Electri. Power Syst. Res.*, *123*, 100–107, 2015, https://doi.org/10.1016/j.epsr.2015.02.001.
28. Chang, S., Cho, J., Heo, J., Kang, J., Kobashi, T., Energy infrastructure transitions with PV and EV combined systems using techno-economic analyses for decarbonization in cities. *Appl. Energy*, *319*, 119254, 2022b, https://doi.org/10.1016/j.apenergy.2022.119254.
29. Sharma, P. and Chinnappa Naidu, R., Optimization techniques for grid-connected PV with retired EV batteries in centralized charging station with

challenges and future possibilities: A review. *Ain Shams Eng. J.*, 14, 7, 101985, 2022, https://doi.org/10.1016/j.asej.2022.101985.
30. Yilmaz, M. and Krein, P.T., Review of the impact of vehicle-to-grid technologies on distribution systems and utility interfaces. *IEEE Trans. Power Electron.*, 28, 5673–5689, 2013, http://dx.doi.org/10.1109/TPEL.2012.2227500.
31. https://www.niti.gov.in/sites/default/files/2021-08/Handbook for EV Charging Infrastructure Implementation 081221.pdf

Appendix

14.A1 PSO-Optimized MPPT Codings

```
function V= PSO(Vpv, Ipv)
persistent u;
persistent dcurrent;
persistent pbest;
persistent p;
persistent dc;
persistent v;
persistent counter;
persistent gbest;
if(isempty(counter))
   counter=0
end
if(isempty(p))
   p=zeros(4, 1);
end
if(isempty(v))
   v=zeros(4, 1);
end
if(isempty(pbest))
   pbest=zeros(4, 1);
end
if(isempty(u))
   u=0;
end
if(isempty(dc))
```

```
    dc=zeros(4, 1);
    dc(1)=0;
    dc(2)=0.3;
    dc(3)=0.5;
    dc(4)=0.9;
end
if(counter>=1 && counter<300)
V=dcurrent;
counter=counter+1;
return;
end
if(u>=1 && u<=4)
    if(Vpv*Ipv)>p(u))
    p(u)=Vpv*Ipv;
    pbest(u)=dcurrent;
    end
u=u_1;
if(u==6)
u=1;
end
if(u==1)
V=dc(u);
dcurrent=V;
counter=1;
return;
elseif(u==2)
V=dc(u);
elseif(u==5)'
[m.i]=max(p);
gbest=pbest(i);
V=gbest;
dcurrent=V;
counter=1;
%update ve and Optimum Voltage
```

```
v(1)=updatevelocity(v(1), pbest(1), dc(1), gbest)
v(2)=updatevelocity(v(2), pbest(2), dc(2), gbest)v(2)
v(3)=updatevelocity(v(3), pbest(3), dc(3), gbest)v(3)
v(4)=updatevelocity(v(4), pbest(4), dc(4), gbest)
%update opt_voltage
dc(1)=updateVopt(dc(1), v(1))
dc(2)=updateVopt(dc(2), v(2))dc(3)=updateVopt(dc(3_, v(3))dc(4)=
updateVopt(dc(4_, v(4))
return;
else
u y='hello'
V=0.1
end
end
function vfinal=updatevelocity(velocity, pobest, pobest, v, gvbest)
w=0.4;
c1=1.2;
c2=2;
fuctiondfinal=update(d,velocity)
dup=v+velocity;
if(dup>1)
dfinal=1;
   elseif(dup<o)
dfinal=0;
else
dfinal=dup;
end
end
vfinal=(w*ve
End
```

14.A2 Overall System in MATLAB/Simulink

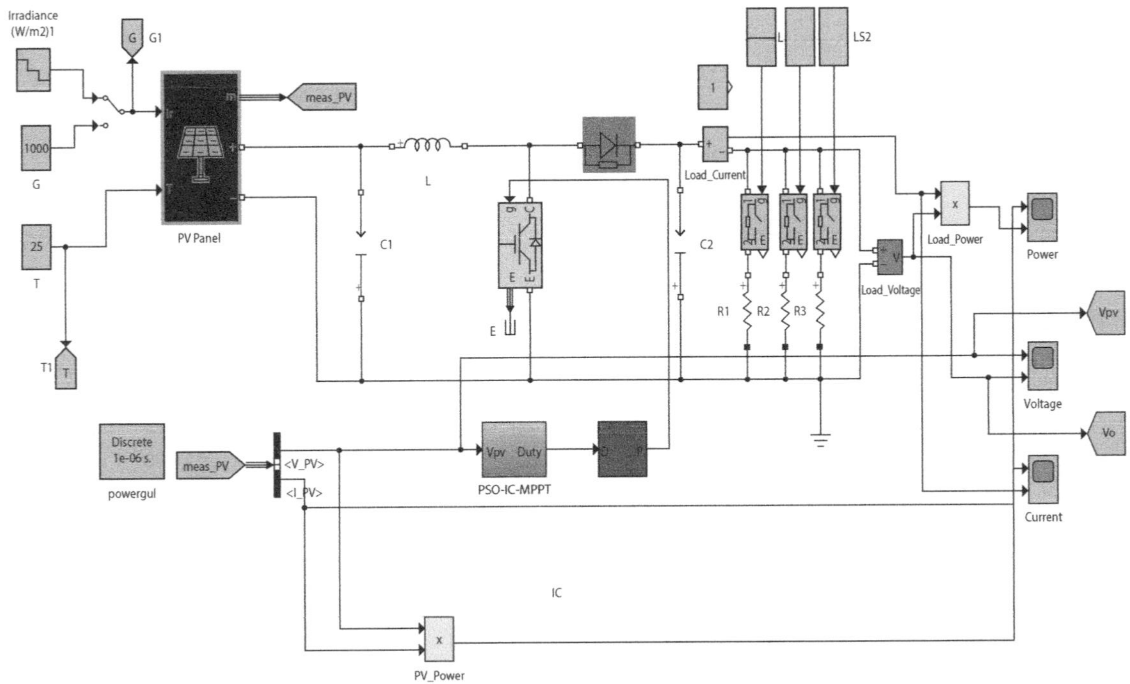

Figure 14.A2 Overall system in MATLAB/Simulink.

15

Closed-Loop Control of Microgrids With Wind and Battery Storage System in Islanding Mode

Shruti Sahu[1], Jyoti Shukla[2] and Basanta K. Panigrahi[3]*

[1]University College of Engineering, Kota, India
[2]Department of Electrical Engineering, SKIT, Jaipur, India
[3]Department of Electrical Engineering, SOA University, Bhubaneswar, India

Abstract
The following chapter gives an overview of a stand-alone permanent magnet synchronous generator (PMSG) based wind-battery energy conversion system which accords the supremacy for employing a battery energy storage system with a PMSG. Due to the intermittent nature of wind, varying-speed wind turbines are brought into play with changing wind speeds for harnessing the maximum power from the different wind patterns taken for analysis, which are step, staircase, and random wind speed. For the stable operation of converters in a microgrid system, control schemes are implemented. The distinctive feature of the battery energy source is to modulate the DC link voltage via a controller and also plays an important role in synchronizing the load voltage and frequency during sudden load change. The control schemes involved in the closed loop control of DC–DC bidirectional converter for charging and discharging of DC link capacitor and closed loop vector-oriented control of load-side converter for critical loading conditions are analysed. The above-mentioned problem statement has been simulated in the MATLAB/Simulink domain via the application of control strategies.

***Keywords*:** Doubly fed induction generator (DFIG), permanent magnet synchronous generator (PMSG), distributed generation (DG), pulse width modulation (PWM)

**Corresponding author*: basanta1983@gmail.com

P. Sivaraman, C. Sharmeela and P. Sanjeevikumar (eds.) Microgrids for Commercial Systems, (409–454) © 2024 Scrivener Publishing LLC

15.1 Introduction

In remote areas where grid integration is not possible due to many reasons like geographical conditions or uneven terrains, microgrids or stand-alone systems proved to be a boon to meet the load demand in critical conditions. As stand-alone systems are dependent on renewable sources which depend on locally available natural resources according to geographical region, an abrupt change in load or wind speed can lead to crucial deviations in frequency and can also lead to voltage fluctuations in the system [1]. The burgeoning in the domain of power electronics and control techniques has become absolutely necessary in coalescing distributed generation which not only minimizes the cost of extraction and eventually minimizes the overall cost of power generation with the advantage of minimizing the emission of greenhouse gases which are responsible for climate change [2]. Nowadays the choice of variable generation is very much in demand and compatible for the varying nature of wind speed; three types of generators that are most commonly used are induction generator, doubly fed induction generator (DFIG), and permanent magnet synchronous generator (PMSG). PMSG, due to its many features, is one of the most desirable choices for wind power generation as there is no need for a gear box for the connection with the wind turbines, and it also does not require any external excitation due to which the cost of maintenance is low compared with others [3]. As PMSG incorporates a permanent magnet as a rotor, there is no need of excitation current; therefore, stator current will only be required for torque production. Hence, the PMSG, for alike output, will operate at high power factor due to no excitation current required, which makes PMSG better than other machines [4].

For the maximum power extraction from intermittent sources which is varying throughout the day, a variable speed wind turbine with a control technique is required for the generator.

The proposed system incorporates the controlling strategies required to manage the fluctuation occurring in DC link voltage due to change in wind speed and sudden change in load change. A pulse width modulation control technique is used to provide pulses to the load side converter, and a converter is used for the speed control of PMSG. Bi-directional power is transferred between the grid side and the generator rotor side under a constant DC link voltage, and the input power from the grid side should be equal to the generator input power [16–24].

Because the change in the speed of the wind will also affect the wind torque of the wind turbine generator, this will give a fluctuating AC voltage which is converted to DC voltage by using an uncontrolled rectifier, and to convert the variable DC voltage to constant DC voltage, a bi-directional converter along with a control structure is used to produce the pulses for the switches of the bi-directional converter to operate in buck and boost mode, and a Li-ion battery storage system is implemented, which charges and discharges the DC link voltage. The voltage dip occurred because of a sudden change in the wind speed which is adjusted by the battery, and the DC–DC bidirectional controller employed maintains the constant DC link voltage to the reference value. The voltage is kept constant to the reference value during the change in wind speed and load demand [29–34].

The control technique utilized in microgrids, particularly in islanding mode, needs the inverter to maintain the load voltage in relation to the reference load voltage intended for the load requiring constant amplitude and frequency voltage. For voltage and frequency control on the load side, the inverter on the supply side employs a complicated vector control approach. The stand-alone system is implemented with a voltage and frequency controller which is capable of controlling the variable load. The vector-controlled technique used offers high dynamics and static performance through inner control loops. This control process consists of a voltage controller which controls the output voltage; the DC link voltage controller stabilizes the voltage of DC link. The load analysis is performed, and the controller is working efficiently and maintains the DC link voltage to regulate the load voltage constant in terms of amplitude and frequency during sudden load change [35–40].

For the removal of harmonics from the inverter output, an LC filter is employed between VOC VSI and the load due to the injection of harmonics introduced by non-linear component employed for the power generation in stand-alone systems as the PWM inverter used produces excessive harmonics in the output voltage which is not suitable for the designed load parameters given to the consumer, so a suitable filter needs to be configured in order to take out the high-frequency harmonics and to enhance the power quality. Usually, filters are required to separate the fundamental component from the rest of the harmonics present in the inverter output; the prime objective is to obtain the pure sinusoidal voltage desired for the designed load.

The proposed system consists of a distributed generation (DG) source and a battery employed for the storage of unused power. The DG source is a wind turbine's generally variable-speed wind turbine system along with a battery system that serves as a storage device and also charges the DC link when the voltage of the DC link is less than the reference voltage. The first stage is when a wind turbine is coupled to a PMSG and a DC–DC buck–boost converter is employed, which also acts as a maximum power point tracker for wind turbine, and the stand-alone configuration needs additional energy storage devices for storing excess energy, so a battery storage system is used along with a DC–DC bi-directional converter to operate in charging and discharging modes depending upon the wind and load scenarios [5–9]. The stand-alone DG systems bring along with it a number of technical challenges such as the following:

- Voltage regulation
- Frequency regulation

Due to change in wind speeds, it is difficult to obtain the fixed frequency voltage from a fluctuating source as our loads are designed to operate at a specific frequency, and also when the load demand changes from time to time, the controller should maintain the voltage and settles the fluctuation as soon as possible during load change and to sustain the load demanded by the loads during unavailability of the main source a battery is employed. To obtain the specific fixed frequency, a control methodology is required to maintain the desired voltage of DC link, and to maintain the flow of power between source and load demanded, a vector-oriented control technique is employed for PWM technique for inverter control [10–13]. Figure 15.1 shows the block diagram of the proposed wind energy conversion system in standalone mode.

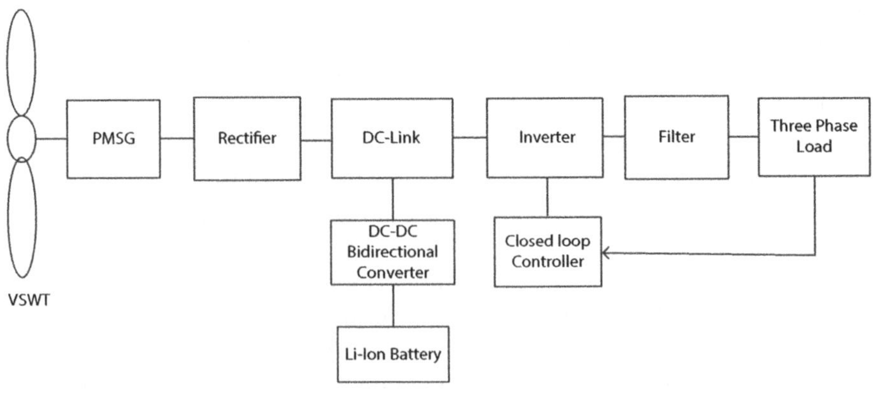

Figure 15.1 Block diagram of VSWT stand-alone wind battery system with controller [1–3].

15.2 Wind Turbine Modeling

Extracting power from the wind is among the innocuous and eco-friendly methods for electricity generation as it does not have any carbon emissions due to uneven heating of the Earth's surface, causing the flow of wind in different regions. The wind turbines are a source of extracting electric power from wind; it converts kinetic energy into rotating mechanical energy [14, 15]. The rotor of the wind turbine is coupled to the PMSG; generally, a horizontal axis wind turbine is preferred in which the rotor shaft is directed towards the wind for maximum power extraction, and mechanical input torque *Tw* and wind speed V_W are obtained from aerodynamics.

$$T_w = \frac{\rho\lambda C_p(\lambda,\beta)V_W R^3}{\lambda} \tag{15.1}$$

$$\lambda = \frac{R\omega_r}{V_w} \tag{15.2}$$

$$C_p(\lambda,\beta) = c_1\left(c_2/\lambda_i - c_3\beta - c_4\right)e^{\frac{-c_5}{\lambda_i}} + c_6\beta \tag{15.3}$$

$$c_1 = 0.5176;\ c_2 = 116;\ c_3 = 0.4;\ c_4 = 5;\ c_5 = 21;\ c_6 = 0.0068 \tag{15.4}$$

$$\frac{1}{\lambda_i} = \frac{1}{\lambda + 0.08\beta} + \frac{0.035}{\beta^3 + 1} \tag{15.5}$$

As the wind speed is not constant throughout the day and the value of C_p maximum power will depend on wind speed (V_W) and pitch angle (β) as per the calculation, maximum power will be obtained at $\beta = 0$, but to obtain maximum power extraction at any speed, maximum power point tracking is required.

15.3 Designing a Permanent Magnet Synchronous Generator

The use of a permanent magnet synchronous generator has proved to be the most suitable generator for wind turbines because of the following features:

- Low maintenance
- Less losses
- Smaller in size

It has a magnetic rotor which does not require any winding in the rotor circuit in other DC machines, and in synchronous machines, an external supply is given for flux generation, i.e., there are no rotor copper losses in PMSG [25, 26]. The wind turbines based on PMSG offers several advantages as follows:

- High efficiency
- Lower losses and maintenance cost
- High reliability
- No slip rings are required
- Do not require excitation

The mathematical modeling of PMSG is based on dq0 theory transformation where d-axis is predominantly lined up with the magnetic axis and q-axis is orthogonal to d-axis. The PMSG-based wind turbine can be defined in the rotor reference frame as follows:

$$V_{dg} = -\left(R_s + \frac{dL_q}{dt}\right)I_q - \omega_r L_d I_d + \omega_r \lambda_m \tag{15.6}$$

$$V_{qg} = -\left(R_s + \frac{dL_d}{dt}\right)I_d + \omega_r L_q I_q \tag{15.7}$$

$$T_e = \frac{3P}{4}[I_d I_q (L_q - L_d) + \lambda_m I_q] \tag{15.8}$$

$$\frac{d\omega_r}{dt} = \frac{P}{2J}(T_m - T_e) \tag{15.9}$$

The correlation between the rotor angular velocity of the generator ω_r and the mechanical angular velocity ω_t of the turbine rotor is as follows:

$$\omega_t = \frac{2\omega_r}{PG} \tag{15.10}$$

Similarly, the turbine torque T_w and the input torque of the generator T_m are given by:

$$T_m = \frac{T_w}{G} \tag{15.11}$$

The d–q components of the generator currents are designed during feeding. The rectifier can be studied as follows:

$$\frac{di_q}{dt} = \frac{1}{L_{sr} + L_q}[-R_s I_q - \omega_r I_d (L_{sr} + L_d) + \omega_r \lambda_m] \tag{15.12}$$

$$\frac{di_d}{dt} = \frac{1}{L_{sr} + L_d}[-R_s I_d + \omega_r I_q (L_{sr} + L_d) + V_{dr}] \tag{15.13}$$

As the wind speed is changing throughout the day, the generator output will similarly vacillate in respect to amplitude and frequency because of variable amplitude and frequency. It is not suitable for the loads which are operated at a fixed frequency. In this condition, we employ control structure and a PWM (pulse width modulation) technique to fulfil the load demand in an islanding microgrid system.

(i) Diode bridge rectifier
(ii) DC–DC boost converter
(iii) DC–DC buck–boost converter
(iv) Three-phase VSI inverter

15.4 Three-Phase Rectifier

The three-phase rectifier converts the generator's alternating AC voltage to DC voltage, and the three-phase rectifier's switches work like a binary switch in that they are open circuit and block the reverse voltage when the switch is OFF or short-circuited or when zero voltage appears across the switch when it is ON. Here in Figure 15.2 shows the diagram of AC to DC conversion from PMSG to three phase uncontrolled rectifier.

Here R_s and L_s represent the stator inductance and resistance of the generator. The instantaneous three-phase voltages are represented as follows:

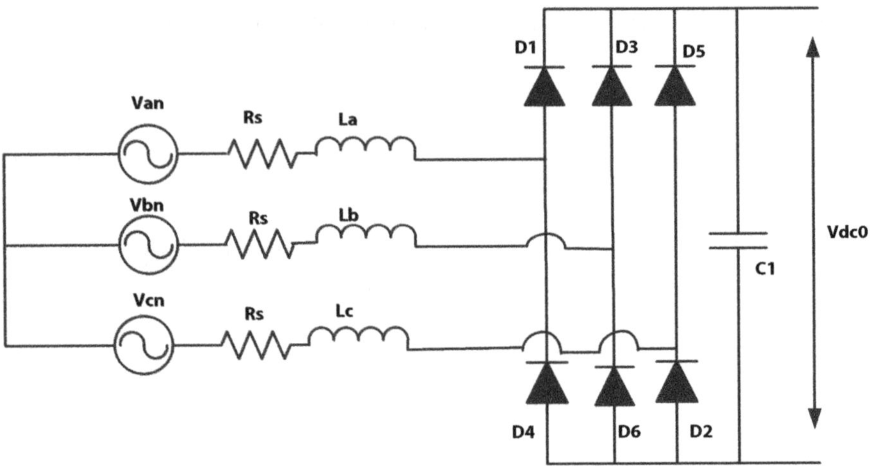

Figure 15.2 PMSG with three-phase diode rectifier [3–7].

$$V_{an} = V_m Sin\ \omega t \tag{15.14}$$

$$V_{bn} = V_m Sin\ (\omega t - 120^0) \tag{15.15}$$

$$V_{cn} = V_m Sin\ (\omega t - 240^0) \tag{15.16}$$

where V_m represents the peak value of the phase voltage.

The basic function of the rectifier is to convert the alternating voltage into DC voltage. V_{dc0} is computed as follows:

$$V_{dc_0} = \frac{3}{\pi} \int_{\frac{\pi}{3}}^{\frac{2\pi}{3}} V_m Sin\ \omega t\ dt \tag{15.17}$$

$$V_{dc_0} = \frac{3\sqrt{3}}{\pi} V_m \tag{15.18}$$

The converted DC voltage has to be maintained constant. Therefore, MPPT control has to be employed to obtain maximum power.

15.5 Modeling of a Boost Converter With MPPT

Maximum power point tracking is used with the help of power electronics and control to track the maximum power point from variable renewables. Not only can maximum power be retrieved, but it can also be modified by controlling the duty cycle of the switch used in the boost converter to monitor and control the rotor speed. Optimally, the tip speed ratio (TSR) λopt has to be optimal to extract the maximum power to get the maximum Cp at any wind speed [27, 28]. Figure 15.3 shows the block diagram of maximum power point tracking using tip speed ratio.

The DC output voltage is applied across the boost converter after the generator output voltage has been rectified by the full-wave diode bridge rectifier in terms of enhancing the voltage of the DC link Vdc. The boost converter is designed in such a way that it can be controlled during varying wind speed conditions. The DC–DC boost converter is used to extract as much power as possible. The DC–DC boost converter's output voltage is calculated as follows:

$$V_{\text{dc actual}} = \frac{V_{dc_0}}{1-D} \tag{15.19}$$

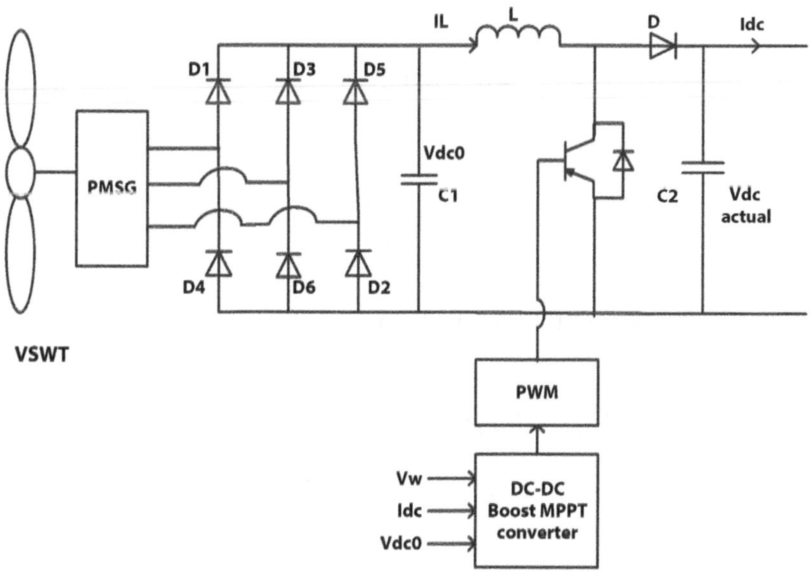

Figure 15.3 Block diagram of boost converter [8–13].

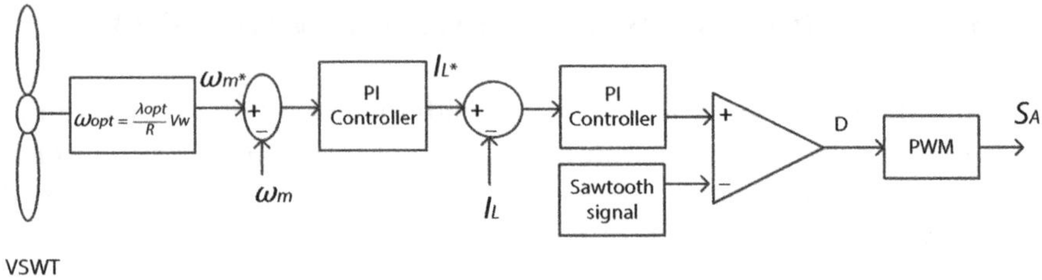

Figure 15.4 Control structure for MPPT [13].

where duty cycle is represented as D of the boost converter. Figure 15.3 shows the block diagram of the boost converter tied up to the wind turbine, and a MPPT controller is employed for controlling the duty cycle of the boost converter.

Variable-speed wind turbines are used for extracting the maximum power operating at any speed using the maximum power point control algorithm. The maximum power that can be extricated from wind turbine is when it is working at maximum C_p, i.e., power coefficient. Therefore, it is necessary that the turbine should operate at optimal tip speed ratio (λ). As the wind speed is vacillating, the speed of the rotor should alter and go along with the change. The block diagram for MPPT control is shown in Figure 15.4.

The optimal power can be calculated from Equation 15.1.

$$P_{m\,opt} = 0.5\rho AC_{p\,opt}\left(\frac{\omega_{opt}R}{\lambda_{opt}}\right)^3 \tag{15.20}$$

$$\omega_{opt} = \frac{\lambda_{opt}}{R}V_w \tag{15.21}$$

Therefore, optimum torque can be given as follows:

$$T_{m\,opt} = 0.5\rho AC_{p\,opt}\left(\frac{R}{\lambda_{opt}}\right)^3(\omega_{opt})^2 \tag{15.22}$$

A control algorithm is required, which incorporates the following steps:

- Measuring the wind speed
- Calculation of the reference speed for the maximum power extraction

$$\omega_{opt} = \frac{\lambda_{opt}}{R} V_w \tag{15.23}$$

- The speed controller output will give reference current I_L*.
- Measuring the input inductor current of the boost converter I_L.
- The error between the two currents is fed to the proportional–integral (PI) to produce duty cycle, *D*.
- The duty cycle D is compared with the sawtooth signal to produce pulses for the IGBT switch used in the boost converter.

15.6 Battery Energy Storage System

Batteries render a dominant part in renewable energy systems as these sources are intermittent in nature, so the main purpose of using a battery is for storing the excess extracted power from the source so that it can be used during the unavailability of the source or in exigencies. Battery storage forms like lithium-ion batteries or supercapacitors are employed in renewable sources to maintain the energy balance between sources via bi-directional converters. Because the lithium-ion battery in concern has a graphite anode and a lithium metal oxide cathode, the electrolyte is made up of a solution of lithium salt (e.g., LiPF6) in a mixed organic solvent (e.g., ethylene carbonate-dimethyl carbonate) (EC-DMC). A layered structure, such as lithium cobalt oxide ($LiCoO_2$), is utilized as the positive electrode or cathode. In between the cathode and anode, a separator is present to prevent direct contact between the two electrodes. The separator allows the transfer of lithium ions by diffusion during charge and discharge. The electrolyte and the dissolved lithium salt provide the medium for the lithium-ion transport.

15.7 Modeling of Lithium-Ion Battery

The modeling of actual batteries is a bit complex as a thorough knowledge is needed of the electromechanical reactions taking place in it. The mathematical modeling of the battery is simulated in MATLAB/Simulink. The most crucial part of the sub-model for an electrical system study is the varying voltage, i.e., based on state of charge (SOC), which is only used as a state variable. A battery's equivalent model is shown in Figure 15.5. The battery implemented is a controlled voltage source model:

$$E = E_0 + K\frac{Q_0}{Q - Q_0} + A.e^{-BQ} \tag{15.24}$$

The battery charge is modeled by the integration of the battery current and also adding the initial charge:

$$Q = Q_{\text{int}} + \int_0^t -I_b.dt \tag{15.25}$$

The modeling of lithium-ion batteries contains two modes: charging mode and discharging mode. It is modeled in MATLAB/SIMULINK, and it has good accuracy. The battery electrical equivalent algorithm using equations are used and shown in Figure 15.5.

The discharging mode is given as follows:

$$V_{batt} = E_0 - \frac{K.Q}{Q - i.t}i^* - \frac{K.Q}{Q - i.t}i.t + A.e^{-B.i.t} \tag{15.26}$$

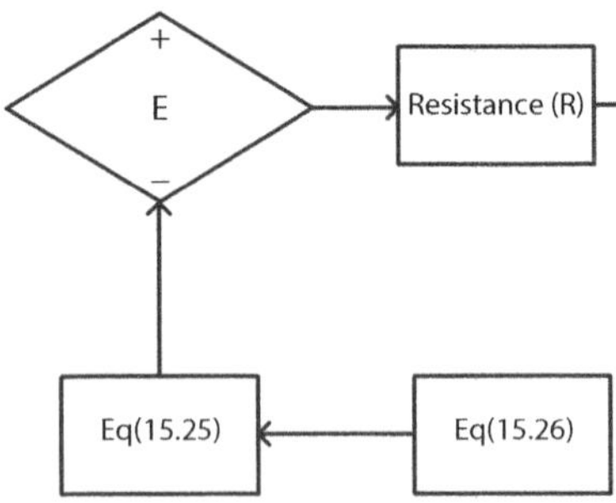

Figure 15.5 Electrical equivalent of battery [14–17].

The charging mode is given as follows:

$$V_{batt} = E_0 - \frac{K.Q}{i.t + 0.1Q} i^* - \frac{K.Q}{Q - i.t} i.t + A.e^{-B.i.t} \qquad (15.27)$$

In the equation above, E_0 is the constant battery voltage, K is the polarization constant, and Q represents the maximum battery capacity as it denotes the extracted capacity and is described by $i(t) = \int i(t)dt\ i^*$ as the filtered battery current. The nominal voltage and nominal capacity of the lithium-ion battery used are 240 V and 300 Ah, respectively. Hence, A and B indicate the exponential voltage and capacity.

15.8 Modeling of a Bidirectional DC–DC Converter

The illustrative diagram of bi-directional DC–DC converters is shown in Figure 15.6, where the converter is supposed to be operating in a continuous conduction mode.

The DC–DC bidirectional converter is assumed to operate continuously. In a certain configuration, the feature of step-up and step-down of voltage is implemented so that it can operate in buck and boost mode. It also has an additional feature of bi-directional power flow.

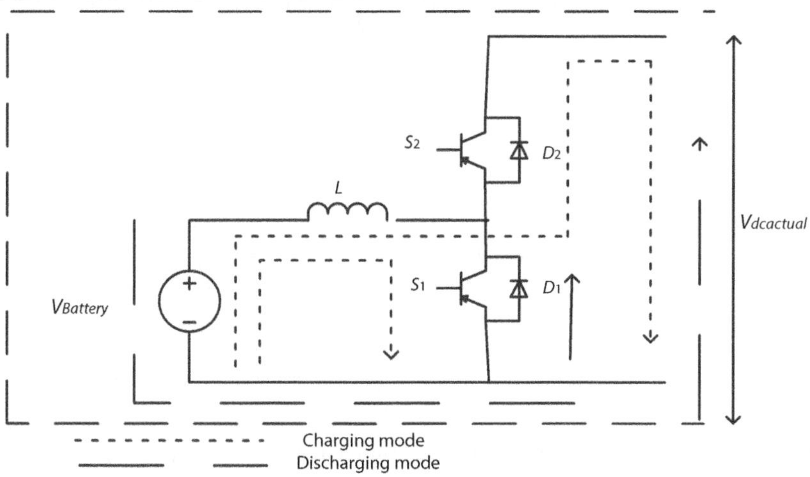

Figure 15.6 Schematic diagram of DC–DC bidirectional converter [17–22].

The operation of the above-mentioned circuit is described below.

Mode I: In this mode, charging of the DC link is described when switching S1. It is turned on, and the battery will energize the inductor, and the inductor current will keep on increasing until the gate pulses are terminated from the switch. The current flowing in the inductor will not change instantaneously, but the voltage polarity will reverse, and it will add up to the battery voltage. Therefore, the diode D2 starts conducting due to forward bias, and the current through the inductor starts charging the DC link. Hence, the power is transported through the battery to the DC link.

Mode II: In this mode, discharging of the DC link or charging of the battery is described when the switch S2 is turned on. The DC link voltage will be greater than the reference value, so it will energize the inductor. The inductor current will not change instantaneously, but the voltage polarity across it changes, and it will forward-bias the diode D1. The power is transferred from the DC link to the battery. The control algorithm schematic diagram is shown in Figure 15.7.

- Bi-directional Converter Voltage Regulation

In order to synchronize the lithium-ion battery charging voltage and maintain a constant DC bus voltage, a proportional integral (PI) controller is utilized to regulate the bidirectional converter voltages for equivalence. The equations below are used to describe the controller.

For boost mode:

$$u_1(t) = K_{p1}[V_{dcref} - V_{batt}(t)] + \int K_{I1}[V_{dcref} - V_{batt}(t)]dt \quad (15.28)$$

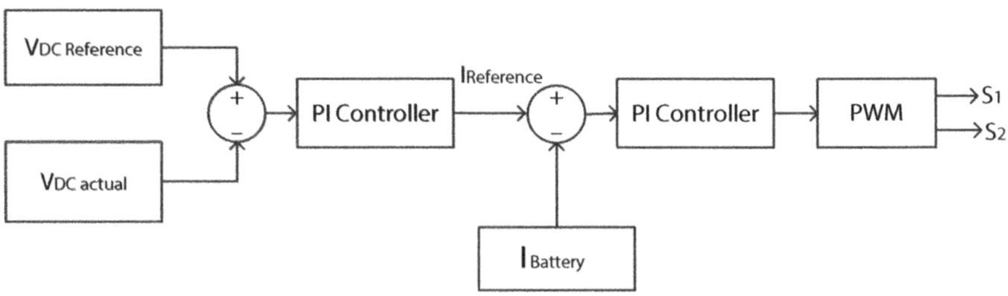

Figure 15.7 Schematic diagram of DC–DC bi-directional converter [16, 17].

For buck mode:

$$u_2(t) = K_{p2}[V_{dcref} - V_{batt}(t)] + \int K_{I2}[V_{dcref} - V_{batt}(t)]dt \quad (15.29)$$

Battery bank is a very important part of the stand-alone system to catch up with the load demand. When a variable-speed wind turbine generates more power than the load requires, then the remaining power has to be fed to the battery bank via a DC–DC bidirectional converter working in charging mode. If the wind energy produces less power than the remaining power, it has to be supplied through the battery. The battery system is connected to a DC–DC bidirectional converter that can switch between buck and boost modes.

- $V_{DC\,ref} > V_{DC\,act}$ The battery system will charge the DC link, and the DC–DC bidirectional converter will operate in boost mode.
- $V_{DC\,ref} < V_{DC\,act}$ The DC link will charge the battery system, and the DC–DC bidirectional converter will operate in buck mode.

A flow chart is shown in Figure 15.8, which describes the control structure of the battery storage system.

The operation of the system comprises four different modes, which are as follows:

- Mode 1: When the power provided by the wind source exceeds the power required by the load, the extra power is stored in the battery system using a bidirectional converter.
- Mode 2: When power generated through the wind energy system is not ample to deliver the power needed for the loads, then the battery storage system will come into play for serving the power needed by the loads via working in the discharging mode through the bidirectional converter.
- Mode 3: In a scenario where the wind system generates no power, then the battery storage systems come into action to fill in the load required to be fed.
- Mode 4: In a situation when the power generated by WECS is not adequate to feed the load and the battery's SOC is less

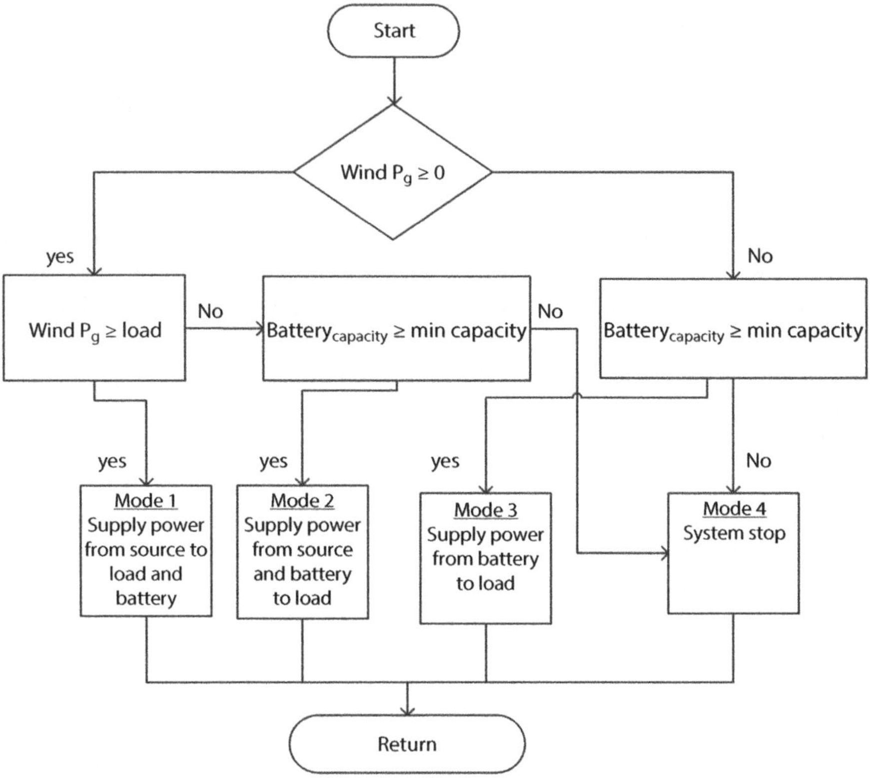

Figure 15.8 Flow chart of control strategy for battery storage system [19–27].

than 40%, in this scene, the system should shut down in order to save the equipment and battery from damage.

15.9 Modeling a Three-Phase Voltage Source Inverter

A voltage source inverter is necessary for the conversion of a fixed DC voltage in a DC link capacitor; hence, the suggested system uses a three-phase two-level voltage source inverter to provide an alternating voltage as an output of the VSI.

The VSI inverter's schematic circuit diagram is depicted in Figure 15.9.

The inverter has six IGBT switches—S_1 to S_6—with two switches for each phase, and the operation of each switch is based on the change in time of operation. The inverter employed in this work was 180° degree VSI mode; each switch will operate for 180° in a cycle, with switches present in every

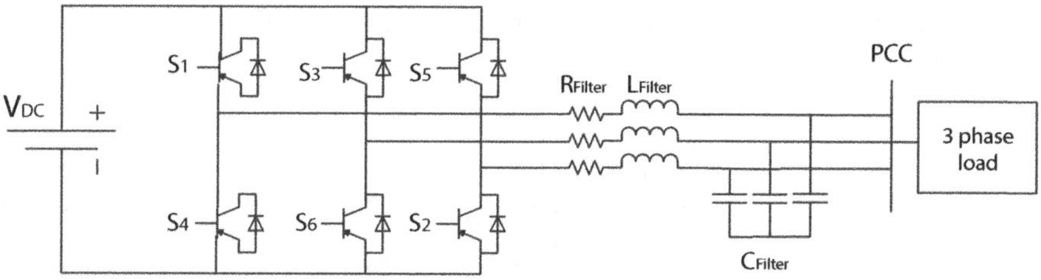

Figure 15.9 Equivalent circuit of two-level VSI [28, 29].

leg—S1, S4; S3, S6; and S5, S2—switched ON in a time span of 180°. In this way, S1 turned ON for 180°, and switch S4 conducts for the upcoming 180° of a cycle. The switches in the upper leg—S1, S3, and S5—are turned ON at a time span of 120°. This means that if S1 is operated at ωt = 0⁰, then S3 should be operated at ωt = 120°. Similarly, S5 at ωt = 240°, the same thing applies for the lower leg of switches. The switching states of the six switches in a three-phase PWM VSI in a stand-alone system are shown in Table 15.1, and an inductance L having an internal resistance R is employed as a filter linking the load and VSI. The VSI phase voltages are signified as V_a, V_b, and V_c, and the load voltages are signified as V_{al}, V_{bl}, and V_{cl} and ω_{ref} is the load angular frequency. The DC link voltage is denoted by V_{dc}, and i_{dc}.

The switching states of the VSI are described in Table 15.1.

For the modeling and control of PWM VSI, mathematical tools are required, which are requisite for analyzing the dynamic implementation and constancy limits of distinct control laws and variables of the system.

Table 15.1 Switching states [28, 29, 37].

State no.	State	Switch state	V_{ab}	V_{bc}	V_{ca}
1	S1, S2, and S6 are on	1 0 0	V_s	0	$-V_s$
2	S2, S3, and S1 are on	1 1 0	0	V_s	$-V_s$
3	S3, S4, and S2 are on	0 1 0	$-V_s$	V_s	0
4	S4, S5, and S3 are on	0 1 1	$-V_s$	0	V_s
5	S5, S6, and S4 are on	0 0 1	0	$-V_s$	V_s
6	S6, S1, and S5 are on	1 0 1	V_s	$-V_s$	0
7	S1, S3, and S5 are on	1 1 1	0	0	0
8	S4, S6, and S2 are on	0 0 0	0	0	0

The three-phase voltage and current are represented in the following equations.

$$\begin{bmatrix} V_{al} \\ V_{bl} \\ V_{cl} \end{bmatrix} = V_m \begin{bmatrix} Sin\,\omega t \\ Sin\,(\omega t - 120) \\ Sin\,(\omega t - 240) \end{bmatrix} \tag{15.30}$$

$$\begin{bmatrix} I_{al} \\ I_{bl} \\ I_{cl} \end{bmatrix} = I_m \begin{bmatrix} Sin\,\omega t \\ Sin\,(\omega t - 120) \\ Sin\,(\omega t - 240) \end{bmatrix} \tag{15.31}$$

Where I_m is the peak value of current and V_m is the peak value of voltage. Applying Kirchhoff's voltage and current laws, the following equations are derived:

$$\begin{bmatrix} V_{al} \\ V_{bl} \\ V_{cl} \end{bmatrix} = R_f \begin{bmatrix} I_{al} \\ I_{bl} \\ I_{cl} \end{bmatrix} + L_f \frac{\partial}{\partial t} \begin{bmatrix} I_{al} \\ I_{bl} \\ I_{cl} \end{bmatrix} + \begin{bmatrix} V_a \\ V_b \\ V_c \end{bmatrix} \tag{15.32}$$

The voltage control technique using PI controller for inverter control in which the rotating reference frame voltage is first converted into stationary reference frame using Park's transformation to convert them into DC as PI controller is a linear controller. By using the vector transform, the equation above can be formulated in the dq0 rotating reference frame synchronizing with the load voltage via Park's transformation.

The decoupled equation can be formulated in state space as follows:

$$\frac{dX}{dt} = A.X - B.Y \tag{15.33}$$

Here the state vector and the input vector are described as follows:

$$X = [I_d\, I_q]^T \tag{15.34}$$

$$Y = [V_d\, V_q\, V_{dl}\, V_{ql}]^T \tag{15.35}$$

Here the system matrix and the input matrix are stated as follows:

$$A = \begin{pmatrix} \frac{-R}{L} & \omega ref \\ -\omega ref & \frac{-R}{L} \end{pmatrix} \tag{15.36}$$

$$B = \begin{bmatrix} \frac{1}{L} & 0 & \frac{-1}{L} & 0 \\ 0 & \frac{1}{L} & 0 & \frac{-1}{L} \end{bmatrix} \tag{15.37}$$

The V_{dl} and V_{ql} components are compared with the reference V_d and V_q, the error is fed to the PI controller, and then the reference dq0 signal is converted

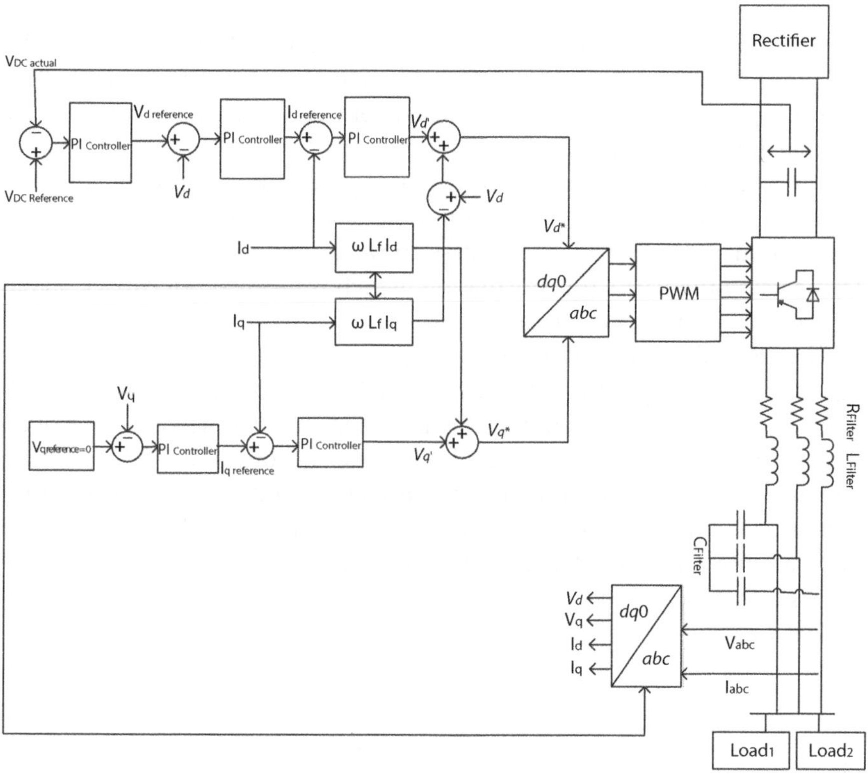

Figure 15.10 Vector control structure for inverter control [1, 2, 25, 26, 32].

into the *abc* reference frame using inverse Park's transformation. The signal is fed to the PWM block to generate the gate pulses for the VSI inverter.

15.10 Vector Control Structure for Inverter Control

The aim of the vector-oriented inverter control in a stand-alone system is mainly implemented for generating the PWM pulses for the inverter; PWM techniques are mostly involved to reduce the harmonics from the fundamental component, so different types of modulation techniques are used. The main objective of the controller is to balance the voltage and the load frequency of the output voltage.

The control structure comprises two loops, namely:

- The outer control loops
- The inner control loops

The outer control loop is a DC link voltage controller which regulates the DC link voltage. The PI controller for the DC link maintains the DC link voltage to the reference value in case of load change. The inner control loop consists of voltage control and current control, which controls the voltage during load change.

For a stationary frame, the time-varying quantities represent the stationary in three-phase reference frame signals. The three-phase filter voltage and current V_{abc} and I_{abc} are first converted into *abc* to *dq0* reference frame using the Park's transformation, which is taken as a feedback signal for the voltage and current controller [41].

The DC link voltage controller is employed to regulate the DC link voltage as shown in Figure 15.10. The actual DC link voltage is compared with $V_{D\ reference}$, and the error output voltage is fed into the next PI controller. This error output voltage of the PI controller becomes the $V_{reference}$ *d-axis* which is regulated with the actual *d-axis* voltage, and then the error output voltage is fed to the next PI controller. The output of this PI controller will generate the reference *d-axis* current which is compared with the actual *d-axis* current.

For *d-axis*:

$$V_{dref} = K_p[V_{dc\,ref} - V_{dc\,actual}] + K_i \int [V_{dc\,ref} - V_{dc\,actual}]dt \quad (15.38)$$

$$I_{dref} = K_p[V_{d\,ref} - V_d] + K_i\int[V_{d\,ref} - V_d]dt \tag{15.39}$$

$$V'_{d\,ref} = K_p[I_{d\,ref} - I_d] + K_i\int[I_{d\,ref} - I_d]dt \tag{15.40}$$

$$V^{*}_{d\,ref} = V'_{dref} + [V_d - \omega\, L_f\, I_d] \tag{15.41}$$

For q-axis:

$$I_{dref} = K_p[V_{d\,ref} - V_d] + K_i\int[V_{d\,ref} - V_d]dt \tag{15.42}$$

$$V'_{q\,ref} = K_p[I_{qref} - I_q] + K_i\int[I_{qref} - I_q]dt \tag{15.43}$$

$$V^{*}_{q\,ref} = V'_{q\,ref} + \omega\, L_f\, I_d \tag{15.44}$$

Via *dq0* transformation, the active and reactive power are formulated as follows:

$$P = \frac{3}{2}(V_d I_d + V_d I_q) \tag{15.45}$$

$$Q = \frac{3}{2}(V_d I_q + V_q I_d) \tag{15.46}$$

According to the reference frame *as* $V_q = 0$ and $V_d = |V|$, the equations of active and reactive power is given as follows:

$$P = \frac{3}{2} V_d I_d \tag{15.47}$$

$$Q = \frac{3}{2} V_d I_q \tag{15.48}$$

Hence, via controlling the direct and quadrature components, respectively, active and reactive power can be controlled.

As shown in Equations (15.42) and (15.45), the d-axis and q-axis components of currents are linked to the variables of the d-axis and q-axis components, indicating cross-coupling of the system. As a result, minor changes in the q-axis variable will influence the d-axis current, and correspondingly, a transition in the d-axis variable will impact the q-axis current.

15.11 Modeling a Passive Filter

A three-phase LC filter is employed in this system to eliminate the harmonics from the signal when the non-linear electronics components are used; it introduces the other harmonics in the fundamental component in the signal [42] when the load terminals are linked to the filter. Because the inverter contributes high-frequency components into the fundamental component of the output voltage, an LC filter is designed to improve the power quality using the following equations, which calculate the values of the filter's L and C parameters. The block diagram of LC filter is represented in Figure 15.11.

$$K = \left[\left(m^2 - \frac{15}{4} m^4 + \frac{65}{5\pi} m^5 - \frac{5}{4} m^6 \right) \frac{1}{1440} \right]^{\frac{1}{2}} \tag{15.49}$$

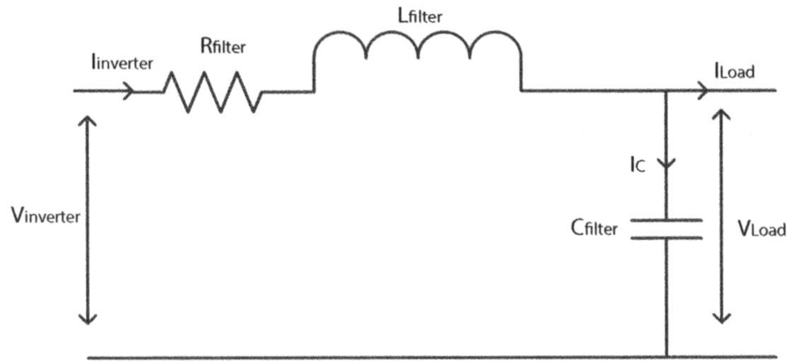

Figure 15.11 Block diagram of LC filter [40, 41].

$$L_f = \frac{V_{rms}}{I_0 f_s}\left\{K\frac{V_{dc\,actual}}{V_0}\left[1+4\pi^2\left(\frac{f_r}{f_s}\right)^2 K\frac{V_{dc\,actual}}{V_0}\right]\right\}^{\frac{1}{2}} \quad (15.50)$$

$$C_f = \left\{K\frac{V_{dc\,actual}}{L_f}\left(\frac{1}{f_s}\right)^2\frac{1}{V_0}\right\} \quad (15.51)$$

Here, m represents the modulation index, V_{rms} is the *rms* voltage of the load, I_0 is the load current, switching frequency is denoted as f_s, f_r is the reference frequency, V_0 is the harmonic voltage, $V_{dc\,act}$ is the actual DC link voltage, L_f is the filter inductance, and C_f is the filter capacitance.

15.12 Simulation Results

A wind and battery energy conversion system using a variable-speed wind turbine along with a controller is modeled in MATLAB/SIMULINK. To check the effectiveness of the controller, some tests have been carried out in the system. The system is designed with three different wind speeds along with the test conducted at the load side. The system parameters tabulated

Table 15.2 System parameters for variable-speed wind energy system [28–34].

System parameters	Value	System parameters	Value
Wind turbine	9.5 kW	DC link capacitance	4,998 μF
PMSG	9.5 kW	Filter inductance (L_f)	4.09 mH
Rated wind speed	12.5 m/s	Filter capacitance (C_f)	40 μF
DC link voltage	790 V	Inverter switching frequency	2.5 kHz
Rated load-side voltage	399 V	DC link voltage controller gain (k_p, k_i)	41,390
Nominal voltage of battery system	230 V	Voltage controller gain (k_p, k_i)	0.13, 1.9
Operating frequency	50 Hz	Current controller gain (k_p, k_i)	0.03, 0.4

for variable speed wind energy conversion system is presented in Table 15.2 [28–34].

For the operation at different wind speeds, a variable speed wind turbine is tested with three varying wind speeds taken, namely:

i. Step change variation in the speed of the wind
ii. Staircase variation in the speed of the wind
iii. Random variation in the speed of the wind

a) Different Wind Speed

(i) Step change variation in the speed of the wind

The operation of a variable-speed wind turbine is analyzed with step wind speed. The initial speed of the wind is 11.05 m/s at t = 29 s. The speed of the wind is changed to 11.95 m/s. As the speed of the wind will change, the rotor speed of the generator will also change. The results are shown below for step change in wind speed. In Figure 15.12 below shows the variable wind speed in the form of step change is taken to check the proposed system variable speed wind turbine [1, 3, 5, 7].

(ii) Staircase wind speed

The wind turbine working on variable speed is analyzed under another wind speed change, i.e., staircase wind speed. In this, the initial wind speed is 11 m/s and at t = 30 s the speed is changed to 11.95 m/s, and at t = 59s the speed is varied to 13.3 m/s. The analysis of the above-mentioned speed change is shown below.

(iii) Random wind speed

The wind turbine is now operated at randomly varying speed, so the system is analyzed with random speed change. The results are shown below.

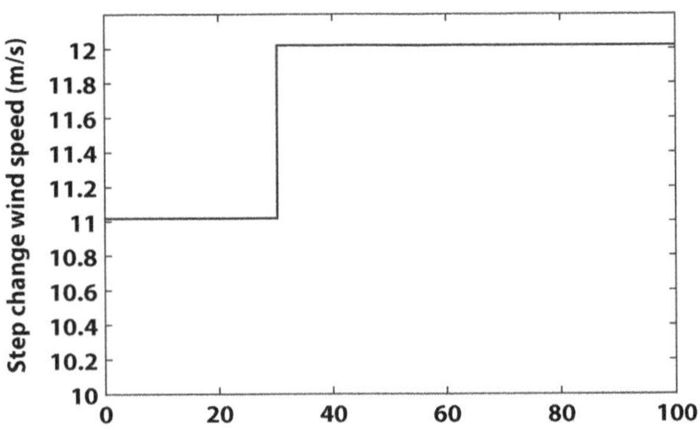

Figure 15.12 Step change variation in wind speed (m/s) [1, 3, 5, 7].

In Figure 15.13 shows below the speed of rotor due to the change in step change wind speed taken to check and analyse the proposed system derived from the references shown in the Figure 15.13. In Figure 15.14 shows below the mechanical torque of PMSG due to the change in step change wind speed taken to check and analyse the proposed system derived from the references shown in the Figure 15.14. In Figure 15.15 shows below the DC current due to the change in step change wind speed taken to check and analyse the proposed system derived from the references shown in the Figure 15.15. In Figure 15.16 shows below the DC output power due to the change in step change wind speed taken to check and analyse the proposed system derived from the references as shown in the Figure 15.16. The wind turbine working on variable speed is analysed under another wind speed change i.e., staircase wind speed in this the initial wind speed is 11 m/s and at t=30s the speed is changed to 11.95 m/s and at t=59s the speed is varied to 13.3 m/s the analysis of the above-mentioned speed change is shown below in Figure 15.17. Here in Figure 15.18 shows below the response of rotor speed of wind turbine due to the application of staircase wind speed across variable speed wind turbine system. Here in Figure 15.19 shows below the response of mechanical torque obtained due to the application of staircase wind speed across variable speed wind turbine system [1, 2, 4]. Here in Figure 15.20 shows below the response of DC current obtained due to the application of staircase wind speed across variable speed wind turbine system [1, 2, 4]. Here in Figure 15.21 shows below the response of DC output power obtained due to the application of staircase wind speed across variable speed wind turbine system [1, 2, 4]. The variable wind speed turbine system is now operated at randomly varying speed so the system is analysed with random wind speed change the results are shown below in Figure 15.22 [8–13]. In Figure 15.23 shown below is the rotor speed of variable wind speed turbine obtained due to the application of random wind speed applied across the variable speed wind turbine system [8–13]. In Figure 15.24 shown below is the Mechanical torque of variable wind speed turbine obtained due to the application of random wind speed applied across the variable speed wind turbine system [8–13]. In Figure 15.25 shown below is the DC output current of variable wind speed turbine obtained due to the application of random wind speed applied across the variable speed wind turbine system [8–13]. In Figure 15.26 shown below is the DC output power of variable wind speed turbine obtained due to the application of random wind speed applied across the variable speed wind turbine system [8-13].

b) Different Load Condition

The proposed system configuration is subjected to different load scenarios, and the controller performance is validated in MATLAB/SIMULINK.

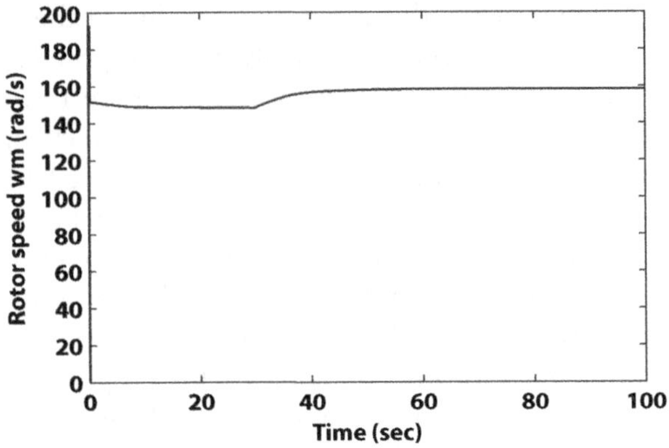

Figure 15.13 Rotor speed for step change wind speed (rad/s) [5–7].

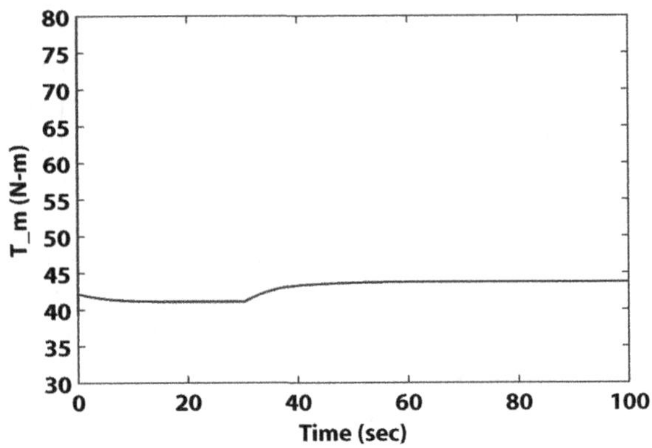

Figure 15.14 Mechanical torque for step change wind speed (N-m) [5–7].

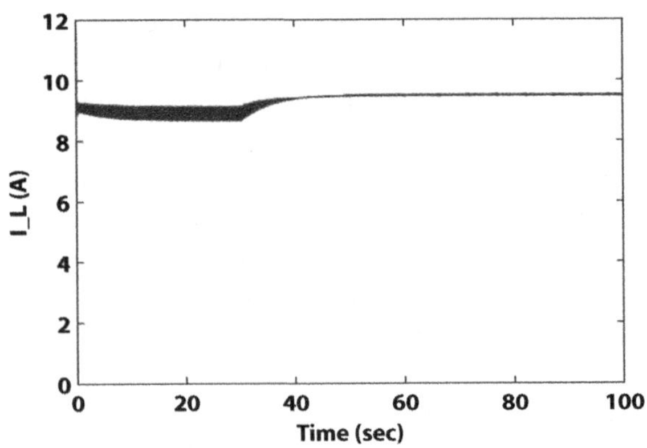

Figure 15.15 DC current for step change wind speed (A) [5–7].

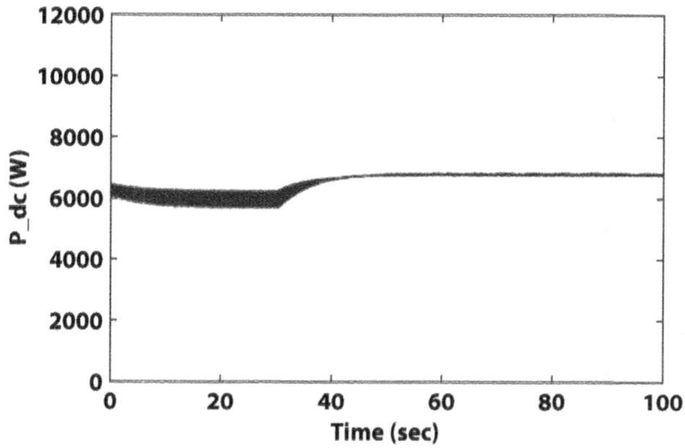

Figure 15.16 DC power output for step change wind speed (W) [5–7].

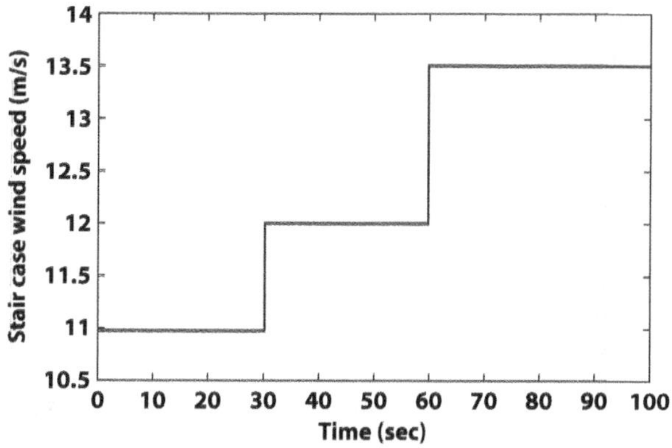

Figure 15.17 Staircase variation in wind speed (m/s) [1, 2, 4].

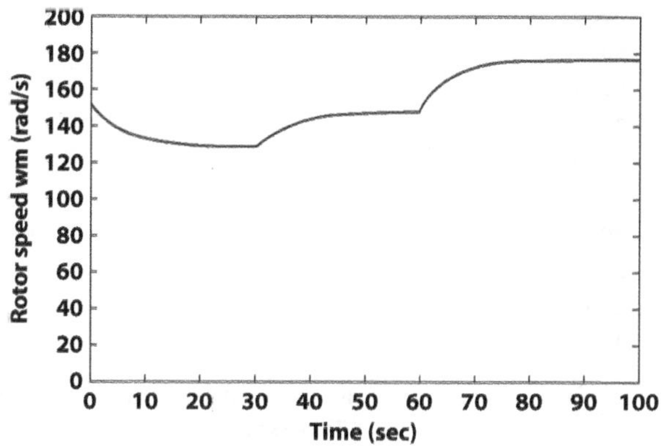

Figure 15.18 Staircase rotor speed (rad/s) [1, 2, 4].

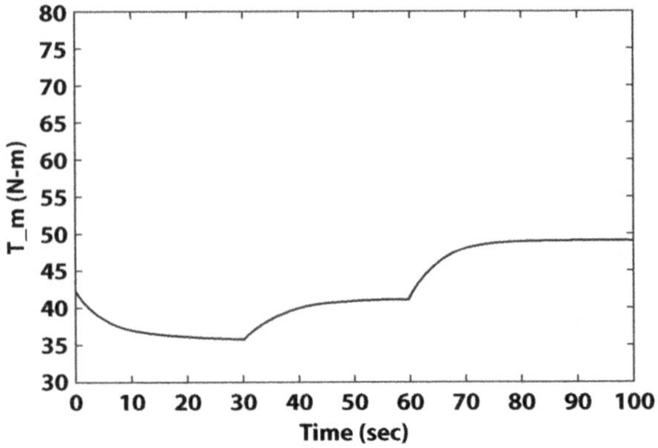

Figure 15.19 Mechanical torque for staircase wind speed (N-m) [1, 2, 4].

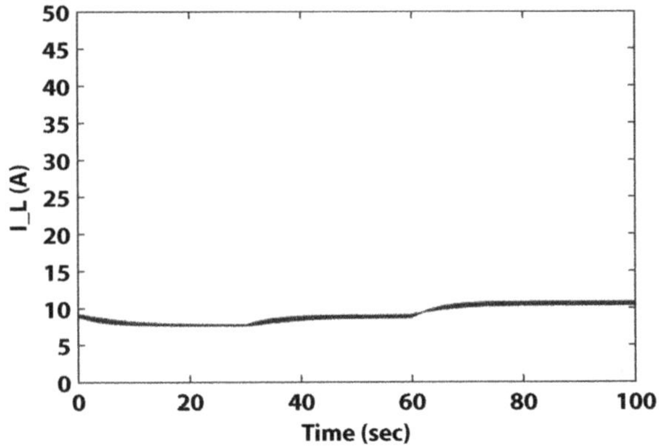

Figure 15.20 DC current for staircase wind speed (A) [1, 2, 4].

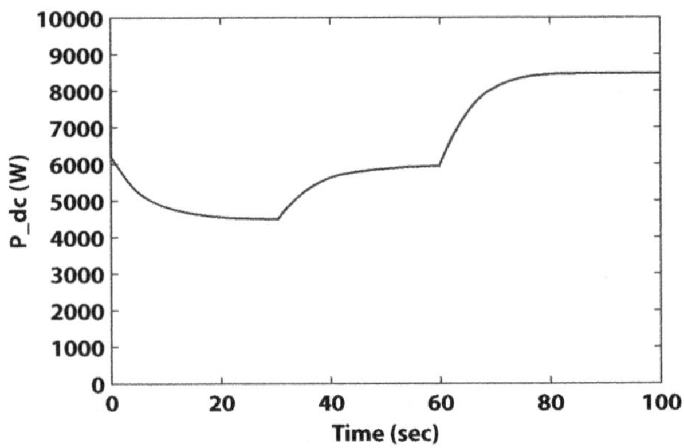

Figure 15.21 DC power output for staircase wind speed (W) [1, 2, 4].

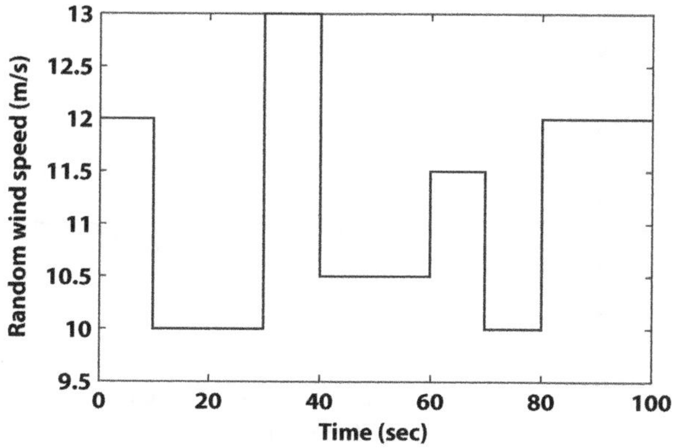

Figure 15.22 Random variation in speed of the wind (m/s) [8–13].

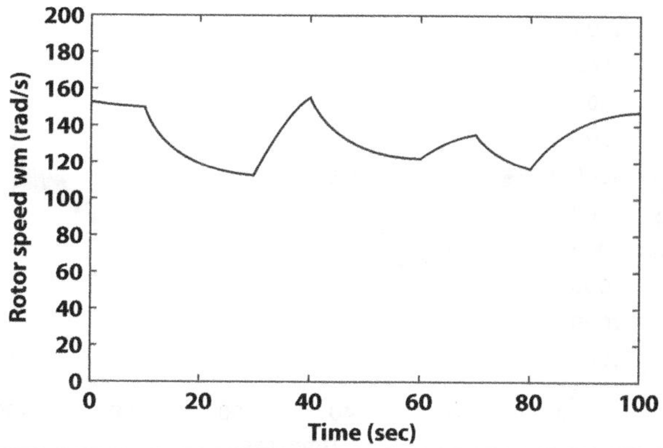

Figure 15.23 Rotor speed for random wind speed change (rad/s) [8–13].

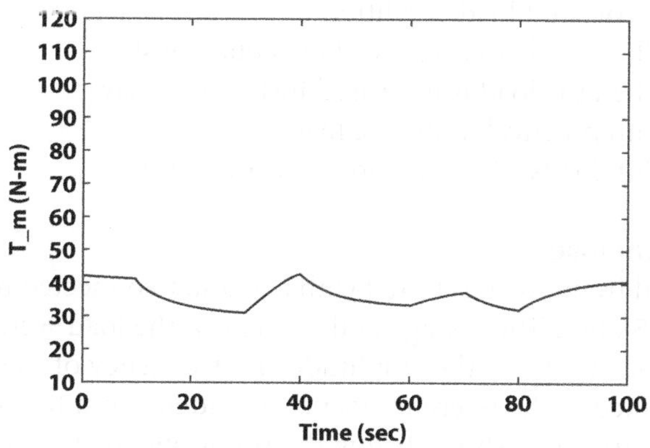

Figure 15.24 Mechanical torque for random wind speed change (N-m) [8–13].

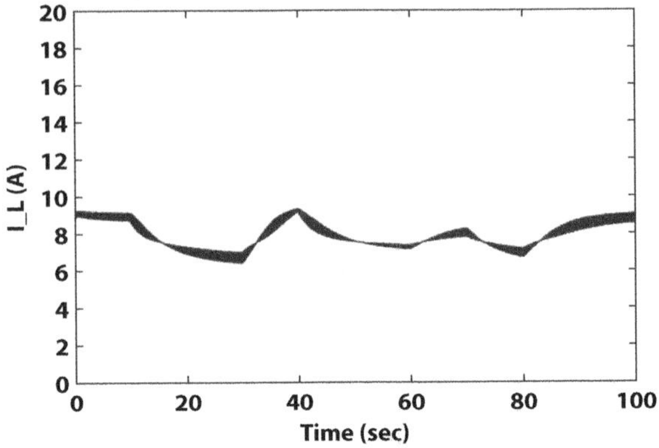

Figure 15.25 DC current for random wind speed change (A) [8–13].

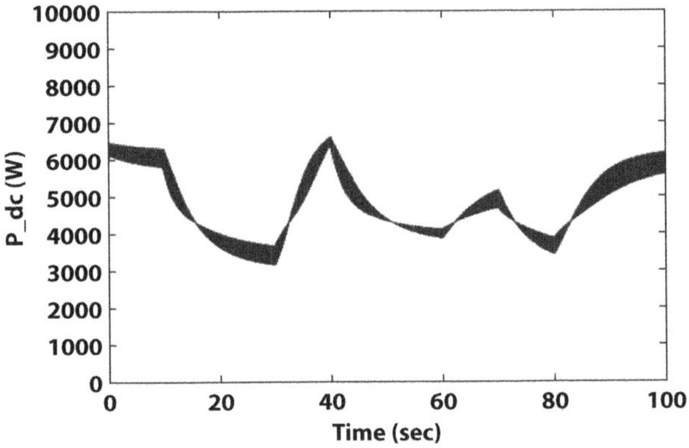

Figure 15.26 DC power for random wind speed change (W) [8–13].

(i) At constant load condition
(ii) When the load is reduced instantaneously
(iii) When the load is increased instantaneously
(iv) With practical inductive load
(v) With loads which are non-linear in nature

(i) At constant load

The proposed framework is tested with a 3-ϕ star connected resistive load, where R = 55 ohm. The voltage and current of the load were maintained at the reference value, or the amplitude and frequency of the load voltage were maintained at the reference value, as evident from the results shown below. In Figure 15.27 shows below the result obtained for DC-link voltage maintained constant fluctuation in DC-link voltage via DC-link controller due to the application of random wind speed change in variable wind speed turbine system [14–21]. In Figure 15.28 shows below the result obtained for the inverter output voltage maintained constant and fluctuation in voltage is controlled via inverter closed loop control system due to the application of random wind speed change in variable wind speed turbine system [23–27]. In Figure 15.29 shows below the result obtained for the inverter output current maintained constant and fluctuation in voltage is controlled via inverter closed loop control system due to the application of random wind speed change in variable wind speed turbine system [23–27]. In Figure 15.30 shows below the result obtained for the load output voltage under constant load the fluctuation in voltage is controlled via inverter closed loop control system due to the application of random wind speed change in variable wind speed turbine system [23–27]. In Figure 15.31 shows below the result obtained for the load output current under constant load the fluctuation in current is controlled via inverter closed loop control system due to the application of random wind speed change in variable wind speed turbine system [23–27]. In Figure 15.32 shows below the result for active power obtained under constant load condition in the variable speed wind turbine system [23–27]. In Figure 15.33 shows below the result for reactive power obtained under constant load condition in the variable speed turbine system [23–27].

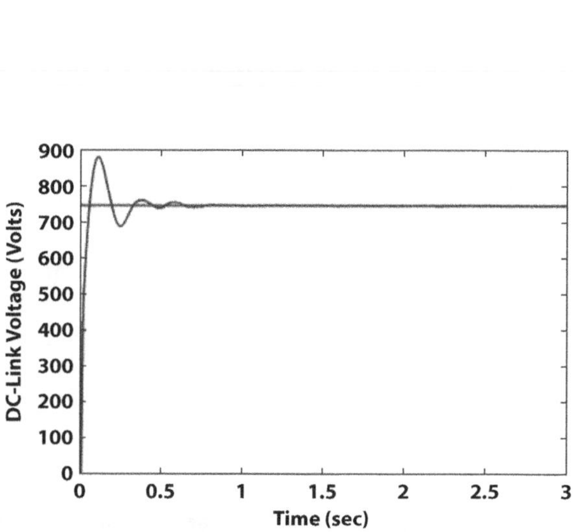

Figure 15.27 DC link voltage at constant load (volts) [14–21].

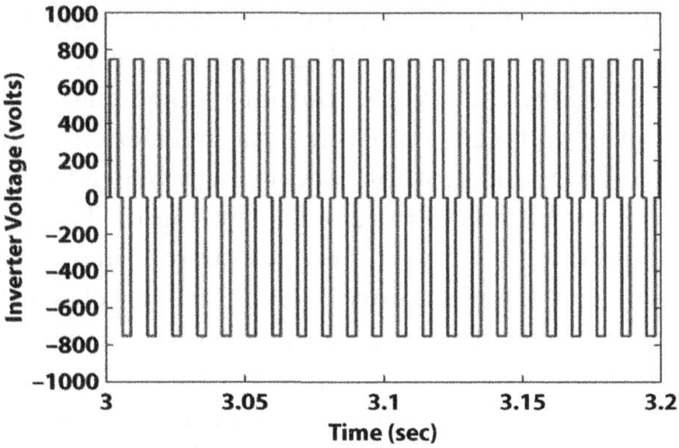

Figure 15.28 Inverter output voltage (volts) [22–27].

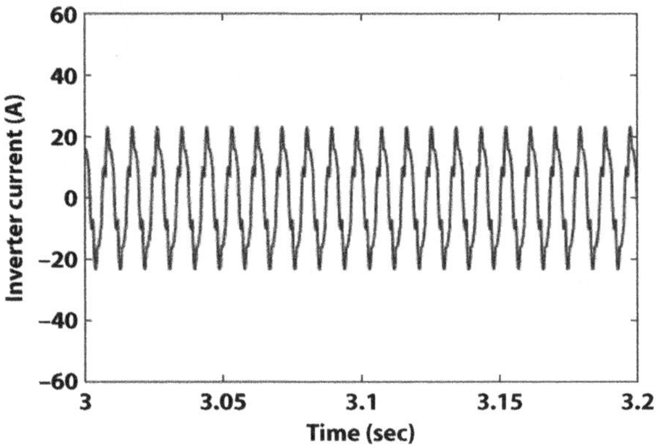

Figure 15.29 Inverter output current (A) [22–27].

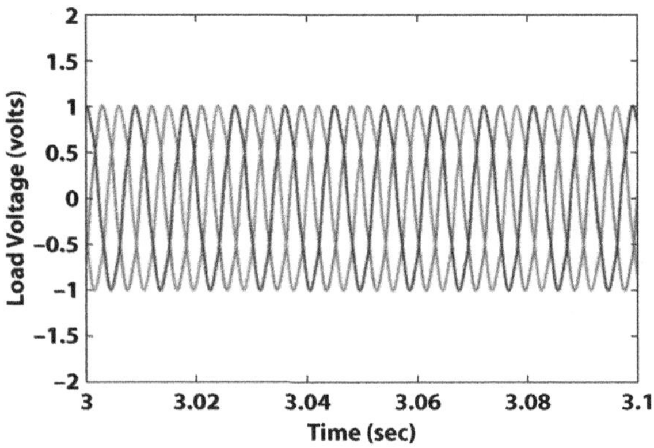

Figure 15.30 Load voltage at constant load (volts) [22–27].

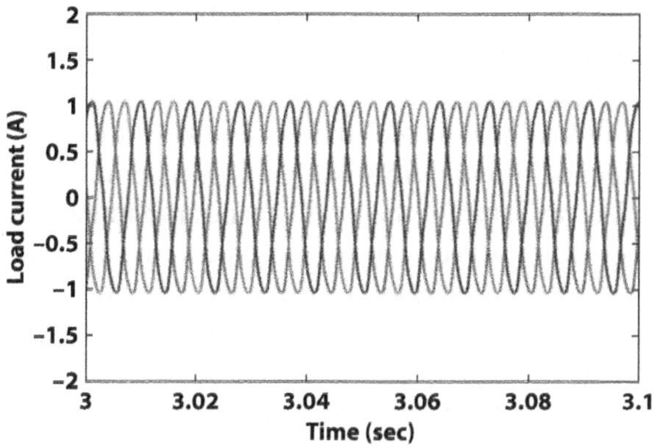

Figure 15.31 Load current at constant load (A) [22–27].

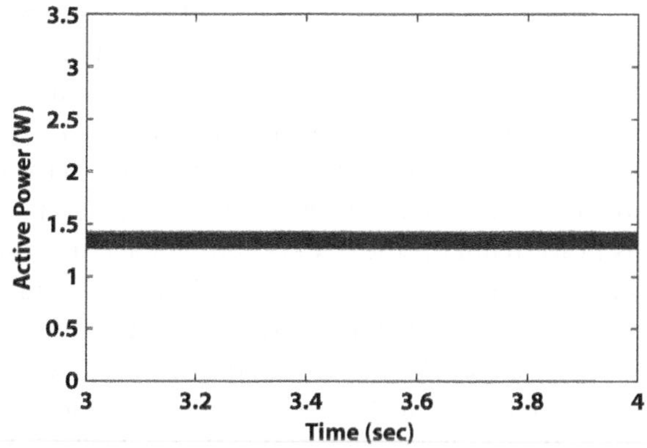

Figure 15.32 Active power at constant load (W) [22–27].

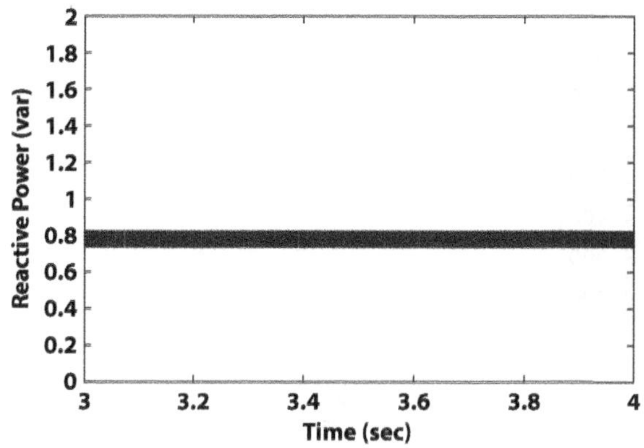

Figure 15.33 Reactive power at constant load (W) [22–27].

(ii) When the load is reduced instantaneously; load shedding (when the load is suddenly reduced to half)

The proposed system is subjected to a load reduction at t = 3.8 s. When a certain amount of load is reduced in a system, then there will be a voltage swell. Thus, in a stand-alone system, the controller should be modeled, such that the controller must sustain the load voltage amplitude and frequency. In this scenario, 3-ϕ star connected load is considered, where R = 102 'Ω. The load was reduced from the system, so the main work of the controller is to maintain the load voltage amplitude and frequency during load variation, and it is controlled within t = 4.15 s. As can be observed, the results show that the controller is working satisfactorily. In Figure 15.34 below shows the DC-Link voltage obtained due to the instantaneously change in load here the load is reduced to half the fluctuation occurred because of sudden load change is controlled via vector oriented closed loop control of three phase inverter from the result it is evident that the controller is fit to maintain the DC-Link voltage constant under sudden load variation [25–30]. In Figure 15.35 below shows the load output voltage obtained due to the instantaneously change in load here the load is reduced to half the fluctuation occurred because of sudden load change is controlled via vector oriented closed loop control of three phase inverter from the result it is evident that the controller is fit to maintain the load output voltage magnitude and frequency constant under sudden load variation [25–30]. In Figure 15.36 below shows the output current obtained due to the instantaneously change in load here the load is reduced to half the fluctuation occurred because of sudden load change is controlled via vector oriented closed loop control of three phase inverter from the result it is evident that the controller is fit to maintain

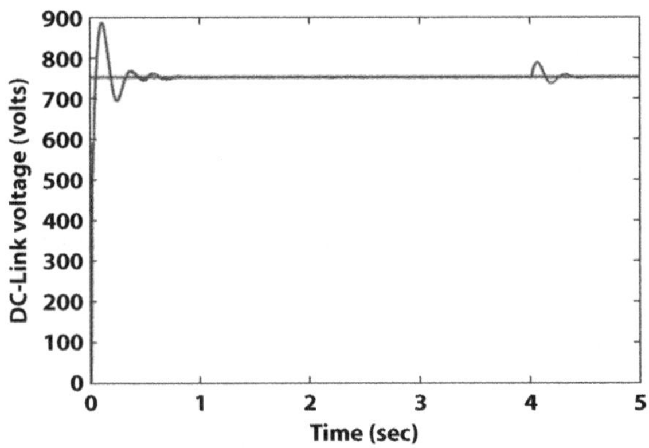

Figure 15.34 DC link voltage when the load is reduced to half (volts) [25–30].

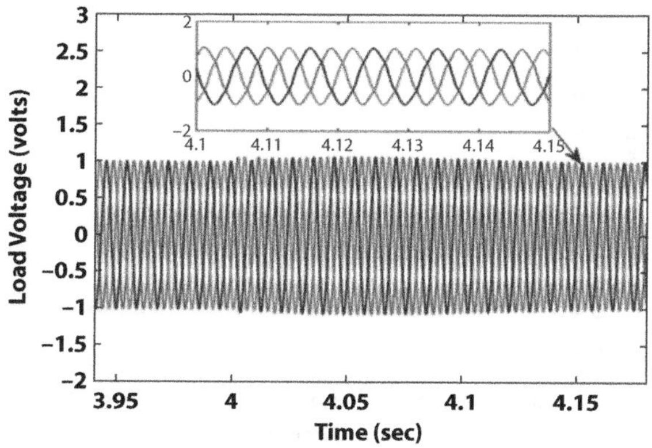

Figure 15.35 Load voltage when the load is reduced to half (volts) [25–30].

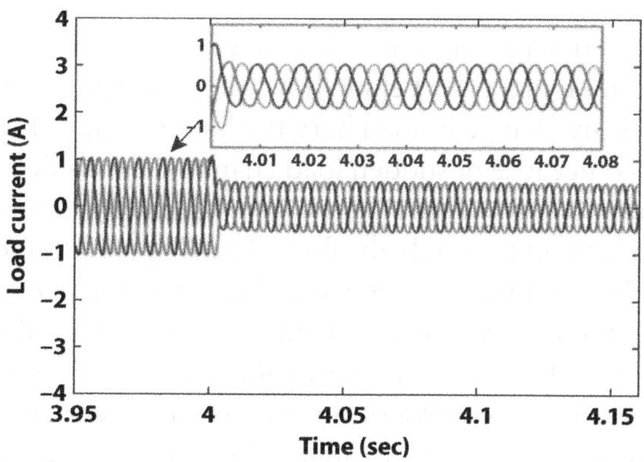

Figure 15.36 Load current when the load is reduced to half (A) [25–30].

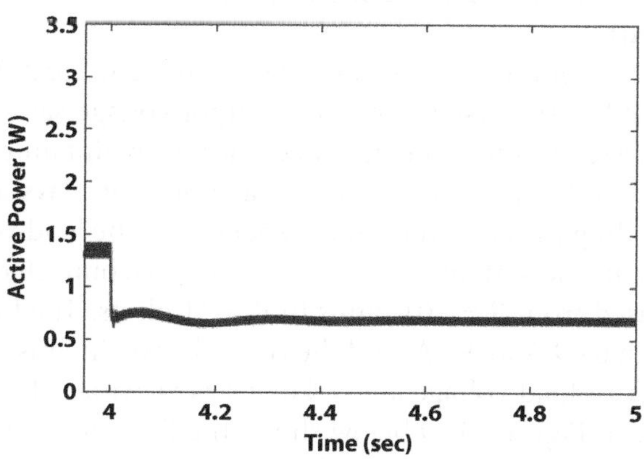

Figure 15.37 Active power when the load is reduced to half (W) [25–30].

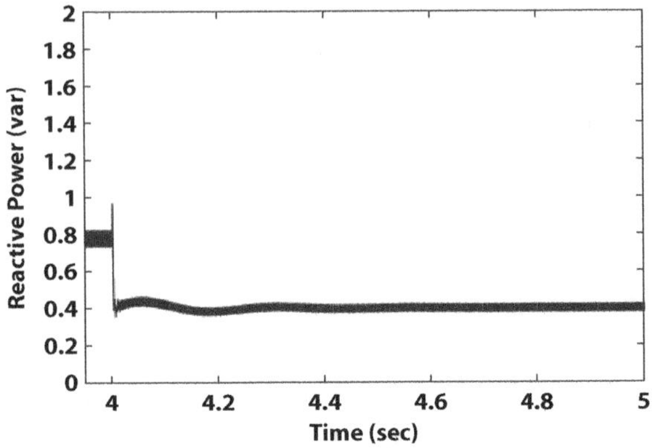

Figure 15.38 Reactive power when the load is reduced to half (VAR) [25–30].

the load output current magnitude and frequency constant under sudden load variation. In Figure 15.37 below shows the active power obtained due to the instantaneously change in load here the load is reduced to half the fluctuation occurred because of sudden load change is controlled via vector oriented closed loop control of three phase inverter from the result it is evident that the controller is fit to supply the flow of active power under sudden load variation [25–30]. In Figure 15.38 below shows the reactive power obtained due to the instantaneously change in load here the load is reduced to half the fluctuation occurred because of sudden load change is controlled via vector oriented closed loop control of three phase inverter from the result it is evident that the controller is fit to supply the flow of reactive power under sudden load variation [25–30].

(iii) When the load is increased instantaneously (when the load is suddenly doubled)

The proposed system undergone a sudden increase in load. When load is suddenly added in the system, then the output voltage sag will appear in stand-alone systems. Therefore, it is necessary to maintain the amplitude and frequency of the load voltage where a controller plays a very crucial part in controlling the amplitude and frequency of the load voltage during a load change in the system at t = 5.48 s. The star-connected three-phase load is increased from R = 101 ohms to R = 51 ohms. The load current is increased from 0.49 A to .99 A, and the controller maintains the amplitude and frequency of load voltage at t = 4.649 s as is evident from the results shown below. In Figure 15.39 below shows the DC-Link voltage obtained due to the instantaneously change in load here the load is increased to double the fluctuation occurred because of sudden load change is controlled via vector oriented closed loop control of three phase inverter from the result it is evident that the controller is fit to maintain the DC-Link voltage constant under sudden load variation [31–37]. In Figure 15.40 below shows the load output voltage obtained due to the instantaneously change in load here the load is increased to the doubled fluctuation occurred because of sudden load change is controlled via vector oriented closed loop control of three phase inverter from the result it is evident that the controller is fit to maintain the load output voltage magnitude and frequency constant under sudden load variation [31–37]. In Figure 15.41 below shows the output current obtained due to the instantaneously change in load here the load is increased to double the fluctuation occurred because of sudden load change is controlled via vector oriented closed loop control of three phase inverter from the result it is evident that the controller is fit to maintain the load output current magnitude and frequency constant under sudden load variation [31–37]. In Figure 15.42 below shows the active power obtained due to the instantaneously change in load here the load is increased to double the fluctuation occurred because of sudden load change is controlled via vector oriented closed loop control of three phase inverter from the result it is evident that the controller is fit to supply the flow of active power under sudden load variation [31–37]. In Figure 15.43 below shows the reactive power obtained due to the instantaneously change in load here the load is reduced to half the fluctuation occurred because of sudden load change is controlled via vector oriented closed loop control of three phase inverter from the result it is evident that the controller is fit to supply the flow of reactive power under sudden load variation [31–37].

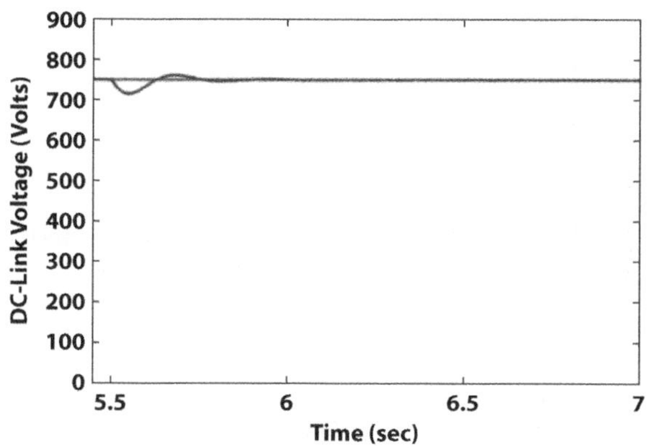

Figure 15.39 Voltage of DC link when the load is increased to half (volts) [31–37].

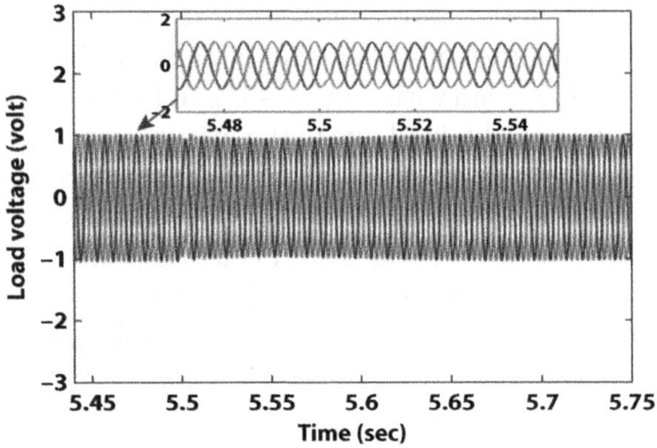

Figure 15.40 Output voltage across load when the load is increased to half (volts) [31–37].

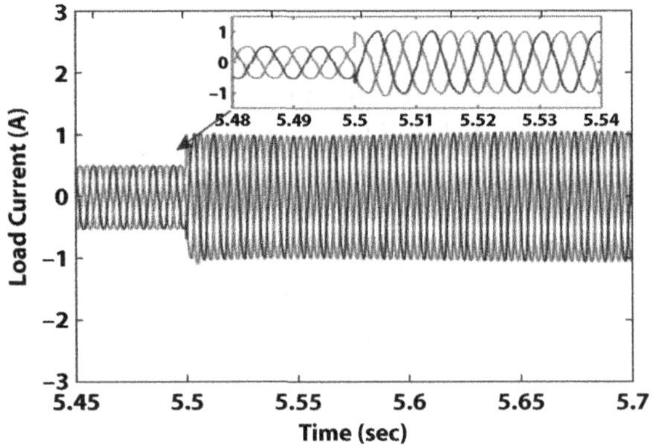

Figure 15.41 Output current across load when the load is increased to half (A) [31–37].

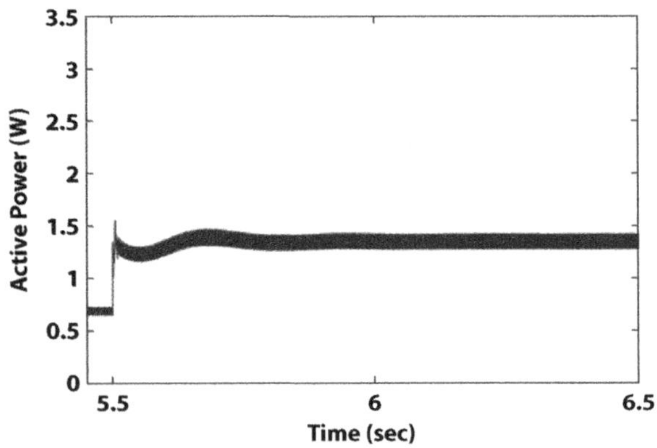

Figure 15.42 Active power when the load is increased to half (W) [31–37].

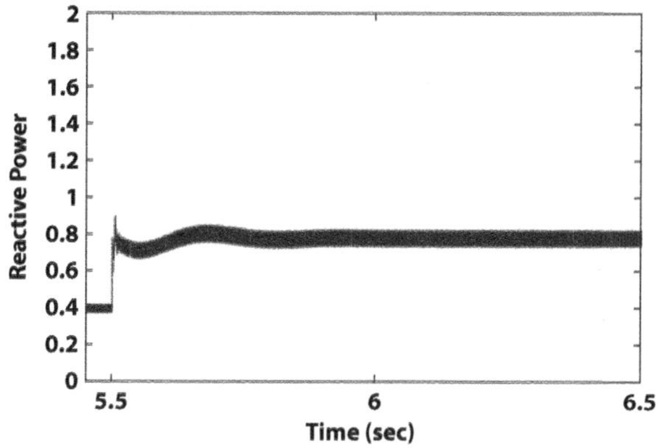

Figure 15.43 Reactive power when the load is increased to half (VAR) [31–37].

(iv) With RL load

The proposed model is validated with a variation in load. The model is tested with a constant 3ϕ star connected resistive load where R = 52 Ω/phase. Then, the system is subjected to a load change at t = 3.9 s, and a practical inductive load—RL load—of R = 102 Ω/phase and L = 14.5 mH is added into the system. It is analyzed that when the load changes, the voltage across the load is constant in terms of amplitude and frequency and is regulated within 0.149 s, as is evident from the results shown below. In Figure 15.44 below shows the output voltage across the load when a three-phase resistive load is applied and after few seconds another three phase RL Load is connected and it is observed from the result obtained that within few seconds the controller perfectly maintains the amplitude and frequency of the output voltage across the load [36–42]. In Figure 15.45 below shows the output current across the load when a three-phase resistive load is applied and after few seconds another three phase RL Load is connected and it is observed from the result obtained that within few seconds the controller perfectly maintains the amplitude and frequency of the output current across the load [36–42]. In Figure 15.46 below shows the active power obtained due to the instantaneously change in load here another RL load is added in the load the fluctuation occurred because of sudden load change is controlled via vector oriented closed loop control of three phase inverter from the result it is evident that the controller is fit to supply the flow of active power under sudden load variation [36–42]. In Figure 15.47 below shows the active power obtained due to the instantaneously change in load here another RL load is added in the load the fluctuation occurred because of sudden load change is controlled via vector oriented closed loop control of three phase inverter from the result it is evident that the controller is fit to supply the flow of active power under sudden load variation [36–42].

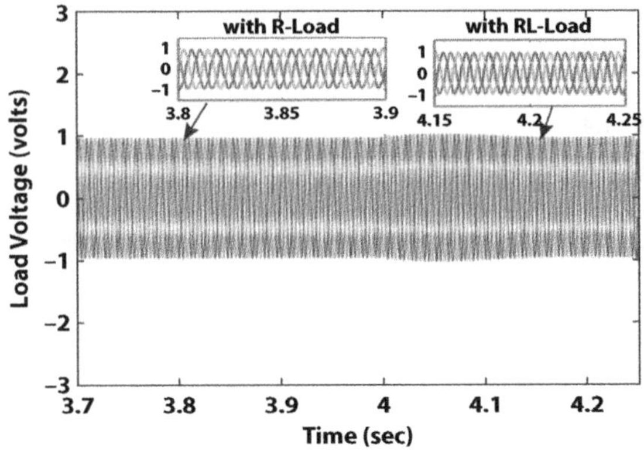

Figure 15.44 Voltage across resistive and inductive load under load switching [36–42].

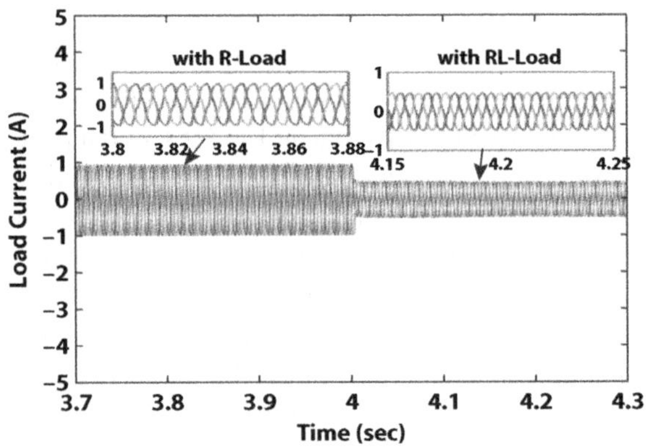

Figure 15.45 Load current across resistive and inductive load under load switching [36–42].

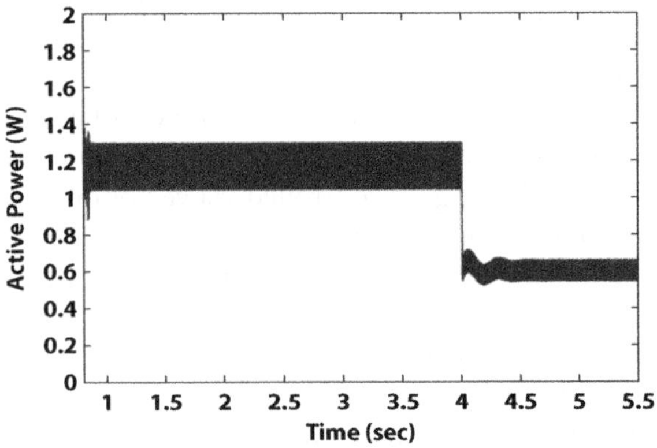

Figure 15.46 Active power across resistive and inductive load under load switching [36–42].

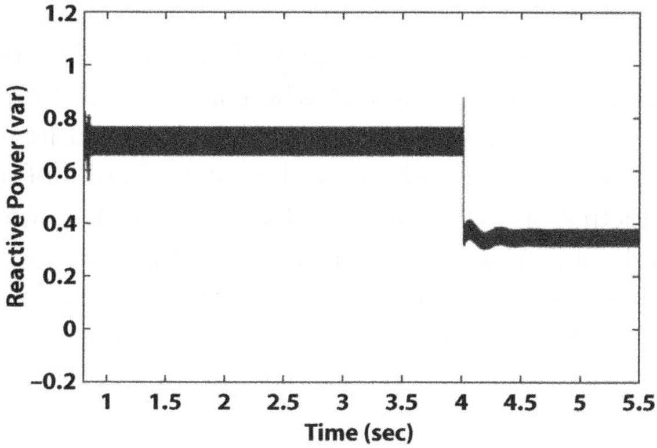

Figure 15.47 Reactive power across resistive and inductive load under load switching [36–42].

(v) With non-linear load

The proposed system is subjected to different load scenarios to validate the controller performance. One such load condition is to test the system with non-linear load at t = 1.9 s. A non-linear load is added into the system here. A three-phase rectifier with RL load R = 102/phase and L = 14.9 mH is considered for the load analysis, and it is shown in the results below that the controller is operating efficiently as it is regulating the load voltage within 0.148 s. In Figure 15.48 below shows the output voltage across the load where after the addition of RL Load another load is added to test the system under non-linear load condition as it is observed from the results obtained that after addition of the non-linear load there is fluctuation in the output voltage but after few seconds the magnitude of voltage and frequency is maintained constant hence, it is evident from the result that the inverter control algorithm is working efficiently [36–42]. In Figure 15.49 below shows the output current across the load where after the addition of RL Load another load is added to test the system under non-linear load condition as it is observed from the results obtained that after addition of the non-linear load there is fluctuation in the output voltage but after few seconds the current its frequency is maintained constant hence, it is evident from the result that the inverter control algorithm is working efficiently [36–42]. In Figure 15.50 below shows the active power across load where after the addition of RL Load another load is added to test the system under non-linear load condition as it is observed from the results obtained that after addition of the non-linear load there is fluctuation in the output voltage but after few seconds the voltage across the load is maintained constant and the filter and the controller improves the voltage profile efficiently current across the load and its frequency is maintained constant hence, it is evident from the result that the inverter control algorithm is working efficiently. In Figure 15.51 below shows the reactive power across load where after the addition of RL Load another load is added to test the system under non-linear load condition as it is observed from the results obtained that after addition of the non-linear load there is fluctuation in the output voltage but after few seconds the voltage across the load is maintained constant and the filter and the controller improves the voltage profile efficiently current across the load and its frequency is maintained constant hence, it is evident from the result that the inverter control algorithm is working efficiently [36–42].

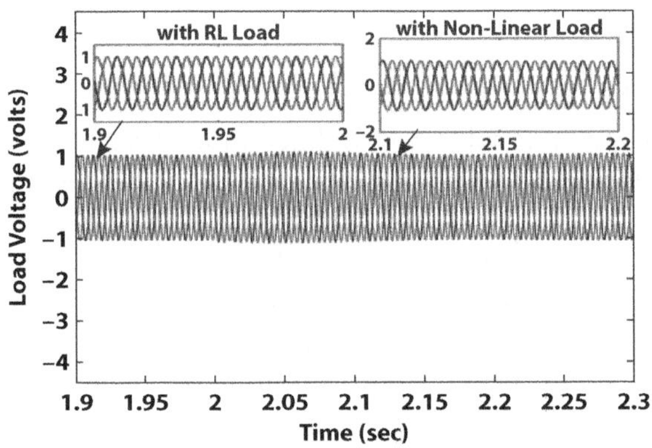

Figure 15.48 Voltage across practical inductive and non-linear load switching [36–42].

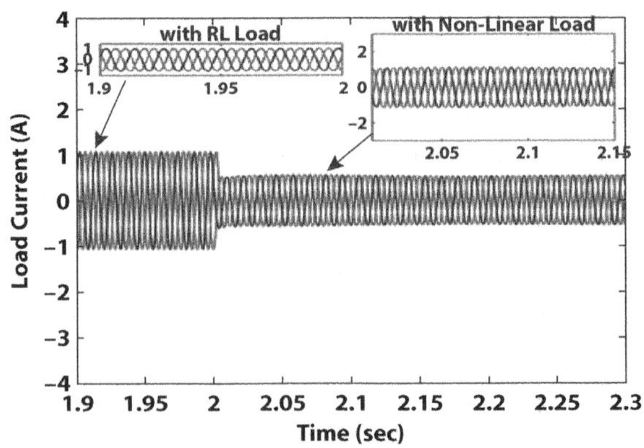

Figure 15.49 Current across practical inductive and non-linear load switching [36–42].

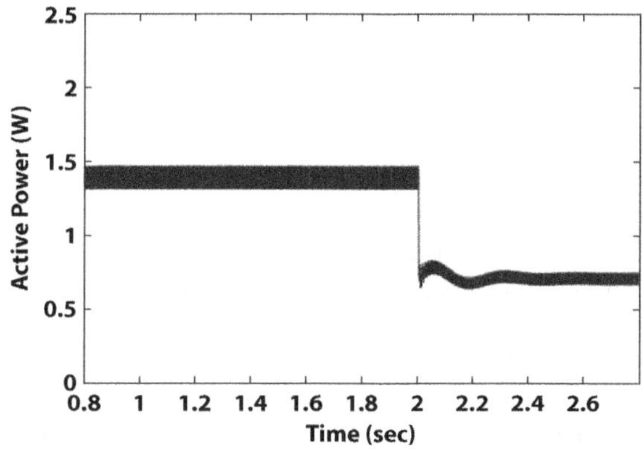

Figure 15.50 Active power across practical inductive and non-linear load switching [36–42].

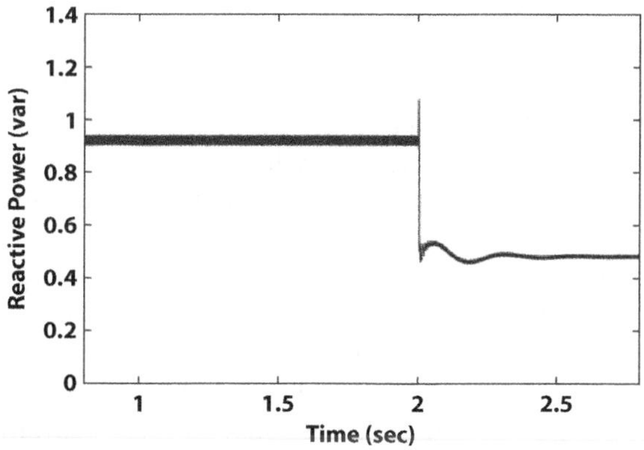

Figure 15.51 Reactive power across practical inductive and non-linear load switching [36–42].

15.13 Conclusion

This chapter provides an overview of a stand-alone PMSG-based wind battery energy conversion system that shows why using a battery energy storage system with a permanent magnet synchronous generator is superior. The control approaches include closed-loop vector-oriented control of the load-side converter for critical loading conditions and closed-loop vector-oriented control of the DC–DC bidirectional converter for charging and discharging of the DC link capacitor. Control techniques were used to simulate the above-mentioned issue statement in the MATLAB/Simulink domain. The wind battery conversion system running at an indecisive speed is modeled with an enhanced controller that successfully manages the frequency and amplitude of the load voltage at various load and speed circumstances, as demonstrated in the above-mentioned results.

References

1. Amin, M.M. and Mohammed, O.A., Development of high-performance grid-connected wind energy systems for optimum utilization of variable speed wind turbines. *IEEE Trans. Sustain. Energy*, 2, 3, 235–245, July 2011.
2. Raju, A.B., Fernandes, B.G., Chatterjee, K., A UPF power conditioner with maximum power point tracker for grid connected variable speed wind energy conversion system, in: *Proc. Ist International conference on Power Electronics*, November 2006.
3. Venkatraman, A., Maswood, A., Sarangam, N., Gabriel, O.H.P., An efficient UPF rectifier stand-alone wind energy conversion system. *IEEE Trans. Ind. Appl.*, 50, 2, 1421–1431, April 2014.
4. Bhende, C.N., Mishra, S., Malla, S.G., Permanent-magnet synchronous generator based stand-alone wind energy supply system. *IEEE Trans. Sustain. Energy*, 2, 4, 725–733, October 2011.
5. Polinder, H., Van der Pijl, F.F.A., de Vilder, G.J., Tavner, P.J., "Comparison of direct-drive and geared generator concepts for wind turbines. *IEEE Trans. Energy Convers.*, 3, 21, 725–733, Sep. 2006.
6. Tan, K. and Islam, S., Optimal control strategies in energy conversion of PMSG wind turbine system without mechanical sensors. *IEEE Trans. Energy Convers.*, 19, 2, 392–399, Jun. 2004.
7. Chan, T.F. and Lai, L.L., Permanent-magnet machines for distributed generation: A review, in: *Proc. IEEE Power Eng. Annu. Meeting*, pp. 1–6, 2007.
8. De Broe, M., Drouilhet, S., Gevorgian, V., A peak power tracker for small wind turbines in battery charging applications. *IEEE Trans. Energy Convers.*, 14, 4, 1630–1635, 1999.
9. Barote, L. and Marinescu, L., PMSG wind turbine system for residential applications. *Proc. of IEEE International Synposium on Power Electronics, Electrical Drives, Automation and Motion, SPEEDAM 2010*, Pisa, Italy, 14–16 June, 2010.
10. Pena, R., Clare, J.C., Asher, G.M., Doubly fed induction generator based back-to-back PWM converters and its application to variable speed wind energy generation. *IEEE Proc. Electr. Power Appl., 143, 3*, 231–241, May 1996.
11. Elmorshedy, M.F., Allam Ahmed, S.M., Rashad, M., Voltage and frequency control of a stand-alone wind energy system based on PMSG. *4th International Conference on Electric Power and Energy Conversion Systems, (EPECS'15)*, 2015.
12. Perreault, D.J. and Caliskan, V., Automotive power generation and control. *IEEE Trans. Power Electron.*, 19, 3, 618–630, May 2004.

13. Amei, K., Takayasu, Y., Ohji, T., Sakui, M., A maximum power control of wind generator system using a permanent magnet synchronous generator and a boost chopper circuit. *Proc Power Conv*, vol. 3, pp. 1447–1452, 2002.
14. Muljadi, E., Drouilhet, S., Holz, R., Gevorgian, V., Analysis of permanent magnet genertaor for wind power battery charging, in: *Proc. IEEE 1996 Industry Applications Society Annual Meeting*, vol. 1, pp. 541–548, 1996, 10.1109/IAS.1996.557087.
15. Müller, S., Deicke, M., De Doncker, R.W., Doubly fed induction genertaor system for wind turbines. *IEEE Ind. Appl. Mag.*, 8, 26–33, May/June, 2002.
16. Xue, Y., Study of single-phase single-stage buck-boost inverters, University of New Brunswick M.Sc, Thesis, Jan. 2004.
17. Xue, Y. and Chang, L., Closed-loop SPWM control for grid-connected buck-boost inverters. *IEEE Power Electronics Specialists Conference 2004*, Aachen, Germany, vol. 5, pp. 3366–3371, June 2004.
18. Blasko, V. and Kaura, V., A new mathematical model and control of a three-phase ac-dc voltage source converter. *IEE Trans. Power Electron.*, 12, 1, 116–123, Jan.1997.
19. Singh, B. *et al.*, A review of three-phase improved power quality AC–DC converters. *IEEE Trans. Ind. Electron.*, 51, 3, 641–660, 2004.
20. Tan, K. and Islam, S., Optimal control strategies in energy conversion of PMSG wind turbine system without mechanical sensors. *IEEE Trans. Energy Convers.*, 19, 2, 392–399, Jun2004.
21. Whaley, D.M., Soong, W.L., Ertugrul, N., Investigation of switched-mode rectifier for control of small-scale wind turbines, in: *Proc. IEEE Industry Applications Society Annual Meeting*, pp. 2849–2856, 2005.
22. Muljadi, E., Drouilhet, S., Holz, R., Gevorgian, V., Analysis of permanent magnet genertaor for wind power battery charging, in: *Proc. IEEE 1996 Industry Applications Society Annual Meeting*, pp. 541–548.
23. Chinchilla, M., Arnaltes, S., Burgos, J.C., Control of permanent-magnet generators applied to variable-speed wind-energy systems connected to the grid. *IEEE Trans. Energy Convers.*, 21, 1, 130–135, March 2006.
24. Haque, M.E. and Rahman, M.F., *Permanent magnet synchronous motor drives analysis, modeling and control*, VDM Verlag, Germany, 2009.
25. Elmorshedy, M.F., Allam Ahmed, S.M., Rashad, M., Voltage and frequency control of a stand-alone wind energy system based on PMSG. *4th International Conference on Electric Power and Energy Conversion Systems, (EPECS'15)*, 2015.
26. Haque, E., Negnevitsky, M., Muttaqi, K.M., A novel control strategy for a variable-speed wind turbine with a permanent-magnet synchronous generator. *IEEE Trans. Industry Appl.*, 46, 1, 1–8, January 2010.
27. Blasko, V. and Kaura, V., A New Mathematical Model and control of a three-phase ac-dc voltage source converter. *IEE Trans. Power Electron.*, 12, 1, 116–123, Jan. 1997.

28. Chen, Z. and Spooner, E., Voltage source inverters for high-power variable power dc sources. *Proc. Inst. Elect. Eng. Gener. Transm. Distrib.*, 148, 439–447, Sep, 2001.
29. Noguchi, T. *et al.*, Direct power control of PWM converter without power source voltage sensors. *IEEE Trans. Ind. Electron.*, 34, 3, 473–479, 1998.
30. Teleke, S., Baran, M.E., Huang, A.Q., Bhattacharya, S., Anderson, L., Control strategies for battery energy storage for wind farm dispatching. *IEEE Trans. Energy Conv.*, 3, 725–732, Sept. 2009.
31. Eriksen, P., Ackermann, T., Abildgaard, H., Smith, P., Winter, W., Garcia, J.R., System operation with high wind penetration. *IEEE Power Energy Mag.*, 3, 6, 65–74, Nov.-Dec. 2005.
32. Kazimierkowski, M.P., Blaabjerg, F., Krishnan, R., *Control in power electronics: Selected problems*, Academic Press, New York, 2002.
33. Lukic, S.M., Jian, C., Bansal, R.C., Rodriguez, F., Emadi, A., Energy storage systems for automotive applications. *IEEE Trans. Ind. Electron.*, 55, 6, 2258–2267, 2008.
34. Wei, L., Geza, J., Jean, B., Real time simulation of a wind turbine generator coupled with a battery supercapacitor energy storage system. *IEEE Trans. Ind. Electron.*, 53, 4, 1137–1145, April 2010.
35. Zhan-li, C., Wei-rong, C., Qi, L., Zhi-ling, J., Zhi-han, Y., Modeling and dynamic simulation of an efficient energy storage component-supercapacitor, in: *Power and Energy Engineering Conference (APPEEC), 2010 Asia-Pacific*, pp. 1–4, 2010.
36. Vazquez, S., Lukic, S.M., Galvan, E., Franquelo, L.G., Carrasco, J.M., Energy storage systems for transport and grid applications. *IEEE Trans. Ind. Electron.*, 57, 12, 38813895, Dec. 2010.
37. Baktash, A. *et al.*, New switching table for improved direct power control of three-phase PWM rectifier. *Aust. J. Electr. Electron. Eng.*, 5, 2, 161–167, 2009.
38. Suryawanshi, H.M., Ramteke, M.R., Thakre, K.L., V.B., Unity-power-factor operation of three-phase ac-dc soft switched converter based on boost active clamp topology in modular approach. *IEEE Trans. Power Electron.*, 23, 1, 229–236, Jan 2008.
39. de Correa, M.B., Jacobina, C.B., da Silva, E.R.C., Lima, A.M.N., A general PWM strategy for four-switch three-phase inverters. *IEEE Trans. Power Electron.*, 21, 6, 1618–1627, Nov. 2006.
40. Steike, J.K., Use of an LC filter to achieve motor friendly performance of the PWM voltage source inverter. *IEEE Trans. Energy Convers.*, 14, 3, 649–654, Sep. 2009.
41. Dahono, P.A., Purwadi, A., Qamaruzzaman, An LC filter design method for single-phase PWM inverters, in: *Proc. Int. Conf Power Electronics and Drive Systems*, vol. 2, pp. 541–548, 1995.
42. Zhang, N., Tang, H., Yao, C., A systematic method for designing a PR controller and active damping of the LC filter for single-phase grid-connected PV inverters. *Energies*, 7, 6, 3934–3954, 2014.

Index

Also of Interest

Check out these other related titles from Scrivener Publishing

ENERGY STORAGE TECHNOLOGIES IN GRID MODERNIZATION, Edited by Sandeep Dhundhara, Yajvender Pal Verma, and Ashwani Kumar, ISBN: 9781119872115. Written and edited by a team of experts, this exciting new volume discusses the various types of energy storage technologies, the applications of energy storage systems, the performance improvement of modern power systems, their role in the real-time operation of power markets, and the operational issues of modern power systems, including renewable-based generating sources.

SMART GRIDS FOR SMART CITIES VOLUME 1, Edited by O.V. Gnana Swathika, K. Karthikeyan, and Sanjeevikumar Padmanaban, ISBN: 9781119872078. Written and edited by a team of experts in the field, this first volume in a two-volume set focuses on an interdisciplinary perspective on the financial, environmental, and other benefits of smart grid technologies and solutions for smart cities.

SMART GRIDS FOR SMART CITIES VOLUME 2: Real-Time Applications in Smart Cities, Edited by O.V. Gnana Swathika, K. Karthikeyan, and Sanjeevikumar Padmanaban, ISBN: 9781394215874. Written and edited by a team of experts in the field, this second volume in a two-volume set focuses on an interdisciplinary perspective on the financial, environmental, and other benefits of smart grid technologies and solutions for smart cities.

SMART GRIDS AND INTERNET OF THINGS, Edited by Sanjeevikumar Padmanaban, Jens Bo Holm-Nielsen, Rajesh Kumar Dhanaraj, Malathy Sathyamoorthy, and Balamurugan Balusamy, ISBN: 9781119812449. Written and edited by a team of international professionals, this groundbreaking new volume covers the latest technologies in automation, tracking, energy distribution and consumption of Internet of Things (IoT) devices with smart grids.

SMART GRIDS AND GREEN ENERGY SYSTEMS, Edited by A. Chitra, V. Indragandhi and W. Razia Sultana, ISBN: 9781119872030. Presenting the concepts and advances of smart grids within the context of "green" energy systems, this volume, written and edited by a global team of experts, goes into the practical applications that can be utilized across multiple disciplines and industries, for both the engineer and the student.

SMART GRIDS AND MICROGRIDS: Concepts and Applications, Edited by P. Prajof, S. Mohan Krishna, J. L. Febin Daya, Umashankar Subramaniam, and P. V. Brijesh, ISBN: 9781119760559. Written and edited by a team of experts in the field, this is the most comprehensive and up to date study of smart grids and microgrids for engineers, scientists, students, and other professionals.

MICROGRID TECHNOLOGIES, Edited by C. Sharmeela, P. Sivaraman, P. Sanjeevikumar, and Jens Bo Holm-Nielsen, ISBN 9781119710790. Covering the concepts and fundamentals of microgrid technologies, this volume, written and edited by a global team of experts, also goes into the practical applications that can be utilized across multiple industries, for both the engineer and the student.

INTEGRATION OF RENEWABLE ENERGY SOURCES WITH SMART GRIDS, Edited by A. Mahaboob Subahani, M. Kathiresh and G. R. Kanagachidambaresan, ISBN: 9781119750420. Provides comprehensive coverage of renewable energy and its integration with smart grid technologies.

Encyclopedia of Renewable Energy, by James G. Speight, ISBN 9781119363675. Written by a highly respected engineer and prolific author in the energy sector, this is the single most comprehensive, thorough, and up to date reference work on renewable energy.

Green Energy: Solar Energy, Photovoltaics, and Smart Cities, edited by Suman Lata Tripathi and Sanjeevikumar Padmanaban, ISBN 9781119760764. Covering the concepts and fundamentals of green energy, this volume, written and edited by a global team of experts, also goes into the practical applications that can be utilized across multiple industries, for both the engineer and the student.

Energy Storage, edited by Umakanta Sahoo, ISBN 9781119555513. Written and edited by a team of well-known and respected experts in the field, this new volume on energy storage presents the state-of-the-art developments and challenges in the field of renewable energy systems for sustainability and scalability for engineers, researchers, academicians, industry.

Energy Storage 2nd Edition, by Ralph Zito and Haleh Ardebili, ISBN 9781119083597. A revision of the groundbreaking study of methods for storing energy on a massive scale to be used in wind, solar, and other renewable energy systems.

Hybrid Renewable Energy Systems, edited by Umakanta Sahoo, ISBN 9781119555575. Edited and written by some of the world's top experts in renewable energy, this is the most comprehensive and in-depth volume on hybrid renewable energy systems available, a must-have for any engineer, scientist, or student.

Progress in Solar Energy Technology and Applications, edited by Umakanta Sahoo, ISBN 9781119555605. This first volume in the new groundbreaking series, Advances in Renewable Energy, covers the latest concepts, trends, techniques, processes, and materials in solar energy, focusing on the state-of-the-art for the field and written by a group of world-renowned experts.

A Polygeneration Process Concept for Hybrid Solar and Biomass Power Plants: Simulation, Modeling, and Optimization, by Umakanta Sahoo, ISBN 9781119536093. This is the most comprehensive and in-depth study of the theory and practical applications of a new and groundbreaking method for the energy industry to "go green" with renewable and alternative energy sources.

DESIGN AND DEVELOPMENT OF EFFICIENT ENERGY SYSTEMS, edited by Suman Lata Tripathi, Dushyant Kumar Singh, Sanjeevikumar Padmanaban, and P. Raja, ISBN 9781119761631. Covering the concepts and fundamentals of efficient energy systems, this volume, written and edited by a global team of experts, also goes into the practical applications that can be utilized across multiple industries, for both the engineer and the student.

INTELLIGENT RENEWABLE ENERGY SYSTEMS: Integrating Artificial Intelligence Techniques and Optimization Algorithms, edited by Neeraj Priyadarshi, Akash Kumar Bhoi, Sanjeevikumar Padmanaban, S. Balamurugan, and Jens Bo Holm-Nielsen, ISBN 9781119786276. This collection of papers on artificial intelligence and other methods for improving renewable energy systems, written by industry experts, is a reflection of the state of the art, a must-have for engineers, maintenance personnel, students, and anyone else wanting to stay abreast with current energy systems concepts and technology.

SMART CHARGING SOLUTIONS FOR HYBRID AND ELECTRIC VEHICLES, edited by Sulabh Sachan, Sanjeevikumar Padmanaban, and Sanchari Deb, ISBN 9781119768951. Written and edited by a team of experts in the field, this is the most comprehensive and up to date study of smart charging solutions for hybrid and electric vehicles for engineers, scientists, students, and other professionals.

Printed and bound by CPI Group (UK) Ltd, Croydon, CR0 4YY

17/09/2026

14964276-0008